Lactose

Lactose

Evolutionary Role, Health Effects, and Applications

Edited by

Marcel Paques

FrieslandCampina Innovation Centre, Wageningen, The Netherlands

Cordula Lindner

FrieslandCampina Innovation Centre, Wageningen, The Netherlands

ELSEVIER

ACADEMIC PRESS

An imprint of Elsevier

Academic Press is an imprint of Elsevier
125 London Wall, London EC2Y 5AS, United Kingdom
525 B Street, Suite 1650, San Diego, CA 92101, United States
50 Hampshire Street, 5th Floor, Cambridge, MA 02139, United States
The Boulevard, Langford Lane, Kidlington, Oxford OX5 1GB, United Kingdom

British Library Cataloguing-in-Publication Data
A catalogue record for this book is available from the British Library

Library of Congress Cataloging-in-Publication Data
A catalog record for this book is available from the Library of Congress

ISBN: 978-0-12-811720-0

For Information on all Academic Press publications
visit our website at https://www.elsevier.com/books-and-journals

Publisher: Mica H. Haley
Acquisition Editor: Charlotte Cockle
Editorial Project Manager: Devlin Person
Production Project Manager: Vignesh Tamilselvvan
Cover Designer: Victoria Pearson

Typeset by MPS Limited, Chennai, India

Contents

List of contributors

Bastiaan H.J. Dickhoff
DFE Pharma, Wageningen, The Netherlands

Michael G. Gänzle
Department of Agricultural, Food and Nutritional Science, University of Alberta, Edmonton, AB, Canada; College of Bioengineering and Food Science, Hubei University of Technology, Wuhan, P.R. China

Gerald A. Hebbink
DFE Pharma, Wageningen, The Netherlands

Kasper A. Hettinga
Dairy Science & Technology, Food Quality & Design Group, Wageningen University & Research, Wageningen, The Netherlands

Andrew Szilagyi
Jewish General Hospital, McGill University School of Medicine, Division of Gastroenterology, Montreal, QC, Canada

Mark G. Thomas
Research Department of Genetics, Evolution and Environment (GEE), University College London, London, United Kingdom

Catherine Walker
Research Department of Genetics, Evolution and Environment (GEE), University College London, London, United Kingdom

Foreword

The American Politician Robert Kennedy coined this famous quote Some men see things as they are, and ask why. I dream of things that never were, and ask why not. This visionary statement is ironically appropriate as the metaphor for *Lactose: Evolutionary Role, Health Effects, and Applications*, the long overdue treatment of perhaps the most underappreciated molecule in biology: lactose. Such scientific indifference is perplexing. No molecule has been so central to the competitive success of higher species and yet so ignored by the scientific community.

The ascendance of mammals over their animate competitors attests to the genius of lactation. The quintessential mammalian construct, the mother–infant dyad, is built around lactation and milk. As lactation is the remarkable bioproduction system of the mammary gland and milk, the biofluid that as diet provides nourishment, protection, fuel, education, communication, and delight to the infant. As a biological process fit for research, is there an equal? In virtually all mammals, the most abundant chemical component within milk (apart from water) is lactose. Among all mammals, human milk contains almost twice as much lactose as any other milk. And yet, we know astonishingly little about this simple molecule that is virtually unique to mammalian milk. It is hard to imagine such scientific disinterest for any other common biomaterial. After all, Charles Darwin spent a lifetime absorbed by the marvels of beetles. Stephen Jay Gould invested a career in the pursuit of land snails! Though the history may be perplexing, finally the scientific community has begun to turn its attention to milk and its charming little disaccharide, lactose.

Lactose: Evolutionary Role, Health Effects, and Applications takes the reader along a path to correcting the imbalance of scientific attention. This book is not a narrow view of a chemical. It is brilliantly diverse in its perspectives. From the evolution of mammals and the role of lactose in their selection to the biochemistry of lactose synthesis and its unique pathways of secretion and metabolism, the chapters of the book are inspiring and exciting. The authors give the reader a view into the fascinating world of lactose and through this window, a glimpse into glycobiology as a field. Not surprisingly the book is fundamentally food for the soul of discovery. While some of the secrets underlying the emergence of lactose as a nutrient are emerging, many more remain to be revealed. It is a book full of wonderful information and teasing motivation to go forward. Students who are approaching the fields of life science and considering diverse career paths will be inspired by this book. They will be inspired both by what we now know about the genius of lactation as a biological process and by the scientific secrets that are tantalizing but still hidden within its structures and functions. Can there be a higher aspiration for a written work than to inspire and guide the next generation of students into a field for a career.

When reading a good book, two sentiments can occasionally torment a reader: (1) I wish I had read this earlier and (2) I wish I had written this book. I defy the readers of *Lactose: Evolutionary Role, Health Effects, and Applications* not to experience at least one of these torments.

Prof. J. Bruce German

Preface

An increasing professional and also widening interest in lactose can be attributed in part to the growing attention in different kinds of media, popular science discussions, and polarized opinions. What is lacking is a concise reference book that brings together the knowledge and latest research from the fields of evolutionary biology, nutritional science, and food science to consolidate current research in this widening field. Therefore the editors initiated the creation of this book, working together with key experts who share their knowledge, gained by years of study and experience from practice, in the present book chapters.

Lactose plays a role in multiple science areas including evolutionary biology, nutritional science, food science, and pharmaceutical sciences, and also in more applied areas such as dietetics, medicine, and pediatrics. Consequently, a handbook as reference is expected to serve a wide range of professionals.

The presence of lactose as the major disaccharide in mammalian milk has long been taken for granted. However, scientists and technologists are beginning to discover the unique properties of lactose and its complex interactions with human physiology and the gut microbiome. The book provides an overview of the biochemistry of lactose, starting with the intriguing evolutionary role of lactose and the biological advantages linked to lactose digestion in humans. The mechanisms involved in digestion, absorption, and the metabolism are discussed next, together with the physiological effects of lactose. Mechanistic explanations for lactose intolerance symptoms are complex and, therefore, addressed in a separate chapter. Complex causal factors are involved, requiring a range of clinical methods for individual diagnosis and therapeutic nutritional and dietary recommendations. Undigested lactose will reach the colon where it is fermented by gut microbiota, which is discussed in detail. In addition, the role of lactose in the manufacturing of many products and as an ingredient in consumer end products, as well as its use as a pharmaceutical excipient, is discussed in two separate chapters.

In principle, each chapter follows a similar structure: starting with an introduction, followed by the main scientific discussion of the relevant and current literature, and closing with main conclusion. It is also indicated where knowledge gaps exist with recommendations for future research.

Multiple cross-referencing is used between the chapters to integrate all the various aspects of lactose and provide the full versatile picture. This required full dedication of the author teams and their interchapter collaboration. To obtain full alignment between the chapters, a one-day, face2face, workshop was held, with participation of representatives of each author team and the editors, supported by a professional facilitator.

Further alignment, unity, and increased readability of the book chapters have been achieved by aligning the definitions of the terms used throughout the book. The creation of a shared language between the chapters and the aligned definitions in the glossary was the result of a dedicated collaboration between the author teams.

In the following text some important aspects and prominent messages about lactose—as captured from the six book chapters—are depicted.

Chapter 1, The evolution of lactose digestion, provides a fascinating view of the environmental factors and genetic responses that confer lactose digestion into adulthood. On the evolutionary timescale, milk and milk products are relatively recent additions to adult human diets that have had profound impacts on our culture, biology, genetics, and behavior. Lactase persistence—the continued expression of the gut enzyme lactase into adulthood—is the most strongly selected single-gene trait to have evolved in Europeans and some African groups over the last 10,000 years, and is the strongest and most often-cited example of recent natural selection, gene–culture coevolution, and convergent evolution in humans.

Chapter 2, Digestion, Absorption, Metabolism, and Physiological Effects of Lactose, deals with the fate of swallowed lactose in the body and resulting physiological effects in different population groups. Emphasis is placed on the role of lactose in neonatal nutrition and development, its contribution to energy supply, and specific structural and developmental roles of the galactose moiety in immune and nervous system, for example, as an important building block for the synthesis of galactocerebrosides, among others precursors for myelination in the maturing nervous system. The activity and regulation of the key enzyme in lactose digestion, intestinal lactase, and the interesting—yet still largely unexplored—phenomenon of the decline of lactase activity with progressing age are described.

Furthermore, the metabolic and physiological consequences of the evolutionary division of the adult world population initially into two Mendelian phenotypes, those who can (lactase persistent) and those who cannot digest lactose (lactase nonpersistent) depending on their ability to express efficient amounts of lactase, are discussed. In the latter, excess of lactose is efficiently metabolized by intestinal bacteria. The fermentation products salvage calories and may confer tolerance in those who can consume large quantities without digestive symptoms.

These phenomena are further described in Chapter 3, Lactose intolerance and other related food sensitivities. Although lactose intolerance or tolerance in adults is often due to the genetically determined lactase persistence (and nonpersistence), this is not a one reason–one consequence relationship. Multiple factors including intestinal integrity and microbial adaptation and activity can influence and even reverse the clinical outcome. Individuals who are lactase nonpersistent can adapt to dairy consumption and be symptom-free with reasonable single dosages, whereas complete dairy avoidance may have a greater negative impact on their health status. Despite our deeper understanding of food sensitivities and their overlap with irritable bowel syndrome, misattribution of gastrointestinal symptoms to lactose ingestion in self-reporting lactose-intolerant individuals remains common and a heavily discussed subject in media-driven popular science. Therefore the different forms and causes of lactose intolerance, its symptoms, diagnostic tests and management, the interrelationships with other ailments, its perception and food sensitivities, and its distinction from cow's milk allergy are thoroughly discussed.

Also the question "Lactose—a conditional prebiotic?" (in lactase nonpersisters) was subjected to a profound analysis and captured in Chapter 4 carrying the same title. Despite a number of supporting aspects, lactose can currently not be regarded as a conditional prebiotic in lactase nonpersisters. Lactose bypassing upper intestinal digestion and reaching the lower intestine can have double-edged effects mediated via microbial fermentation in the gut. Both possible effects, positive [production of beneficial fermentation products, such as SCFAs (short chain fatty acids) and favorable microbiota changes] and negative (excessive gas production contributing to gastrointestinal discomfort), evolve from microbial activity and, therefore, can be regarded as two sides of the same coin. However, the activated bacteria, extend and type of metabolic activity are dependent on many factors, such as genetic factors and health status, (history of) diet and microbiota adaptation, medication influencing the function of the microbiome, and thus are subject even to temporal variability within the same individual.

The use of lactose in food application needs to recognize that lactose has multiple and strongly dose-dependent and host factor–dependent roles as digestible and glycemic carbohydrate, as a FODMAP (fermentable oligo-, di-, monosaccharide, and polyol) with possible adverse health effects, and also as a potential health beneficial prebiotic carbohydrate.

Chapter 5 is entitled "Application of lactose in the pharmaceutical industry." A majority of pharmaceutical dosage forms contain lactose, and in volume, lactose is one of the biggest pharmaceutical excipients. Lactose is a versatile excipient that is safe to use, relatively cheap, and widely available in many forms, for example, as a filler to provide bulk to, for instance, tablets, as a binder to provide the strength to a dosage form to keep it together, and it can assist in delivering the drug to the place of action like in dry powder inhalers. Lactose also provides the flow to a formulation which is required for producing it. The different manufacturing processes available in the pharmaceutical industry all require different grades of lactose with specific properties to cope with the formulation challenges. Concerns on intake of pharmaceutical lactose with medication in case of lactose intolerance are groundless. It is proven practice and generally accepted that a daily intake of 10–12 g lactose does not result in physical problems, which is an amount that is hard to reach by pharmaceutical intake.

Chapter 6 is entitled "Lactose in the dairy production chain." Lactose, the main carbohydrate in milk, is a unique sugar produced in the mammary gland of mammals and is the main source of energy in newborns. The physical and chemical properties of lactose play a major role in the properties and quality of many dairy products and dairy-based ingredients. Lactose can be isolated for direct use, but it can also be chemically converted into many lactose-derived components, which have very different functionalities. For consumer products, lactose is the energy source for the lactic acid bacteria, which play a crucial role in dairy fermentation, and lactose acts as substrate for specific flavor components. As ingredients are usually dried, and lactose is a major component of the dry matter, lactose determines most of the physical and chemical characteristics. Recent technological developments enable the manufacture of more complex oligosaccharides on an industrial scale. This opens the opportunity for application in, for example, future infant formulations, which may be even better at mimicking the functions of human milk oligosaccharides.

These six chapters as a whole represent the current research, industrial applications, and history of lactose as we look forward to ways in which lactose will continue to contribute to and shape our health and well-being.

CHAPTER

The evolution of lactose digestion

1

Catherine Walker and Mark G. Thomas

Research Department of Genetics, Evolution and Environment (GEE), University College London, London, United Kingdom

1.1 Introduction

Milk is a nutrient- and energy-dense food that provides a well-balanced matrix of macro- and micronutrients. Its consumption by neonates is shared across all extant mammals, and for the majority, it is the primary food source at birth, decreasing during and after the weaning period. However, regular milk consumption in adulthood is, as far as we know, a trait unique to humans. The main sugar in most mammalian milks is the disaccharide lactose, and the presence of this nutritional component has a deep evolutionary history and is a key factor in adult human milk consumption over the last 10,000 years (see Chapters 2–4 for more specific discussions of lactose and lactose digestion).

One obvious reason why only humans consume milk as adults is that we are able to utilize domestic animals for its production, but another key factor influencing adult milk consumption is lactose digestion. Lactose is digested to its component monosaccharides—glucose and galactose—by the membrane-bound gut enzyme lactase. In most mammals, including most humans, lactase production is downregulated to low levels, sometimes during or after the weaning period is over. Without lactase, lactose will pass undigested into the colon where it can cause adverse effects such as cramps, bloating, and diarrhea. However, about one-third of the adult humans in the world continue to produce sufficient quantities of lactase to permit lactose digestion, a trait known as lactase persistence (LP). The evolution of adult milk drinking and LP is the strongest, best-evidenced, and most widely cited example of both cultural and genetic adaptations to dietary change (Gerbault et al., 2011; Ingram, Mulcare, Itan, Thomas, & Swallow, 2009; Itan, Powell, Beaumont, Burger, & Thomas, 2009; Nielsen, Hellmann, Hubisz, Bustamante, & Clark, 2007; Tishkoff et al., 2007).

In this chapter we will explore the evolutionary history of lactose in milk, and the cultural and biological adaptations that have led to milk being a major food source for adult humans today.

Lactose. DOI: https://doi.org/10.1016/B978-0-12-811720-0.00001-5

1.2 Evolutionary origins of milk and lactose

Milk production has defined the biological class Mammalia since 1758. Carolus Linnaeus devised a system of classification in his tenth edition of *Systema Natura* which divided the animal kingdom into six classes, the smallest in terms of numbers of species being mammals (Linnaeus, 1758). While other classes such as reptiles, birds, fish, and amphibians have higher numbers of species, mammals include species with the largest average body mass of all known aquatic-dwelling species, in the blue whale, and also of living land-based species, in the African elephant. Although mammals share other unique characteristics such as having a four-chambered heart, three ear bones, and body hair, Linnaeus chose the ability to produce milk as the defining feature. In this sense, producing, ingesting, and digesting milk is a trait shared by all mammals.

Lactose is a disaccharide made up of the monosaccharides—glucose and galactose—which are joined by a β 1–4 glycosidic bond. It is present in varying concentrations in the milk of virtually all mammals (Oftedal, 1995). It is widely known as "milk sugar" as it is only found in milk and its only known natural source is the mammary glands of mammals (Urashima, Fukuda, & Messer, 2012), where it is synthesized in either free or bound forms. Although there have been historical reports of lactose synthesis in forsythia pollen (Curtis & Jones, 1959; Kuhn & Low, 1949) and in Sapotacea fruit (Reithel & Venkataraman, 1956), more recent gas chromatography investigations by Toba, Nagashima, and Adachi (1991) found no evidence of this. While the evolutionary origins of lactose and milk itself remain contested, there is a growing body of literature on its earliest functions and the selective pressures that contributed to its changing composition through time (Lefèvre, Sharp, & Nicholas, 2010; Oftedal, 2011).

Based on the significant similarities and also key differences observed in mammalian milk compositions, milk secretion most likely evolved in ancestral therapsids before the appearance of the first mammals some 166–220 million years ago (ya) (Bininda-Emonds et al., 2007; Lefèvre et al., 2010; Madsen, 2009; Oftedal, 2011; Urashima et al., 2012). Within mammals, there are three infraclasses: eutherian (placental), metatherian (marsupials), collectively known as theria (bearing live young), and prototherian (monotremes). Although eutherian mammals are widely known as placentals, there is evidence in the fossil record of earlier nonplacental eutherians. All living eutherians give birth to live young and are nourished during gestation by a placenta. Placental mammals are the most numerous subclass, with over 4000 species. Metatherians (marsupials) are made up of over 300 species found predominantly on the Australasian and American continents. Marsupials gestate offspring briefly in the presence of a placenta before the young transfer to a pouch with nipples, to provide continued nourishment. The metatherian and eutherian lineages probably diverged from one another some 135–140 million ya, while prototherians, split earlier from theria, probably 166–220 million ya (Bininda-Emonds et al., 2007;

Killian et al., 2001; Madsen, 2009) (Fig. 1.1). Prototherians, or monotremes, are the smallest infraclass of mammals. These egg-laying mammals include just five extant species: the platypus and four species of echidnas. Monotremes have a less advanced system of lactation than the theria (Lefèvre et al., 2010); they do not have nipples, but instead the young suckle from areas with specialized hairs near the mammary glands.

Investigations of the nutrient composition of milk in these three infraclasses (prototherian, metatherian, and eutherian) have provided insights into their evolutionary relationships and the evolution of lactose as the dominant sugar in eutherian milk. While eutherians, marsupials, and monotremes all produce milk in mammary glands to feed their young, the carbohydrate and other nutrient contents vary both between and within infraclasses. With the exception of a few marine mammals, particularly pinnipedia (e.g., seals, sea lions) (Oftedal, Boness, & Tedman, 1987), lactose is present in almost all mammalian milks (Oftedal, 2002a). In most eutherians, lactose is the main milk carbohydrate, although smaller quantities of oligosaccharides can also be present. Exceptions to this are a few

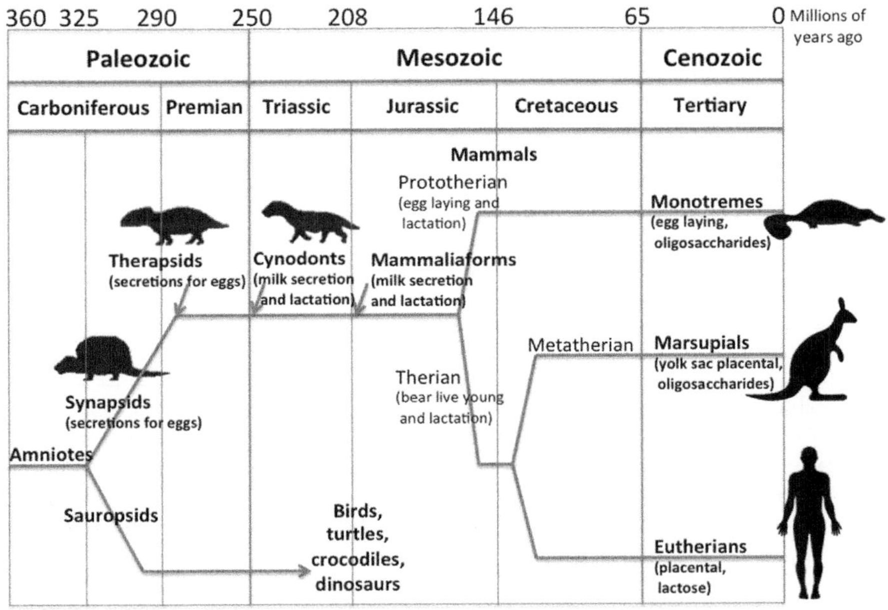

FIGURE 1.1

Mammalian evolution noting reproduction strategy and principal milk carbohydrate in the three infraclasses.

Adapted from Lefèvre, C. M., Sharp, J. A., & Nicholas, K. R. (2010). Evolution of lactation: Ancient origin and extreme adaptations of the lactation system. Annual Review of Genomics and Human Genetics, 11, *219–238.*

species of Canoidea, including bears, mink, coati, giant panda, and some pinnipedia, where oligosaccharides are higher in concentration than lactose. This is also the case in some marsupials and monotremes, with some lactose present (Messer & Kerry, 1973; Oftedal, 2002a; Urashima and Messer, 2007; Urashima et al., 2012). This broad pattern of higher lactose concentrations in eutheria is consistent with the hypothesis that oligosaccharides were the earlier principal milk carbohydrates in species ancestral to eutheria, with lactose becoming the dominant carbohydrate later in most eutherians, in response to changing selective pressures and preferred biochemical pathways to provide accessible glucose to offspring (Oftedal, 2002a) (Table 1.1).

Table 1.1 Lactose concentrations (lactose values expressed as a percentage of total milk components) for a variety of mammals.

Order (family)	Common name	Species	Lactose* %	Reference
Monotremata (Tachyglossidae)	Short-beaked echidna	*Tachyglossus aculeatus*	2.3[a]	Griffiths, Green, Leckie, Messer, and Newgrain (1984)
Monotremata (Ornithorhynchidae)	Platypus	*Ornithorhynchus anatimus*	3.7[b]	Griffiths et al. (1984)
Diprotodontia (Macropodidae)	Eastern grey kangaroo	*Macropus giganteus*	0.2	Messer and Mossop (1977)
Chiroptera (Vespertilionidae)	Cave bat	*Myotis velifer*	4.4[c]	Kunz, Oftedal, Robson, Kretzmann, and Kirk (1995)
Primate (Lemuridae)	Brown lemur	*Eulemur fulvus*	8.5[c]	Tilden and Oftedal (1997)
Primate (Cercopithecidae)	Rhesus macaque	*Macaca mulatta*	7.9[c]	Lönnerdal, Keen, Glazier, and Anderson (1984)
Primate (Hominidae)	Human	*Homo sapiens*	7.0[d]	Jenness (1974); McClellan, Miller, and Hartmann (2008)
Carnivora (Canidae)	Domestic dog	*Canis familiaris*	3.8	Oftedal (1984)
Carnivora (Ursidae)	Brown bear	*Ursus arctos*	2.2	Jenness, Erickson, and Craighead (1972)
Carnivora (Phocidae)	Hooded seal	*Cystophora cristata*	1.0	Oftedal, Boness, and Bowen (1988)
Carnivora (Phocidae)	Grey seal	*Halichoerus grypus*	–	Iverson, Bowen, Boness, and Oftedal (1993)

(Continued)

Table 1.1 Lactose concentrations (lactose values expressed as a percentage of total milk components) for a variety of mammals. *Continued*

Order (family)	Common name	Species	Lactose* %	Reference
Carnivora (Phocidae)	Harp seal	*Phoca groenlandica*	0.8	Oftedal (1995)
Artiodactyla *(Delphinidae)*	Bottlenose dolphin	*Tursiops truncatus*	1.1	Jenness, 1974
Perissodactyla (Equidae)	Ass	*Equus asinus*	5.9	Oftedal and Jenness (1988)
Perissodactyla (Equidae)	Przewalski horse	*Equus przewalskii*	6.7	Oftedal and Jenness (1988)
Proboscidea *(Elephantidae)*	Asian elephant	*Elephas maximus*	4.7	Jenness, 1974
Artiodactyla (Cervidae)	Reindeer	*Rangifer tarandus*	3.4	Luick, White, Gau, and Jenness (1974)
Artiodactyla (Camelidae)	Bactrian camel	*Camelus bactrianus*	4.5	Konuspayeva, Faye, and Loiseau (2009)
Artiodactyla (Bovidae)	Cow	*Bos taurus*	4.8	Jenness (1974)
Artiodactyla *(Suidae)*	Pig	Sus scrofa	5.5	Jenness, 1974
Artiodactyla (Bovidae)	Dall sheep	*Ovis dalli*	5.3	Cook, Pearson, Simmons, and Baker (1970)
Artiodactyla (Bovidae)	Rocky mountain goat	*Oreamnos americanus*	4.5	Carl and Robbins (1988)
Artiodactyla (Bovidae)	Water buffalo	*Bubalus bubalis*	4.8	Jenness (1974)

**Lactose values that also include other sugars are noted.*
[a]*Total sugars, principally the oligosaccharide sialyllactose with the monosaccharide hexose;*
[b]*Total sugars, mainly the oligosaccharide difucosyllactose and hexose;*
[c]*Total sugars, mostly lactose with a small oligosaccharide component;*
[d]*Lactose concentration in human milk is 70 grams per litre (g/l) with oligosaccharides 12.9 g/l.*

The presence of lactose, albeit at varying concentrations, in all three mammalian infraclasses provides evidence that lactose in milk most likely evolved before their common ancestor, rather than arising independently on at least two separate occasions. It has been hypothesized that milk secretion may have its origins in the reproductive strategies of synapsids (sometimes described as mammal-like reptiles), a branch which diverged from sauropsids ~325 million ya and are the direct ancestors of mammals (Oftedal, 2002a, 2002b) (see Fig. 1.1). During the Triassic period some 250–210 million ya, milk secretion and lactation may have evolved in within cynodonts (therapsids), a class ancestral to mammals (Oftedal, 2002a). About 200 million ya, in the Late Triassic period, milk secretion and lactation may also have been established in mammaliaformes with oligosaccharides as the predominant carbohydrate in these early milks (see later for the biochemical explanation of why oligosaccharides predate lactose as the dominant milk sugar).

Oftedal (2002a) hypothesized that the mammary gland evolved from apocrine-like glands, which are typically found near hair follicles. These glands would have secreted fluids absorbed by the porous parchment shells of egg-laying synapsids. Oftedal (2002b) further argued that these secretions would have prevented dehydration and protected against microbial infection in the eggs, gradually becoming more nutrient dense over time, alongside a decrease in egg size (Oftedal, 2002a, 2002b). Milk oligosaccharides have antimicrobial properties in addition to contributing as an energy source in these ancestral species (Messer & Urashima, 2002).

Biochemical investigations by Vorbach, Capecchi, and Penninger (2006) further support the view that early milk secretions played an antimicrobial and/or antiinflammatory role, with nutrition being a minor or later adaptation. Noting that xanthine oxioreductase and lysozyme are antimicrobial enzymes that are also involved in milk carbohydrate synthesis, Vorbach et al. (2006) argued that milk first evolved in therapsids as part of the innate immune system. The α-lactalbumin molecule, which is found in whey protein, catalyzes lactose synthesis in addition to its role in the immune system. Based on Vorbach et al.'s (2006) observation of 40% identity of the amino acid sequence of lysozyme and α-lactalbumin, they hypothesized that a lysozyme gene duplication event generated the precursor from which the α-lactalbumin gene was derived. Blackburn, Hayssen, and Murphy (1989) noted that the biochemical similarities between lysozyme and α-lactalbumin had been identified in the 1960s but only recently has the evolutionary impact of these findings been highlighted (Blackburn et al., 1989). As α-lactalbumin is found only in mammals, but lysozyme is found more widely in animals, it is likely that α-lactalbumin was derived from lysozyme (Messer & Urashima, 2002).

Lactose synthase is an enzyme complex of the milk protein α-lactalbumin and β4galactosyltransferase. Lactose synthase catalyzes the transgalactosylation of UDP-Gal (donor) and glucose (acceptor) to synthesize the disaccharide (Fig. 1.2). Like lactose (Gal(β 1−4)Glc) itself, α-lactalbumin is only found in the mammary glands and milk, and becomes the rate-limiting factor in lactose biosynthesis, whereas β4galactosyltransferase is found in many other tissues. When no α-lactalbumin is present, β4galactosyltransferase synthesizes *N*-acetyllactosamine (Gal(β 1−4)GlcNAc) by transferring galactose UDP-Gal to nonreducing *N*-acetylglucosamine (GlcNAc) residues in glycoconjugates, where *N*-acetylglucosamine, a monosaccharide glucose derivative, is a component in glycoprotein synthesis. However, in the presence of α-lactalbumin, β4galactosyltransferase changes its preferred acceptor to glucose and thereby synthesizes lactose. This preferred biochemical pathway helps to explain why lactose synthesis only occurs in mammals.

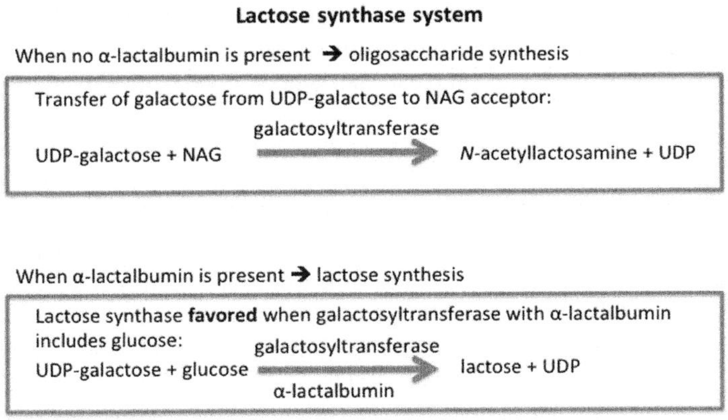

FIGURE 1.2

Lactose synthesis. *UDP*, Uridine diphosphate; *NAG*, *N*-acetylglucosamine.

Adapted from Blackburn, D. G., Hayssen, V., & Murphy, C. J. (1989). The origins of lactation and the evolution of milk: A review with new hypotheses. Mammal Review, 19*(1), 1–26.*

N-acetyllactosamine, as the precursor of milk oligosaccharides, can provide unique insights into the evolutionary history of mammals and the complexities of milk itself. As mentioned earlier, lactose and milk oligosaccharides probably evolved as components of the protolacteal secretions in the early mammary glands of ancestral therapsids. This may have occurred perhaps some 200 million ya as part of the developing innate immune system. These protolacteal fluids would likely have principally contained proteins and lipids, with carbohydrates as a later adaptation. Within the primitive mammary glands of ancestral species, many of the glycosyltransferases seen today in extant mammals were also probably present, as these enzymes are required for basic cell metabolism and maintenance. When α-lactalbumin, derived from lysozyme, initially appeared in the primitive mammalian glands, lactose would have been synthesized at low levels, and in turn the glycosyltransferases would have catalyzed the conversion of lactose into milk oligosaccharides by adding monosaccharides onto lactose. The hypothesized early role of oligosaccharides in milk was to provide offspring with antimicrobial defenses and immunity, and oligosaccharides feature in the milk of almost all mammals, but are in higher concentrations in monotremes, marsupials, and a few exceptions among eutherians (e.g., Canoids).

As the concentration of α-lactalbumin within mammary glands increased over time, this would allow for a higher concentration of oligosaccharides in milk. In monotreme and marsupial milks, these oligosaccharides evolved a dual function of providing both immunological and nutritional benefits. Today, monotreme milk oligosaccharides contain fucose bound to a base of lacto-*N*-neotetraose or lacto-*N*-neohexaose varying from 0.2% to 0.9%, far in excess of glycoprotein requirements, suggesting that fucose provides an energy source for suckling monotremes. While intestinal lactase is absent in neonate monotremes, it is hypothesized that the oligosaccharides are transported by pinocytosis or endocytosis into the lysosomes and supranuclear vacuoles of enterocytes by lysosomal glycosidases, including β-galactosidase, α-fucosidase, and neuraminidase (Urashima, Saito, Nakamura, & Messer, 2001). Similarly, metatherian oligosaccharides vary in concentration from 1% to 13% during the longer periods of lactation, with bound lactose at the reducing ends of those oligosaccharides (mainly galactosyl saccharides and derivatives). Energy is provided to suckling neonates via lysosomal acid β-galactosidase hydrolyzing milk oligosaccharides into monosaccharides which are absorbed in the small intestine (Urashima et al., 2001).

Most eutherian milks, by contrast, are dominated by lactose, with oligosaccharides present in lower concentrations. The faster synthesis of lactose probably led to its increased concentration in early eutherian milks while also providing a more accessible energy source, but oligosaccharides were retained, possibly for their role in protecting neonates from pathogens (Andersson, Porras, Hanson, Lagergard, & Svanborg-Eden, 1986; Cravioto et al., 1991; Meurant, 1995; Newburg, 2000; Urashima et al., 2001), and as possible prebiotics (i.e., substances that encourage the growth of beneficial microorganisms) (see Chapter 4: Lactose—a conditional prebiotic? for a discussion of milk as a prebiotic). Whereas in monotremes and marsupials, the dominant milk oligosaccharides served a dual role, in eutherians, these milk carbohydrates diverged and specialized in their roles, with oligosaccharides providing microbial protection and possibly contributing nutrients required for postnatal brain growth and development (Carlson, 1985; Kunz, Rudloff, Baier, Klein, & Strobel, 2000), lactose principally supplying energy in the form of glucose, and galactose possibly being involved in postnatal cognitive development (Coelho, Berry, & Rubio-Gozalbo, 2015; Newburg & Neubauer, 1995). Unlike monotremes and marsupials, eutherians have not evolved a system of pinocytosis or endocytosis where oligosaccharides can be transported into the lysosomes and supranuclear vacuoles of enterocytes for hydrolysis.

One further hypothesized selective advantage of lactose over milk oligosaccharides may be the dual roles of its component monosaccharides: glucose and galactose. Lactose is cleaved by the enzyme lactase-phlorizin hydrolase, a β-galactosidase, into glucose and galactose in the brush border of the small intestine. Both glucose and galactose are then actively transported across the epithelial cell membrane by the sodium-glucose-linked cotransport SGLT1 and the proximal nephron tubule by SGLT2. The uni-transporters GLUT1 and GLUT2 then transport glucose across plasma membranes in the intestine and kidneys. Renal glucose is reabsorbed into the blood and carried to the liver. In the liver, excess glucose is converted into glycogen, and galactose can be converted into glucose via the Leloir pathway. However, in addition to galacto-oliogsaccharides, some galactose also passes into the colon where it may act as a prebiotic, contributing to the growth and maintenance of a healthy gut microbiome, including bacteria such as bifidobacteria and lactobacilli, which in turn confer a benefit to the host (Cederlund et al., 2013; Coelho et al., 2015; Gibson et al., 2017; Macfarlane &

Cummings, 1999) (see Chapter 4: Lactose—a conditional prebiotic? for further discussion of prebiotics).

While the extent to which galactose maintains healthy gut microbiota requires further investigations, it is possible that evolutionary adaptation has favored lactose twofold: Lactose is preferred over oligosaccharides in eutherians for providing energy to neonates while oligosaccharides are maintained for their antimicrobial, prebiotic and neurodevelopmental roles, and from lactose digestion itself, in which glucose supplies the dominant energy source while galactose may both provide energy (via the Leloir pathway) and be involved in maintaining a favorable gut flora (Coelho et al., 2015). This may provide some explanation for the observed differences in lactose and oligosaccharide concentrations between eutherians, marsupials, and monotremes and possibly within these infraclasses. Marsupials and monotremes may be transporting oligosaccharides across membranes through pinocytosis and/or endocytosis to initiate hydrolysis, hence the higher concentrations of oligosaccharides in these infraclasses. It also may explain the maintenance of oligosaccharides in eutherian milks, despite the majority of energy being derived from lactose and the inability of eutherians to hydrolyze these complex carbohydrates, as these oligosaccharides play a role in supporting the developing microbiome through fermentation in the lower gut.

1.3 Dietary shifts in human prehistory

When considering lactose from an evolutionary perspective, most studies are concerned not with its deep prehistory, reaching back some 300 million years, during which time lactation and lactose production evolved in mammals and mammalian ancestors, but rather focus on recent developments in human dietary prehistory, when milk became an integral part of the subsistence strategy for some human populations. The regular consumption of milk and/or dairy products in adulthood began some 9–10,000 ya, after the domestication of milkable animals in a period known as the Neolithic, when humans began a sedentary way of life, and started producing food rather than hunting and gathering it. The addition of milk and dairy products to the human diet had major consequences for culture, genetics, health, reproduction, and survival rates.

Before considering this Neolithic shift in human diets, which included the introduction of dairying and exogenous milk consumption, it is helpful to see these changes within the wider context of the major dietary shifts in human prehistory (Fig. 1.3).

1.3.1 Lower Paleolithic

The Lower Paleolithic spans from ~3.3 million to 300,000 ya, and is characterized by successive stone tool industries among our hominin ancestors and their

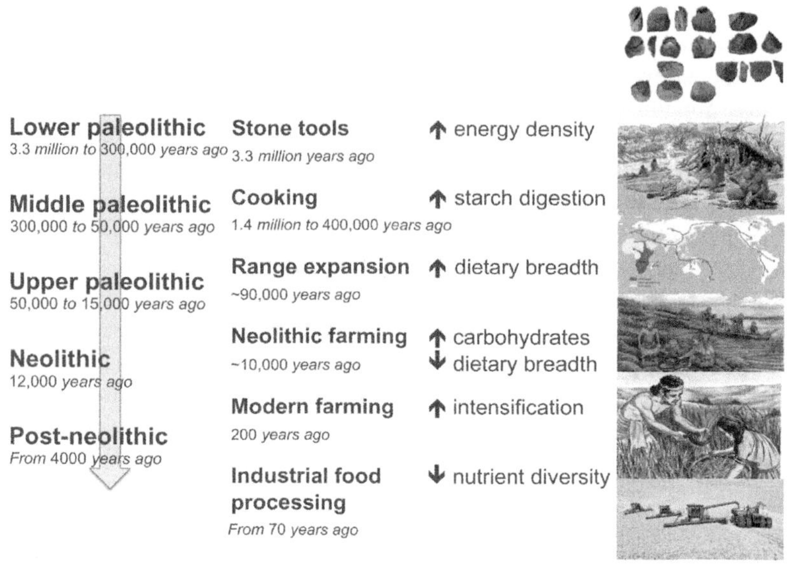

FIGURE 1.3

Dietary shifts in human evolution and the likely effects on nutrition.

close relatives: pre-Mode 1, or Lomekwian (dating to ~3.3 million ya), Mode 1, or Oldowan (starting ~2.6 million ya), and Mode 2, or Acheulean (starting ~1.7 million ya). Prior to 2 million ya, the cranial capacity of our hominin ancestors was ~450 cm^3 (Falk et al., 2000), and only started increasing systematically in the *Homo* lineage after this; modern human brains range from 1300 to 1500 cm^3. Thus the development of stone tools seems to have preceded brain size increase in ancestral hominins. It has been argued that the use of stone tools, particularly from Mode 1 tools onward, led to critical changes in diet that permitted the evolution of increased brain size (Aiello & Wheeler, 1995). In particular, a shift from a diet rich in hard herbivorous foods to one rich in meat (probably scavenged rather than directly hunted in the early stages of stone tool use) would have provided the calories necessary to maintain an energetically expensive organ like a large brain. The addition of meat would also have provided a rich source of iron, protein, zinc, vitamins, and possibly other key nutrients such as long chain fatty acids, supplementing foraged diets of bulbs, corms and large seeds, with possible implications for reproductive capacity.

Another major innovation during the Lower Paleolithic was the controlled use of fire and cooking. Cooking leads to a wide range of chemical and physical changes in food, but three factors have been argued to be particularly advantageous to our ancestors: (1) a reduction in the pathogen load of the ingested foods, particularly scavenged meats, (2) the tenderizing of meats leading to a decrease in required chewing time (Carmody & Wrangham, 2009; Wollstonecroft, 2011;

Wollstonecroft, Ellis, Hillman, Fuller, & Butterworth, 2012; Wrangham & Carmody, 2010; Wrangham & Conklin-Brittain, 2003), and (3) an increase in the digestibility of plant starches by α-amylases (Hardy, Brand-Miller, Brown, Thomas, & Copeland, 2015). However, dates for the widespread use of cooking range from 1.8 million ya, from evidence of charcoal in archeological sites (Boback et al., 2007; Carmody & Wrangham, 2009; Wrangham & Carmody, 2010; Wrangham & Conklin-Brittain, 2003; Wrangham et al., 1999) to 400,000 ya, with the earliest direct evidence of a hearth, excavated at Qesem Cave in Israel (Shahack-Gross et al., 2014).

Starches are polysaccharides made up of the linear glucose polymer amylose and the branched glucose polymer amylopectin, in various proportions. In their raw state most starches are highly crystalline and only poorly accessible to α-amylases. However, cooking in the presence of water gelatinizes the helical amylose, leading to a disordered mix of amylose and amylopectin, which increases accessibility by α-amylases, resulting in increased starch hydrolysis and a dramatic increase in glucose yields from digestion (Butterworth, Warren, & Ellis, 2011; Cooke & Gidley, 1992; Xie et al., 2014). Interestingly, humans, unlike other hominids—including Neanderthals, chimpanzees, and gorillas—have multiple copies and are copy number-variable for the salivary and, to a lesser extent, pancreatic amylase genes (Carpenter et al., 2015). Furthermore, salivary amylase gene copy number positively correlates with salivary amylase production in the mouth, and oral starch digestion rates (Mandel, des Gachons, Plank, Alarcon, & Breslin, 2010). This led Hardy et al. (2015) to propose that cooking, an increase in the consumption of starch-rich plants, and the evolution of amylase gene copy number variation, led to a gene–culture coevolutionary scenario that provided the additional free glucose necessary to permit further increase in human brain size over the last 800,000 years. However, while an evolutionary relationship between starch-rich food consumption, cooking, and salivary amylase gene copy number increase, and increased in human brain size, remains highly plausible, a recent study estimated a more recent date for the origins of salivary amylase gene copy number variation; ~450,000 ya (Inchley et al., 2016). This more recent date is certainly more consistent with secure archeological evidence of cooking and also corresponds well with recent estimates of the timing of accelerated brain size increase in the human lineage (Dunbar, 2017).

1.3.2 Middle Paleolithic

The Middle Paleolithic, a period from ~300,000 to 50,000 ya, is noted for the emergence of anatomically modern human (AMH) ~300,000 ya (Bradshaw, 1997; Hublin et al., 2017) as well as for the emergence of Neanderthal material culture in western Eurasia. With continued range expansions out of Africa during this period, particularly following the interglacial period ~100,000 ya (Watson, 2009), new foods would have been incorporated into diets as small bands of hunter-gatherers exploited local food resources in new ecologies.

1.3.3 Upper Paleolithic and Mesolithic

The Upper Paleolithic, often termed the Late Stone Age in an African context, from 50,000 to 15,000 ya, was characterized by further range expansion together with new hunting and food processing technologies. This period is associated with increasing dietary breadth, as novel food sources were encountered in newly colonized regions, with the use of new materials, including bone, antler, shell, ivory, the development of new hunting technologies including microliths, fine bone needles, harpoons, spears, and projectiles, and the appearance of food pounding and other processing tools (Ambrose, 2001; Powell, Shennan, & Thomas, 2009). A shift to smaller and lower ranking prey is also noted in the archeological record (Stiner, Munro, Surovell, Tchernov, & Bar-Yosef, 1999). Fishing nets and fine spears are also associated with this period, and in addition to stable isotope data from human bones, they provide evidence of increased aquatic foods in the human diet (Richards & Hedges, 1999).

1.3.4 Neolithic

The Neolithic transition first occurs ~12,000 ya in southwestern Asia and is associated with a rapid change in diet and lifestyle as humans transitioned from living in small and mobile hunter gatherer bands to living in sedentary communities. During this transition, human ecology fundamentally changed: from being food-seeking to food-producing (Bocquet-Appel & Bar-Yosef, 2008; Fuller, 2007; Zeder, 2008). The Neolithic is associated with an increased reliance on cereals and other grains, with a concomitant reduction in diet breadth, but an increase in and more regular supply of carbohydrate. This has been proposed to have led to increased fertility whereby, with the introduction of farming, the total fertility increased by two births per female, attributable to a reduction of birth interval with more energy in the diet from high-calorie foods—quickening the return of the maternal postpartum energy balance. In addition, sedentary lifestyles may have added an energy gain from mothers not carrying infants while foraging, which was required in pre-agrarian hunter-gatherer societies (Bocquet-Appel, 2011). However, the Neolithic transition is also associated with a decrease in long-term health (Frayer, 1980; Macintosh, Pinhasi, & Stock, 2016; Stiner, 2001). It has been estimated that foraged foods, fruit and vegetables made up only 20% of the total calorific intake during the Neolithic, with grains and cereals providing the majority of calories (Eaton, Eaton, & Cordain, 2002). As a result, the presumed decreased intake of vitamins, minerals, and micronutrients has been proposed to have led to a decline in health, as inferred from as stature and bone deformities, and increased susceptibility to pathogens (Armelagos & Cohen, 1984; Molleson, 1994; Mummert, Esche, Robinson, & Armelagos, 2011). However, this increased reliance on cereals and other grains in the Neolithic needs to be understood in a wider context. Starch-rich roots, tubers, and underground storage organs may have contributed to the human diet to a greater extent over the last 400,000 years than previously thought (Dunbar, 2017; Hardy et al., 2015; Inchley et al., 2016). Thus the reduced health status of early Neolithic farmers may be a result not of an increase in dietary carbohydrate but because the particular carbohydrate sources were less diverse and micronutrient-balanced, or other factors.

1.4 Domestication of plants and animals leading to dairying

Neolithization is a complex process of cultural, economic, and technological changes which encompassed fundamental shifts in subsistence. These changes included an increasingly sedentary way of life, adherence to some form of property rights (Gallagher, Shennan, & Thomas, 2015), the domestication of plants and animals, the introduction of pottery, specialization of labor, inequality, growth of proto-urban societies, and population growth (Bocquet-Appel, 2011; Diamond, 2002; Leonardi, Gerbault, Thomas, & Burger, 2012; MacHugh, Larson, & Orlando, 2017; Richards, 2002). Domestication is the process of adaptive breeding of plants or animals for human use. The process of plant and animal domestication began ~12,000 ya in southwestern Asia and also arose later, but presumably independently in parts of China and Mesoamerica (Asouti & Fuller, 2013; Diamond, 2002; Fuller, Willcox, & Allaby, 2011; Zeder, 2015). Animal domestication led to the introduction of lactose into the diets of some adult human populations. Milk and dairy products, principally from domesticated goats, sheep, and cattle, may have been important as supplements to the energy-dense but micronutrient-limited grain-based diets during the Neolithic (Leonardi et al., 2012; O'Brien & Bentley, 2015).

1.4.1 Plant domestication

Early cultivation, leading to the domestication of ancient wheats, barleys, lentils, and oats, has been inferred for at least 10 sites from the southern and northern Levant as early as ~12,000 ya (Bar-Yosef & Meadow, 1995; Colledge, 2001; Edwards, Meadows, Sayej, & Westaway, 2004; Fuller et al., 2011; Weiss, Kislev, & Hartmann, 2006). Further work in this region suggests that as many as 24 centers may have contributed to this multiregional process of harnessing and selecting plants for human consumption (Purugganan & Fuller, 2009). This points to multiple origins of plant cultivation and domestication in southwestern Asia, and is consistent with the high genetic differentiation found among humans in the region from early Neolithic ancient DNA data (Broushaki et al., 2016; Hofmanová et al., 2016; Lazaridis et al., 2016). A similar pattern is observed in China and Southeast Asia, with a diffuse process of cultivation of rice (*japonica*) from ~ 8000 ya, and also Chinese millet in northern sites and eastern inner

Mongolia at about the same time or earlier than rice cultivation (Bettinger, Barton, & Morgan, 2010; Fuller et al., 2010). Other regions of plant domestication include Ethiopia, West Africa, Sahel (the area between the Sahara and the Sudan Savannah), the eastern United States, Mesoamerica, the Andes/Amazonia, and Papua New Guinea (Diamond, 2002). As early farmers selected wild plants for desirable traits (larger, nonshattering seeds to maximize harvesting yields, timing of ripening), the investment of time and labor required to manage those plants increased. With management it became possible to rely on cultivated and ultimately domesticated cereals and crops.

1.4.2 Animal domestication

Evidence of early plant domestication in southwestern Asia extends back to ~12,000 ya; animal domestication occurred at a similar time or shortly after (Bar-Yosef & Meadow, 1995; MacHugh et al., 2017), although some wild herd management may have occurred earlier, particularly of goats. The only known exception to this is domesticated dog (*Canis lupus familiaris*), which was probably domesticated from gray wolves (*Canis lupus*) before the earliest crops (Larson et al., 2012). With the domestication of several key animals, principally goats, sheep, pigs, and cattle, human subsistence changed fundamentally (Makarewicz & Tuross, 2012). It has been argued that early domestic animals were initially exploited for meat, hide, and bone. These "primary products" were resources that could be obtained only after slaughter. However, domesticated animals could also be exploited for their "secondary products" while the animal was living, such as milk, wool, and traction. Andrew Sherratt's work on the "Secondary Products Revolution" posited that utilization of these resources occurred later in the Neolithic, and that animals were originally domesticated for their primary products (Sherratt, 1981). However, there is a growing body of evidence to support a revision of Sherratt's theory. While domesticated animals in the early Neolithic were exploited for their primary products, the widespread use of their secondary products—particularly milk—followed immediately or very soon after (Evershed et al., 2008; Makarewicz & Tuross, 2012; Vigne & Helmer, 2007).

The domesticates providing milk in the early Neolithic are goats, sheep, and cattle. Goat (*Capra hircus*) domestication began ~11,000 ya across a large region between the Taurus and Zagros Mountains (Makarewicz & Tuross, 2012; Zeder & Hesse, 2000; Zeder, 2008). Sheep (*Ovis aries*) were domesticated at about the same time as goats, overlapping in some of the Taurus Mountain range with goats (Zeder, 2008). The earliest evidence of cattle (*Bos taurus*) domestication has been found on Cyprus from ~10,500 years before present (Vigne, Carrere, Briois, & Guilaine, 2011) and in southwestern Asia within a few hundred years of this date (Bollongino et al., 2012; MacHugh et al., 2017). Pig (*Sus scrofa*) domestication began in a similar region ~10,500 ya (Helmer, Gourichon, Monchot, Peters, & Segui, 2005). Within a thousand years, these early domesticates had spread out of their original domestication centers into nearby regions of

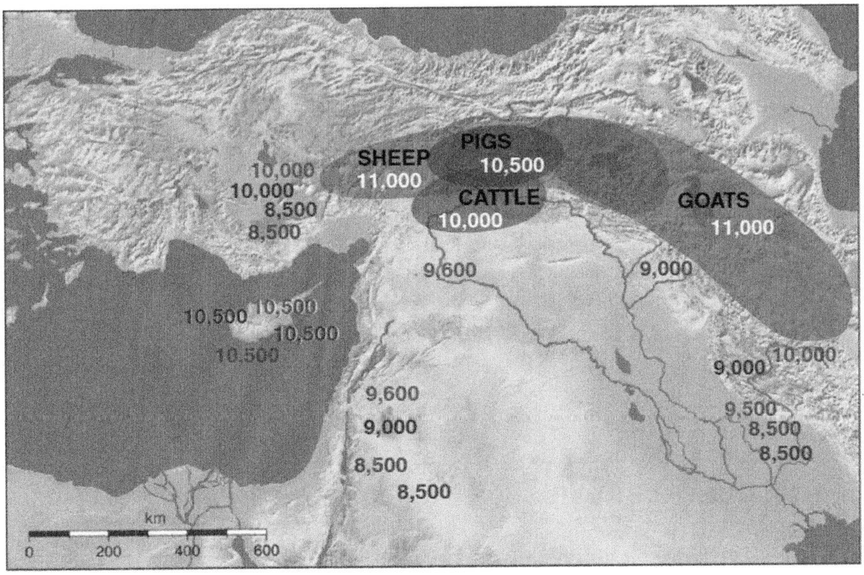

FIGURE 1.4

Approximate regions and dates of domestic livestock dispersal in calibrated years before present.

© 2008 National Academy of Sciences. Reproduced with permission from the Zeder, M. A. (2008). Domestication and early agriculture in the Mediterranean Basin: Origins, diffusion, and impact. Proceedings of the National Academy of Sciences, 105*(33), 11598.*

the southwestern Asia. By 8400 ya, these domesticated animals had reached the Aegean and Balkan regions, transported by early farmers expanding into new territories (Hofmanová et al., 2016; Leonardi et al., 2012) (Fig. 1.4).

Over the next few thousand years, domestic livestock spread into central and northern Europe alongside farming communities, reaching Britain by ~6000 ya (Evershed et al., 2008; Richards, Schulting, & Hedges, 2003). Other land routes radiating out of the Fertile Crescent carried domesticated goats, sheep cattle, and pigs through Iran into India and northwest into Russia. Within 4000 years, these milkable livestock were thriving throughout many parts of Eurasia.

Other domesticated animals such as zebu cattle (*Bos indicus*), water buffalo, and horses also provided milk. Genetic and archeological data indicate an independent domestication of zebu cattle in the Indus Valley of northwest India ~8000 ya. Zebu cattle from India also spread into East Africa, and today African cattle are a mixture of both *B. taurus* and *B. indicus*, to varying degrees. The domesticated water buffalo is classed into two major categories, based broadly on phenotype and geographic distribution. The river water buffalo (*Bubalus arnee bubalis*) was domesticated ~5000 ya (Kumar et al., 2007) in India and the swamp water buffalo (*Bubalus arnee carabanesis*) a thousand years later in China

(Yang, Liu, Chen, & Speller, 2008), likely in an independent domestication event. Although some interbreeding between the river and swamp water buffalo has caused researchers to question the timing, mitochondrial genetic analyses currently support that these are two related subspecies (Kumar et al., 2007). Horses (*Equus ferus caballus*) may also have had several independent domestications, one of which occurred in the Eurasian Steppe of northern Kazakhstan ~5500 ya (Gaunitz et al., 2018; Outram et al., 2009). Other domesticates include llamas (*Lama glama*) and alpacas (*Vicugna pacos*). These camelids were domesticated ~6000 ya in the Peruvian Andes (Kadwell et al., 2001). However, there is little or no evidence of the native pre-Columbian people exploiting camelids for milk until the Spaniards introduced cows, sheep, goats and dairying practices to the indigenous populations in the 16th century (Gade, 1999).

1.5 Archeological evidence of dairying and milk consumption

The earliest evidence of dairying comes from Anatolia and the Levant and dates to ~9000 ya in the form of milk lipids identified in pottery fragments (Evershed et al., 2008; Salque et al., 2013). At the time it is highly unlikely that LP was present at appreciable frequencies, if at all, so the consumption of fresh milk by adults may have been unfavorable. However, milk fermentation to produce yoghurt, butter fat, and, through further processing, cheese leads to products with dramatically reduced lactose content, but which retain many of the nutrients of milk. In addition, fermentation, and in particular cheese making, would have increased the longevity and portability of milk products (Burger, Kirchner, Bramanti, Haak, & Thomas, 2007; Itan et al., 2009) as well as adding key nutrients to a diet that had decreased in breadth, as measured by the number of foods in the diet and the diversity of micronutrients in those food sources (Eaton et al., 2002; Stiner, 2001).

1.5.1 Milk lipids recovered from fragments of pottery

Within 1500 years of the first goat, sheep, and cattle domestications in the Fertile Crescent, there is direct evidence of dairy processing in pottery fragments from Anatolia, the Levant, and southeastern Europe (Evershed et al., 2008; Salque et al., 2013). As pottery surfaces are irregular and to some extent porous, organic residues such as fats and proteins can become embedded in the uneven interior surfaces. Using gas chromatography to separate out different fatty acids (mainly $C_{16:0}$ and $C_{18:0}$), and coupled mass spectrometry to examine stable carbon isotope ratios (principally ^{13}C and ^{12}C), it is possible to determine whether the lipid residues are ruminant adipose, nonruminant adipose, or ruminant dairy in origin. These differences are mainly due to different fatty acid compositions of various fat sources and differential fractionation of stable carbon isotopes in the metabolic processes leading to fatty acid synthesis. Over 2200 potsherds were examined to show that milk fats were the dominant residues (Evershed et al., 2008). These analyses strongly indicate that early Neolithic people in Anatolia, the Levant, and southeastern Europe were deriving nutritional benefit from livestock through dairying without having to slaughter the animals (Bogucki, 1984; Salque et al., 2013).

Early evidence of dairying was not confined to Anatolia, the Levant, and southeastern Europe. Farming and milkable animals reached northern Europe by ~7500 ya, and similar methods to those described earlier have provided evidence of milk processing and cheese making in these populations. Interestingly, early Neolithic ceramics from two sites in the Kuyavia region of Poland that were pierced with small holes—which have been interpreted as cheese strainers or sieves used to separate the semisolid milk curds from the liquid whey—also yielded clear evidence of milk fats (Salque et al., 2013). Carcass fats were not identified on the majority of these sieves, indicating specialization of ceramics for specific dairy processing (Fig. 1.5).

Further archeological investigations of early European Neolithic sites in present-day Romania and Hungary dating from 7900 to 7500 ya provide evidence of the spread of agro-pastoralism northward into Europe, with dairying practices confirmed

FIGURE 1.5

An example of a ceramic pot with perforations, possibly used in milk processing.

From author's own collection, Lluc Sanctuary Museum, Mallorca, 2017.

by milk lipids in potsherds (Craig et al., 2005). Ceramic fragments from two sites in the Danube riverine basin contained ruminant dairy fats. Investigations of the faunal remains from these sites revealed that domesticated sheep and goats dominated, with domestic cattle also present. Also of note was the presence of wild ungulates, suggesting that these early agro-pastoral cultures exploited domesticated animals for secondary products such as milk, wool, and traction but also relied to some extent on wild animals for supplementary meat consumption (Craig et al., 2005).

From potsherd lipid analyses, it is possible to infer the timing of dairying and estimate when dairy products featured in the human diet. Archeological investigations of beakers, rather than fragments of larger vessels, have provided evidence that these cups contained fresh milk (Heron et al., 2015). The Rzucewo Culture, which inhabited the site of Nida in Lithuania from ~5200 to 4400 ya, produced Corded Ware beakers. Dairy residues have been found in some of these beakers, implying that fresh, rather than processed, milk was being consumed by this culture more than 5000 ya.

Analyses of 437 pottery fragments from six Neolithic sites across southern England identified ruminant dairy fats in 26% of the sherds, as well as carcass fats, but also that size of the pot correlated with its use (Copley et al., 2003; Copley, Berstan, Dudd, et al., 2005; Copley, Berstan, Mukherjee, et al., 2005). Smaller pots, with a mean diameter of 172 mm, were associated with dairy processing whereas the larger pots, with a mean diameter of 207 mm, were more likely to contain carcass fats. Experimental investigations of the different degradation profiles of milk fats versus butter fats were conducted by burying milk and butter infused pottery sherds. Over time, the triacylglycerol (TAG) profiles in the milk sherds were degraded by soil microorganisms and leaching, leaving only trace fatty acids, possibly accelerated by the presence of proteins and carbohydrates in addition to fats. By comparison, the butter impregnated sherds, with an absence of protein and carbohydrates to encourage the proliferation of soil microbes, retained a higher proportion of TAGs. These findings helped researchers to conclude that the smaller pots, many with TAG residues, were most likely used in producing butter, not storing milk (Copley, Berstan, Dudd, et al., 2005) and provide evidence of milk processing in the British Isles very soon after the arrival of farming some 6100 ya.

In Ireland, further archeological evidence of milk processing and butter making has been found in peat bogs, in the form of buried caches of butter. Known as "bog butter," stores of this ancient milk product have been found in various locations, probably preserved as a consequence of the cool anaerobic conditions. It is unclear if the butter was buried for preservation, to improve the taste, or as part of a religious offering. The oldest "bog butter" sample from Tullamore, County Offaly, has recently been radiocarbon dated to 3000 ya. Gas chromatography combustion isotope ratio mass spectrometry investigations (GC-C-IRMS) of similarly aged Scottish samples confirmed the presence of dairy fats (Berstan et al., 2004). Investigations of Irish Iron Age samples ranging from 2400 to 1950 ya buried in wooden casks revealed a content (94%–100%) of predominantly milk fats, with little adipose fats detected (Cronin et al., 2007). Thus most of the "bog butter" finds are indeed dairy in origin, not stores of tallow and/or carcass fats.

Evidence of dairying and dairy processing in the archeological record is not confined to Europe. Excavations of mummies at Xiaohe, a Bronze Age cemetery in Xinjiang, China, provide the first direct evidence of kefir cheese making in eastern Eurasia, and by extension the presence of cattle and caprine-keeping culture (Yang et al., 2014). Taklamakan mummies dating to ~3800 ya were buried with necklaces decorated with triangular organic foods. Proteomic analyses indicate that these were cheeses made predominantly from bovine milk, with a few containing mixed milk proteins from cattle, sheep, and goats. Furthermore, evidence of the presence of *Lactobacillus* sp. and various *Saccharomycetaceae* sp., but not of rennet, suggests the production of virtually lactose-free kefir cheese that did not require the slaughter of animals to obtain rennet for processing. Kefir cheese production uses a yeast and bacterial culture to curdle the ruminant milk into solids, whereas rennet cheeses use stomach enzymes obtained from slaughtered animals to curdle the casein in milk. In addition, consuming the liquid whey by-product would have provided a protein-rich drink with a potentially prebiotic effect maintaining a healthy and adaptive gut microbiome for these Taklamakan semi-pastoralists (Yang et al., 2014). Lactoferrin and β-lactoglobulin in the whey protein have antimicrobial properties and possibly help regulate immune system function while α-lactalbumin, in binding to opioid receptors, may contribute to cardiovascular health (Beermann & Hartung, 2013). Producing kefir cheese provided a portable and preserved food source with the added benefit of the liquid whey by-product containing additional nutrients and energy (see Chapter 4: Lactose—a conditional prebiotic? for a discussion of dairy foods as prebiotics).

In contrast to Neolithic Europe and Eurasia, where sedentism and plant and animal domesticates preceded widespread adoption of dairying, pastoralism emerged in Africa before domestic plants were prevalent (Dunne et al., 2012; Marshall & Hildebrand, 2002; Timpson et al., 2018; Gerbault et al., 2016). While currently much of Saharan Africa is inhospitable for grazing and herding, during the African Humid Period starting 10,000 ya, these lands would have been temperate, fertile, and supportive of African pastoralists managing herds of domesticated animals (Ortiz et al., 2000). Archeological evidence for cattle, sheep, and goats dating to 8000 ya in the Saharan savannas suggests that herds were being managed in this environment, and rock art depictions of cattle with full udders, and numerous drawings of cattle together with humans from Saharan sites such as Wadi Teshuinat II record scenes of African pastoralism. While rock art is difficult to date accurately, potsherds from this same rock shelter have been examined for evidence of dairy lipids. These analyses found that more than 50% of the potsherds from the Middle Pastoral Period (from ~7200 ya) contained dairy lipids (Dunne et al., 2012). Of particular interest is the absence of dairy lipids in the early Pastoral Period (from ~8300 ya). This may indicate a rapid intensification of milking and "secondary products" use as the pastoral economy matured within one millennium. These potsherd analyses confirm that dairying and milk processing were widely practiced in Saharan Africa from ~7000 ya, and that dairy products played a substantial role in the diet of these northern African pastoralists.

1.6 Herd mortality profiles

In addition to radiochemical identification of milk lipids in potsherds, it is also possible to infer the targets of herd management strategies (e.g., meat or dairy production) from the faunal remains from larger encampments and settlements (Helmer, Gourichon, & Vila, 2007; Vigne & Helmer, 2007). Using teeth eruption and wear patterns it is possible to estimate the age at death of each animal, and by constructing graphs based on age-at-death profiles, patterns emerge that are consistent with herd management principally to obtain meat, or dairy, or mixed productivity. In herds managed to provide meat, typically calves are slaughtered at an older age when they are fully or near-fully grown, to optimize meat yield. In a dairying herd, different culling strategies are practiced whereby a high proportion of calves are culled soon after birth—particularly male calves—to increase the milk yield available for human consumption. Vigne and Helmer (2007) examined 36 early Neolithic sites ranging from the Levant to the northwest Mediterranean. They reported that 30 of the 36 kill-off profiles were consistent with a mixed strategy for meat and milk exploitation for sheep and goats. Shillourokambos in southern Cyprus provides the earliest evidence of nascent milk exploitation: from ~10,500 to 10,000 ya, within a mixed meat/milk profile. Similarly, kill-off profiles from Çafer Hoyuk in Turkey dating to ~10,000 ya indicate early milk use, with Tell Halula in Syria exploiting sheep predominantly for milk. Milk from cattle also contributed to the diet in the early Neolithic, but because early domesticated cattle only lactated in the presence of a calf, kill-off profiles providing clear signals of milk herd management are more difficult to discern. However, Vigne and Helmer concluded that six of ten cattle kill-off profiles from early Neolithic sites provided evidence of mixed herd management. Examining cattle kill-off profiles from Tell Aswad in Syria, lower proportions of calves appear to have been culled soon after birth in the early Neolithic, but these proportions increased over time, suggesting greater reliance on milk. Overall, herd mortality profiles, in conjunction with other archeological evidence, provide insights not only into the timing of early dairying but also into regional differences of milk exploitation from goats, sheep, and cattle.

1.7 Direct archeological evidence of milk consumption—dental calculus

Dental calculus is a mineralized form of tooth plaque that has the potential to preserve organic molecules—including proteins and DNA—in archeological remains.

β-Lactoglobulin is abundant in ruminant milk, but is not biosynthesized by humans in the oral cavity. Using tandem mass spectrometry on trypsin digested proteins, the presence of the whey protein β-lactoglobulin was identified in a range of western Eurasian dental calculus samples dating from the Bronze Age (~5000 ya) onward (Warinner et al., 2014). Thus while lipid analysis of potsherds confirmed that milk was being processed over a wide geographic region, the presence of β-lactoglobulin in dental calculus provides direct evidence that milk and dairy products were being consumed. Species-specific β-lactoglobulin identification is also possible in some circumstances, due to amino acid residue variation between sheep, goats, cattle, and other milkable animals. One sample from Hungary was shown to contain milk proteins from cattle, sheep, and goats whereas a sample from Russia contained only cattle proteins. However, while β-lactoglobulin peptides were not detected in some dental calculus samples, it should be noted that some dairy processing, including making cheeses and butter, removes the majority of β-lactoglobulin, whereas it remains high in soured milks, kefir, ricotta, and whey (Warinner et al., 2014).

1.8 Summary of archeological data

Archeological data provides strong and consistent evidence that in western Eurasia, dairying began soon after animal domestication, and milk featured in the Neolithic diet from at least 9000 ya. Potsherd lipid analysis indicates that milk was being processed into butter, cheese, yoghurts and soured milks, as well as being consumed fresh, as seen in the Lithuanian Nida beakers. What remains unclear is why so many distinct cultures, spanning a range of geographic and climatic regions, began consuming milk products, and possibly fresh milk. Although they possessed the technology to ferment milk, not only preserving excess dairy product but also rendering it transportable, storable, and more digestible by removing the majority of lactose, there is evidence that some of these cultures may have consumed fresh milk. Mapping the geographic and temporal adoptions of dairying reveals a highly structured pattern of dairy consumption over a diverse landscape. What also remains unclear, given humans had already adapted culturally to consuming dairy by fermenting and/or processing the milk which reduced lactose to negligible levels, is why did humans evolve LP as an adaptation to drinking fresh milk?

1.9 Lactase persistence

In infants, dietary lactose is digested in the brush border of the small intestine by the membrane-bound enzyme lactase-phlorizin hydrolase; the full name of the enzyme is often shortened to lactase. The two primary activities of lactase are the hydrolysis of lactose via β-galactosidase activity (Arribas et al., 2000; Swallow, 2003) and the hydrolysis of plant glucosides via β-glucosidase activity (Day et al., 2000; Jäger & Saaby, 2011). The enzyme lactase-phlorizin hydrolase is encoded by the lactase gene (LCT) located in humans on chromosome 2, position 21.3 (Harvey et al., 1993) (see Chapter 2: Digestion, absorption, metabolism, and physiological effects of lactose and Chapter 3: Lactose intolerance and other related food sensitivities for further discussions of lactose digestion and tolerance).

In most mammals, LCT expression is downregulated sometime after weaning, leading to low levels of lactase in the small intestine. In the absence of lactase, lactose will enter the colon where bacteria ferment it to produce short chain fatty acids (acetate, butyrate, and propionate) as well as gasses (hydrogen, carbon dioxide, and methane), potentially leading to flatulence and cramping. In addition, colonic lactose can have an osmotic effect resulting in reduced fluid absorption and diarrhea (Ingram, Mulcare, et al., 2009). These phenomena are described in more detail in Chapter 2: Digestion, absorption, metabolism, and physiological effects of lactose and Chapter 3: Lactose intolerance and other related food sensitivities.

While LCT downregulation after weaning is seen in nearly all mammals and most humans, it is not the case for all humans. The phenotype of downregulated LCT expression is known as lactase nonpersistence (LNP) which is the ancestral state and the norm in most humans. However, some humans continue to express lactase at higher levels after weaning and into adulthood (Ingram, Mulcare, et al., 2009; Itan, Jones, Ingram, Swallow, & Thomas, 2010). LP is the best-known, and in many populations the strongest example of recent natural selection, convergent evolution and gene–culture coevolution in humans. Within the past ~10,000 years, not only has our culture adapted to the availability of milk from domestic animals, but also our biology has adapted to these changes.

Currently ~35% of the world's adult population is lactase persistent (Ingram, Mulcare, et al., 2009; Itan et al., 2010; Liebert et al., 2017). However, its distribution is highly structured, both globally and, in some cases, locally (Fig. 1.6). Around 90%–97% of northern Europeans are lactase persistent, falling to between 44% and 84% in central Europe and 11%–64% in southern and southeastern Europe (Ingram, Raga, et al., 2009; Itan et al., 2010). Relatively high frequencies of LP are observed in some northwestern Indian subcontinent populations while lower frequencies are found in southern and eastern India (Gallego Romero et al., 2011; Swallow & Hollox, 2000). LP is relatively rare in East and Southeast Asian populations, as well as in indigenous Australasian, Oceanian, and American populations. However, as the current American population is made up largely of immigrants over the past 500 years, the distribution of LP is determined by the ancestry of the incoming population. Approximately 75% of Americans of northern European ancestry are LP whereas those of Mexican descent are 27% LP, Afrocaribbean ancestry under 20% LP, and Native Americans ~1% LP (Swallow & Hollox, 2000).

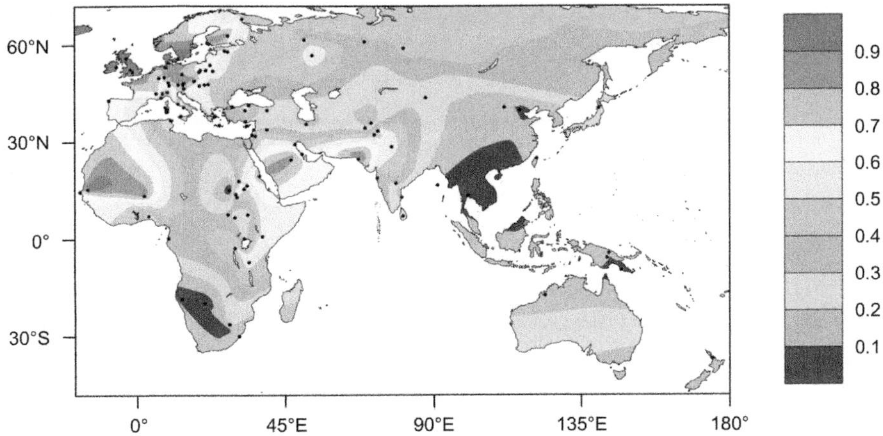

FIGURE 1.6

Interpolated map of the distribution of lactase persistence phenotype showing the distribution of lactose digesters. Areas in red and orange (northern Europe, western Africa, Saudi Arabia and south-western Iran) show the highest concentration of LP, correlating with dairying practice, whereas areas in green and blue (all other areas) show regions where the majority of people do not digest lactose in adulthood (Itan et al., 2010)

© Itan et al.; Licensee BioMed Central Ltd. 2010. Updated maps available from https://www.ucl.ac.uk/mace-lab/resources/glad/LP_maps.

In Africa and the Middle East the distribution of LP is somewhat more complicated. While broad-level geographic patterns do exist with LP being found throughout eastern and northern Africa, and Arabia, it is often the case that large differences are observed between neighboring populations. In these populations, ethnicity and subsistence strategy better explain the distribution of LP than geography (Holden & Mace, 1997; Mulcare et al., 2004; Tishkoff et al., 2007). In particular, high frequencies of LP, sometimes similar to those of northern Europeans, are observed within African and the Middle Eastern dairying or pastoralist populations, while lower frequencies are found in nondairy agriculturalist and hunter-gatherer populations (Holden & Mace, 1997). For example, ~70% of the Bedouin (nomadic pastoralists) of Jordan are LP, whereas <20% of their neighboring nonpastoralist Arab groups are LP. Similarly, over 70% of herders in the Sinai Desert are LP, but only ~15% of nearby urban Egyptians are LP (Swallow, 2003). This high degree of structuring of LP at local levels is most likely due to a combination of differences in subsistence, differing population origins, and limited gene flow between populations due to linguistic and other cultural barriers.

1.9.1 Molecular basis of lactase persistence

LP is an autosomal dominant genetic trait. Despite decades of research, there are still unanswered questions regarding its molecular basis and regulation. There are

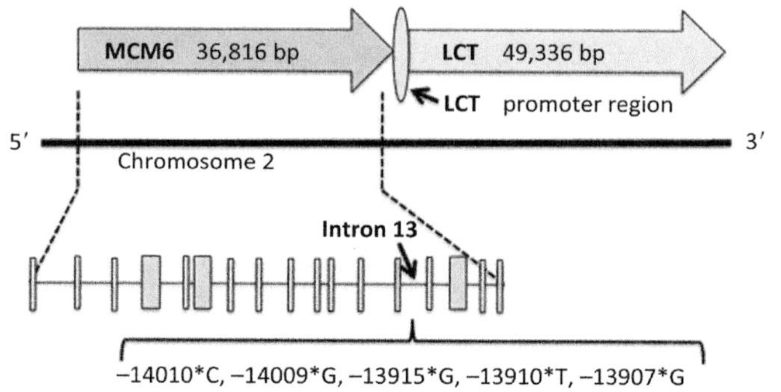

FIGURE 1.7

Schematic drawing of LCT and MCM6 genes with five known LP alleles.

multiple genetic causes of LP with at least five alleles conferring LP, including -13910*T, -13915*G, -13907*G, -14009*G, -14010*C (genome positions relative to the start codon of LCT) (Fig. 1.7). These alleles are located within a 100 base pair region, not in LCT itself but some 1400 bp upstream; within intron 13 of the adjacent gene MCM6 (mini-chromosome maintenance 6) (Enattah et al., 2008; Ingram, Mulcare, et al., 2009; Ranciaro et al., 2014; Tishkoff et al., 2007). One further associated allele identified in intron 9 of MCM6 at position -22018*G is in strong linkage disequilibrium with -13910*T, the major European LP-associated allele. This single nucleotide polymorphism is associated with LP and either is causative or perhaps enhances the expression of lactase, but has no effect on LP without -13910*T (Ranciaro et al., 2014). It is not uncommon in the human genome to find alterations near a gene, usually upstream, affecting transcription. These changes in the DNA sequence can affect whether a gene is transcribed and also the rate of transcription (synthesizing more or less of the associated mRNA). In the case of lactase, the five LP alleles are located in a regulatory region near the LCT promoter. Each allele causes lactase to continue to be expressed into adulthood, whereas the ancestral state is the downregulation of this gene sometime after weaning.

While all five known causative LP variants are found in agropastoralist or pastoralist populations—like LP itself—their geographic distributions are highly structured. The major Eurasian allele is -13910*T (Enattah et al., 2008; Gerbault et al., 2011; Itan et al., 2010; Mulcare et al., 2004; Ranciaro et al., 2014; Swallow, 2003). Functional studies have confirmed that -13910*T enhances lactase expression (Labrie et al., 2016; Lewinsky et al., 2005). The mechanism responsible for this increased lactase expression with this allele is enhanced rates of transcription of LCT. The -13910*T allele has been shown to directly affect LCT promoter activity with the octamer-binding protein 1 (Oct-1) transcription factor. Oct-1 binds more tightly to -13910*T. This results in higher levels of lactase in the small intestine. Oct-1 binds less tightly to the -13910*C allele, resulting in reduced rates of transcription and reduced lactase levels in the small intestine (Lewinsky et al., 2005).

More recent work has indicated that lactase expression can also be influenced by age-related differences in methylation patterns. Methylation can change the expression of a gene without changing the DNA sequence itself. It is a process by which a methyl group (CH_3) covalently bonds, often to a cytosine base in the DNA, and slows or prevents transcription. This modification, particularly in promoter regions, can inhibit the transcription of a gene. The ancestral -13910*C allele accumulates more methylation over time and this correlates with suppressed transcription of LCT, whereas -13910*T remains largely unmethylated, and continued lactase expression is promoted. It has been hypothesized that epigenetic modification may contribute to the observed age-related reduced lactase expression levels (Labrie et al., 2016; Swallow & Troelsen, 2016). Epigenetic modifications, such as methylation, are heritable changes in gene expression, and can occur in response to the environment, including age, lifestyle, diet, and environmental toxins.

There is also variation in the patterns and timings of LCT downregulation in LNP individuals, during or after weaning. In Finnish and Estonian populations downregulation of lactase appears to occur relatively late (8–12 years old), whereas Chinese and Japanese children show evidence of reduced lactase levels as early as 1 year old, and some African infants under a year old may begin downregulating lactase (see Table 1.2 for timings of LCT downregulation). The reduced ability to digest lactose correlates with lower levels of lactase mRNA (Rossi et al., 1997; Swallow, 2003; Wang, Harvey, Rousset, & Swallow, 1994). However, whether this interpopulation reduction in mRNA is driven by genetics or in response to environmental stimuli (including dietary change with weaning) has not yet been determined (Swallow, 2003). Methylation may partially explain lower mRNA levels in humans but it is unlikely that lactase levels fall in response to lower lactose intake and it may be that the timing simply coincides with weaning for other reasons.

1.9.2 The distribution of lactase persistence alleles

The -13910*T allele reaches its highest frequencies in British and Scandinavian populations (69%–95%), then declines southward and eastward toward the Middle East (Ingram, Mulcare, et al., 2009; Itan et al., 2010). It is also the most common LP-associated allele among northern Indian herders and in southern Asia (Gallego Romero et al., 2011) and has also been observed in various northern, eastern, and central Africa pastoralist populations (Bersaglieri et al., 2004; Enattah et al., 2008; Ingram et al., 2007; Ingram, Mulcare, et al., 2009; Jones et al., 2013, 2015; Mulcare et al., 2004; Myles et al., 2005; Tishkoff et al., 2007). A conserved haplotype background indicates that the -13910*T allele in Africa and southern Asia did not arise from independent mutation events, but rather

Table 1.2 Variation across human populations in the estimated age when lactase is downregulated in LNP individuals.

Population	Lactase downregulation age	Reference
Chinese	1–5 years	Sahi (1994)
	7–8 years	Yang, He, Cui, Bian, and Wang (2000)
Hong Kong Chinese	3–10 years	Chang, Hsu, Chen, Lee, and Hsu (1987)
Japanese	1–5 years	Sahi (1994)
Taiwanese	2–6 years	Chang et al. (1987)
Thai	By 5 years	Wang et al. (1998)
	Under 12 months to 3 years	Keusch, Troncale, Miller, Promadhat, and Anderson (1969)
Bangladesh	7–18 months to 3 years	Brown, Parry, Khatun, and Ahmed (1979)
African	Between 1 and 8 years	Northrop-Clewes, Lunn, and Downes (1997); Rasinperä et al. (2004)
Nigerian Yoruban	Under 12 months to 4 years	Kretchmer, Hurwitz, Ransome-Kuti, Dungy, and Alakija (1971)
East African Baganda	Some downregulation observed from 1 week to 5 months	Cook (1967)
Somalian	5–10 years	Rasinperä et al. (2004)
Ugandan	After 3 years	Cook and Kajubi (1966)
Sardinian	3–9 years	Schirru et al. (2007)
Greek	5–12 years	Ladas, Katsiyiannaki-Latoufi, and Raptis (1991)
Finnish	8–12 years	Rasinperä et al. (2004)
Finnish and Estonian	8–12 years	Sahi (1994)
Israeli Jewish	From 6.5 years	Gilat, Dolizky, Gelman-Malachi, and Tamir (1974)
Native American	1–5 years or later	Bose and Welsh (1973); Caskey et al. (1977); Johnson et al. (1977)
American Black ancestry	1–4 years	Bayless et al. (1975)
	3–9 years	Pribila, Hertzler, Martin, Weaver, and Savaiano (2000)
	6–11 years	Huang and Bayless (1967)
American Mexican ancestry	1–4 years	Bayless et al. (1975)
American European ancestry	5 years or later	Bayless et al. (1975)
Mexican	Under 4–13 years	Rosado et al. (1994)

Some of the observed variation in timing may be attributable to different methods used to determine downregulation. For example, studies use different methods to establish lactase persistence status and some studies did not test for infection which can affect lactose digestion. Future work systematically confirming these age estimates of downregulation would contribute significantly to our understanding of these variations.

shares a common origin with European carriers, and probably came to those regions by migration.

Unlike most of Eurasia, where -13910*T is the main cause of LP, in Africa and the Middle East at least four additional alleles are associated with the trait. The -14010*C allele has been observed in Tanzania, Kenya, and southern Africa (Liebert et al., 2017; Ranciaro et al., 2014). Interestingly, this allele is not only found in agropastoralist populations belonging to different language families (Afro-Asiatic, Nilo-Saharan and Niger-Congo) but has also been observed in hunter-gatherer populations such as the South African San and Tanzanian Sandawe, although notably not in the neighboring Hadza (Liebert et al., 2017; Ranciaro et al., 2014). The -14010*C allele haplotypic background is similar in eastern and southern Africa, again indicating a common origin, and it has been hypothesized that gene flow from East to South Africa is responsible for the appearance of this allele in San hunter-gatherers and Xhosa.

While the -13915*G allele is often referred to as the Middle Eastern variant, where it probably arose and reaches frequencies as high as 72%–88%, it is also found in East and North Africa (Imtiaz et al., 2007; Ranciaro et al., 2014). This allele has not been observed in western and southern African populations (Imtiaz et al., 2007). Like the other African variants, the -13907*G allele distribution is highly structured (Liebert et al., 2017; Ranciaro et al., 2014). It is most common in Ethiopia and the horn of Africa, and is particularly associated with Cushitic-speaking populations in Ethiopia, Sudan, and Kenya, but has not been observed in Tanzania. The variant is also present in low frequency in Kenyan Nilo-Saharan speaking populations (Ranciaro et al., 2014). A recently identified functional allele, -14009*G, has been observed in Ethiopian, Somali, and Sudanese populations (Ingram, Raga, et al., 2009; Jones et al., 2013; Liebert et al., 2017). Among the Beja herders of northern Sudan, the frequency of this variant is ~23.5%, although it is only 1.6% in the central Sudanese Baggara. This allele also appears at low frequency (~1.0%) in the Middle East (Fig. 1.8).

1.9.3 When did lactase persistence alleles arise?

By measuring the extended haplotype homozygosity (EHH) of a genomic region surrounding an allele, it is possible to estimate its age, and when this information is combined with present-day frequencies, it is possible to test if positive natural selection has been acting (Sabeti, Reich, Higgins, & Levine, 2002). Under positive natural selection, beneficial traits that increase their carrier's chances of survival and reproduction will increase in frequency in populations over time. The basic principle here is that while allele frequencies change through time by genetic drift (random changes in frequency from generation to generation), if an allele is relatively young but common in a population, such as the -13910*T allele, then it must have risen to high frequency quickly. If that rate of increase in frequency is greater than that expected under genetic drift alone, then positive natural selection can be invoked.

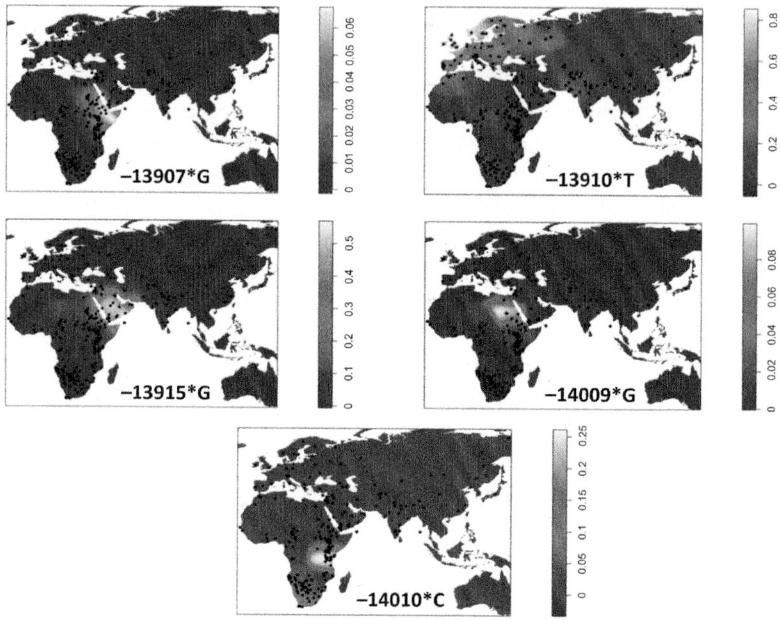

FIGURE 1.8

Interpolated contour maps showing the frequency distribution for the five LP-associated alleles (Liebert et al., 2017).

Maps © Liebert, A., López, S., Jones, B. L., Montalva, N., Gerbault, P., Lau, W., Swallow, D. M. (2017). World-wide distributions of lactase persistence alleles and the complex effects of recombination and selection. Human Genetics, *136(11–12), 1445–1453 and reproduced according to the Creative Commons Attribution 4.0 International License (http://creativecommons.org/licenses/by/4.0/). Updated maps available from https://www.ucl.ac.uk/mace-lab/resources/glad/LP_maps.*

EHH patterns associated with the -13910*T allele have revealed conserved haplotypes in excess of 1 Mb, much longer than haplotypes associated with the -13910*C allele. Using these data the 13910*T was originally estimated to be between 2188 and 20,650 years old (Bersaglieri et al., 2004). Haplotype-based investigations of the other LP-associated alleles yielded similar dates: -13915*G has been estimated to have arisen in Saudi Arabia 4075 ya (2050–6100 95% CI) (Enattah et al., 2008) and age estimates of the -14010*C allele in Tanzanian Afro-Asiatic and Kenyan Nilo-Saharan populations are ~7000 ya (2000–16,000 95% CI) (Tishkoff et al., 2007). However, it has been noted that the LCT region may have restricted rates of recombination—which would lead to elevated EHH and possibly to underestimates of the age of LP-associated alleles (Liebert et al., 2017). Nonetheless, these date estimates broadly bracket those for milkable animal domestication.

Studies of DNA from archeological bones in Europe indicate that while the 13910*T variant may have originated before or after the origins of dairying, it did not rise to appreciable frequencies until somewhat later. Individuals carrying the -13910*T variant have been identified in Scandinavian hunter-gatherers from ~5400 to 3400 ya with a population frequency estimation of 5% (Malmström et al., 2010). The variant has also been found in farming populations in northwest Spain from ~5000 to 4500 ya (Plantinga et al., 2012), although this has not been replicated and it has not yet been identified in other European farming populations from this period (Burger et al., 2007; Gerbault, 2013; Ségurel & Bon, 2017; Sverrisdóttir et al., 2014). In recent years there has been a considerable increase in the amount of genomic data published from ancient Europeans, and while the patterns of -13910*T variant distribution in space and through time are not fully elucidated, it appears that it did not reach appreciable frequencies until the Bronze or Iron Ages (e.g., Mathieson et al., 2015). These data indicate that LP was not common until well-after milk and dairy foods entered the human diet.

1.9.4 Strength of selection estimates

The relatively recent and so rapid rise in frequency of LP-associated alleles is difficult to explain by genetic drift alone. Indeed, when these inferred rates of frequency increase are considered in the context of other variants in the genome, or expected rates of genetic drift, it appears that LP has been favored by strong natural selection. The first formal estimates of the strength of selection—based on EHH—were 1.4%–19% for the -13910*T allele in European populations (Bersaglieri et al., 2004) and 2.6%–14.2% for the -14010*G allele in East African populations (Tishkoff et al., 2007). Subsequent estimates based on EHH (Enattah et al., 2008), spatially explicit modeling (Gerbault, Moret, Currat, & Sanchez-Mazas, 2009; Itan et al., 2009), ancient DNA analyses (Burger et al., 2007; Mathieson et al., 2015; Sverrisdóttir et al., 2014), and by combining a range of population genetic statistics (Peter, Huerta-Sanchez, & Nielsen, 2012) have all fallen into these ranges, and now provide overwhelming evidence that LP is the most strongly selected single-gene trait to have evolved in European and many Middle Eastern, Southern Asian, and African populations over the last 10,000 years. By comparison, selection estimates on multiple alleles involved in skin, hair, and eye pigmentation over a similar timescale were lower, with point estimates of between 2% and 10% (Wilde et al., 2014). A conservative LP selection coefficient estimate of 0.05 would mean that 5% more individuals carrying the LP variant would survive to reproduction *in each generation*. This 5% would have been compounded over generations, leading to the dramatic rises in frequencies observed.

1.10 Positive selection favoring lactase persistence

Considering the varied ecologies in which LP has evolved, it is unlikely that one single hypothesis explains the strong signatures of natural selection inferred across multiple regions. Rather than look for one unifying explanation, perhaps what is now required is a new framework for understanding the variation in selective pressures attributable to cultural and environmental differences. It may be possible to define a common set of hypotheses from which unique combinations explain the regional differences in distribution frequencies seen today. Searching for one unifying reason may oversimplify the complex processes of evolution of at least five separate LP alleles.

1.10.1 The cultural-historical hypothesis

The observed global pattern of LP positively correlates to dairying and milk consumption (Holden & Mace, 2009). This codistribution unifies multiple LP-causative alleles with distinct geographic distributions into a single, coherent pattern. The cultural-historical hypothesis was proposed by two independent researchers in the early 1970s, Simoons and McCracken, who noted that reliance on milk drinking in the diet would likely have created positive selective pressures for LP (McCracken, 1971; Simoons, 1969). This hypothesis is primarily concerned not with the specific advantages(s) of LP, but with when it evolved in relation to the origins of dairying. It posits that dairying came first, and the genetic response followed. In opposition to this, the reverse-cause hypothesis posits that LP arose first and dairying only followed in populations where LP was sufficiently high to confer a benefit (Bayless, Paige, & Ferry, 1971; Nei, 1986; Simoons, 1970). In 2007 Joachim Burger and colleagues (Burger et al., 2007) examined DNA from the bones of early farmers in central Europe with clear archeological evidence of dairying, and showed that the European -13910*T allele was either absent or present at very low frequencies. Indeed, not only were these early farmer populations most likely producing milk, but their ancestors for at least 1500 years had also been doing so (Evershed et al., 2008; Salque et al., 2013). Subsequent ancient DNA studies have confirmed that -13910*T frequencies remained very low until ~4000–5000 ya in Europe (Mathieson et al., 2015). These data overwhelmingly support the cultural-historical model for the origins of LP in Europe, and similar analyses indicate the same scenario in Africa (Salque et al., 2013; Schlebusch, Sjödin, Skoglund, & Jakobsson, 2013) and southern Asia (Gallego Romero et al., 2011).

The cultural-historical hypothesis can account for the distribution of LP in many European, African, Southern Asian, and the Middle Eastern populations, but is difficult to reconcile with the low frequency of LP observed among nomadic herders of the Central Asian Steppe. The estimated LP frequency in the Mongols (~12%) and Kazakhs (24%–30%) is lower than similar dairying populations in Africa. LP in these groups is mostly caused by the -13910*T allele. Based on haplotype structure surrounding the -13910*T allele, the variant is most likely to come to the Central Asian Steppe via gene flow from Europe (Gallego Romero et al., 2011). The amount of lactose in camel milk is similar to cow's milk, so the differences in frequencies are unlikely attributable to lactose concentration (Ségurel & Bon, 2017). However, these nomads differ significantly from Europeans and Africans in that they seldom drink fresh milk. They process milk into butter, yoghurts, and cheeses to preserve it and make it more transportable, in the processing removing most of the lactose. In this regard, the cultural-historical hypothesis is best applied to groups drinking fresh milk, rather than dairying in general.

LP individuals could derive up to 30% more calories from fresh milk by lactose digestion than by not being able to digest lactose (Brüssow, 2013), with the additional advantage of an absence of side effects such as diarrhea, cramps, and excessive colonic gasses. This up to 30% energetic benefit is unlikely to explain strong natural selection on LP alone, although susceptibility to the side effects may have prevented non-LP individuals from accessing the other beneficial nutrients in milk, such as fats, protein, vitamins, and minerals. Indeed, milk is a highly nutrient-dense, nutrient-diverse, and relatively nutrient-balanced food, and these factors alone may explain the strong natural selection favoring LP. However, a number of more specific advantages to drinking fresh milk, and therefore LP, have been proposed, and these are described below.

1.10.2 Calcium assimilation hypothesis

In Europe the distribution of LP is positively correlated with latitude. Noting this relationship, Flatz and Rotthauwe (1973) proposed the calcium assimilation hypothesis, whereby milk provided an important source of calcium and to a lesser extent vitamin D in northern Europe. Mesolithic hunter-gatherer diets were high in aquatic resources, and so in vitamin D (Richards & Hedges, 1999). However, Neolithic cereal-based diets were relatively deficient in vitamin D, which is required for calcium absorption. In addition, wheat is high in phytate, which has been demonstrated to impair calcium absorption by chelation (Cordain, 1999). Vitamin D can be made in the skin by exposure to UVB light, which photoconverts 7-dehydrocholesterol into cholecalciferol (vitamin D_3). However, at higher latitudes, incident UVB levels are often too low for adequate vitamin D production (Jablonski & Chaplin, 2010), and without an adequate dietary source this can lead to rickets and cause pelvic flattening. It has been suggested that the resultant narrowing of the birth canal could have increased maternal mortality in childbirth (Cordain, Hickey, & Kim, 2012). Milk would have supplemented dietary vitamin D and provided calcium in agrarian diets of northern Europe. However, high levels of LP are observed in African and Middle Eastern groups, where UVB levels are high resulting in sufficient vitamin D biosynthesis. The calcium assimilation hypothesis thus may explain some of the LP distribution seen in northern Europe, but is unlikely to be relevant at lower latitudes.

1.10.3 Climate aridity hypothesis

The climate aridity hypothesis (Cook & Al-Torki, 1975; Cook, 1978) posits that fresh milk provided a source of relatively pathogen-free fluid, which would have been particularly advantageous in arid regions or times of drought. Non-LP individuals would suffer maldigestion and diarrhea, exacerbating dehydration, with potentially fatal consequences. This hypothesis may explain the LP distribution in arid African and Middle Eastern regions, where dairy animals act like walking canteens. In Europe, the climate aridity hypothesis is less relevant as severe drought is rare, but milk would still have provided a relatively pathogen-free fluid, and perhaps a vital bridging food during intermittent crop failures.

1.10.4 Famine diarrhea hypothesis

In populations that depend on both dairying and agriculture, crop failure can be common. Under such circumstances milk or milk products can act as important fallback foods. Dairy products would have had a range of lactose contents and it is possible that under famine conditions, the low lactose content products (e.g., cheeses) would have been consumed rapidly, leaving the higher lactose content foods to be consumed later. High lactose content foods would benefit malnourished LP individuals, but consumption by malnourished non-LP individuals could be catastrophic as diarrhea can be fatal under starvation conditions. This would lead to episodic but very high selection differentials between LP and non-LP individuals (Ingram, Mulcare, et al., 2009; Sverrisdóttir et al., 2014).

1.10.5 Malaria resistance hypothesis

Some have hypothesized that reliance on milk consumption confers protection against malarial symptoms by restricting *p*-aminobenzoic acid (PABA) availability; the suppression of malarial symptoms is abrogated when PABA is added to milk or PABA-deficient diets of infected animals (Cordain et al., 2012; Lokki et al., 2011). The Fulani are nomadic milk-drinking pastoralists found throughout West and central Africa. With 68% LP (Swallow, 2003) and lower levels of malarial infection than neighboring nonpastoralist groups such as the Dogon (Bereczky et al., 2006), researchers investigated whether LP was associated with resistance to malaria. While the Fulani are less parasitized than neighboring groups, the results were not statistically significant (Lokki et al., 2011). Other studies of LP and malaria in Sardinia, where infection rates are moderate, have also concluded that LP does not seem to protect against malarial symptoms (Meloni, Colombo, Ruggiu, Dessena, & Meloni, 1998). However, LP and lower rates of malaria correlate in West Africa, with no convincing cause yet identified.

1.10.6 Phlorizin digestion hypothesis

The LCT (lactase-phlorizin hydrolase) cleaves lactose into its component monosaccharides and also hydrolyzes phlorizin as well as a number of other plant glycosides. Phlorizin is a glucoside found in the stems, roots, and bark of plants in the Rosaceae family. These wide-ranging fruit trees include apple, cherry, and pear, and are found in both northern Europe and Africa where LP levels are high. The Hadza of Tanzania have no history of dairying or herding, and yet 47% of these hunter-gatherers are lactase persistent based on the results of lactose tolerance tests measuring blood glucose levels (Ranciaro et al., 2014). Genetic investigations have not found any of the five known LP variants in the Hadza, suggesting the possibility that unknown LP variant(s) may exist in the Hadza or microbiome adaptations may ferment lactose to raise blood glucose levels. As phlorizin is used in traditional medicine to treat malaria and other infectious diseases, it is possible that the LP seen in the Hadza is in response to positive selective pressures to metabolize phlorizin, rather than lactose (Németh et al., 2003; Ranciaro et al., 2014).

1.10.7 Galactose gut microbiome health hypothesis

In the early evolution of milk, oligosaccharides likely played a major role in antimicrobial protection, with milk carbohydrates acquiring a nutritional role later in the evolution of mammals. However, colostrum—which is very high in milk oligosaccharides—has retained some of that early function by providing immunity to newborns (Brand Miller & McVeagh, 1999). Once cleaved, lactose yields glucose, which is rapidly absorbed, and galactose. While the majority of galactose is also absorbed in the small intestine and transported to the liver for conversion to glucose (and storage in the form of glycogen), some passes into the colon where it may feed certain colonic bacteria, possibly acting as a beneficial prebiotic (Cederlund et al., 2013; Gibson, 1998; Gibson et al., 2017; Macfarlane & Cummings, 1999; Wahlqvist, 2015). Recent studies also showed that galacto-oligosaccharides added to formula milk acted as a prebiotic in newborns by stimulating the growth of bifidobacteria and increased the intestinal metabolic activity of the microbiome (Knol et al., 2005) which may lend support to the role of galactose as a prebiotic. With increasing population density in the Neolithic and the resulting rise in infectious disease, the ability to cleave lactose into adulthood may have not only provided additional vital calories, but also played a lifelong prebiotic role in maintaining a healthy gut microbiome. This hypothesis requires further examination.

1.11 Conclusion

At the timescale of the evolution of mammals some 150 million ya, and on the more recent timescale of human evolution over the past ~10,000 years, lactose has been a key player. This simple and yet highly milk-specific disaccharide has had a strong influence on nutrition, immunity, and ultimately survival. Lactose digestion continues to capture public interest, but despite considerable scientific research into its nutritional, health, and evolutionary roles, there is still much to learn.

Milk has been, and continues to be, an integral part of mammalian evolution. The earliest milk-like secretions probably arose some 300 million ya, with lactose gradually becoming the dominant carbohydrate in eutherians. Milk then entered the adult human diet some 10,000 ya. Despite its long history, there remain many unanswered questions about the evolution of lactose and the human cultural and genetic adaptions to its consumption. While there are some fundamental, perhaps accidental, biochemical reasons why lactose became the preferential disaccharide from which the oligosaccharides in the earliest milks were synthesized, we have not fully answered the question of why lactose was favored over other possible disaccharides. Other questions surrounding the timing and reasons for lactase downregulation also merit further research: why do mammals downregulate lactase during or after weaning, when the energetic costs of maintaining enzyme production are likely to be low? The extent to which environmental factors influence the genetic control of LP and the precise mechanism of this interaction also require further investigation.

The current pattern of LP seen globally, with ~2.45 billion individuals able to digest lactose into adulthood, may well persist into the future in the absence of ongoing selective pressures. If food security challenges dietary sufficiency in some populations or geographic regions, it is possible that LP frequencies could be driven up in these areas. A recent study has found evidence of selection for LP in Chilean populations since European colonization (Montalva et al., 2019). However, in the absence of selective pressures, it is unlikely that the allelic frequency will change much by drift alone, given the size of the world's population. However, we may observe shifts in LP frequency. Because LP is a dominant trait, with the advent of global travel and increased mobility, admixture and gene flow could potentially lead to an increase in the numbers of LP individuals. This would occur by increasing the number of LP heterozygotes. As one allele produces sufficient lactase to hydrolyze lactose, it may be that through gene flow areas currently largely LNP may change over time.

However, if there are benefits or costs to being lactase persistent in the current landscape (perhaps immunity related via galactose metabolism or toxin stress related via phlorizin digestion), then there may be selective pressures at work which have not yet been quantified (Montalva et al., 2019). Selective pressures change over time and seldom, if ever, unidirectional. The positive selective pressure that drove LP-associated allelic frequencies up to those observed today are unlikely to still be acting, but that does not mean there are no selective pressures at work (Montalva et al., 2019). Selection may be different, perhaps weaker, but probably not entirely absent.

What is apparent is that genetic diversity confers adaptive flexibility in the face of changing environments. Understanding the genetic mechanisms at work and their interactions with the environment—within the cell, within the organism, between organisms, and with their surroundings—is critical to our understanding of LP as a genetic adaptation that has impacted human culture, health, and demography. In understanding the historical, cultural, and biological factors that shaped our modern state, we may better understand our adaptive flexibility in face of other, as yet unforeseen, challenges. Milk is the gift that keeps on giving.

References

Aiello, L. C., & Wheeler, P. (1995). The expensive-tissue hypothesis—the brain and the digestive-system in human and primate evolution. *Current Anthropology*, *36*(2), 199–221. Available from https://doi.org/10.1086/204350.

Ambrose, S. H. (2001). Paleolithic technology and human evolution. *Science*, *291*(5509), 1748–1753.

Andersson, B., Porras, O., Hanson, L. A., Lagergard, T., & Svanborg-Eden, C. (1986). Inhibition of attachment of *Streptococcus pneumoniae* and *Haemophilus influenzae* by human milk and receptor oligosaccharides. *Journal of Infectious Diseases*, *153*(2), 232–237.

Armelagos, G. J., & Cohen, M. N. (1984). *Paleopathology at the origins of agriculture*. Orlando, FL: Academic Press.

Arribas, J. C., Herrero, A. G., Martin-Lomas, M., Canada, F. J., He, S., & Withers, S. G. (2000). Differential mechanism-based labeling and unequivocal activity assignment of the two active sites of intestinal lactase/phlorizin hydrolase. *European Journal of Biochemistry*, *267*(24), 6996–7005.

Asouti, E., & Fuller, D. Q. (2013). A contextual approach to the emergence of agriculture in Southwest Asia reconstructing early Neolithic plant-food production. *Current Anthropology*, *54*(3), 299–345. Available from https://doi.org/10.1086/670679.

Bar-Yosef, O., & Meadow, R. H. (1995). The origins of agriculture in the Near East. In T. D. Price, & A. B. Gerbauer (Eds.), *Last hunters, first farmers: New perspectives on the prehistoric transition to agriculture* (pp. 39–94). Sante Fe, NM: School of American Research Press.

Bayless, T. M., Paige, D. M., & Ferry, G. D. (1971). Lactose intolerance and milk drinking habits. *Gastroenterology*, *60*(4), 605–608.

Bayless, T. M., Rothfeld, B., Massa, C., Wise, L., Paige, D., & Bedine, M. S. (1975). Lactose and milk intolerance: Clinical implications. *New England Journal of Medicine*, *292*(22), 1156–1159. Available from https://doi.org/10.1056/NEJM197505292922205.

Beermann, C., & Hartung, J. (2013). Physiological properties of milk ingredients released by fermentation. *Food and Function*, *4*(2), 185–199. Available from https://doi.org/10.1039/c2fo30153a.

Bereczky, S., Dolo, A., Maiga, B., Hayano, M., Granath, F., Montgomery, S. M., & Doumbo, O. (2006). Spleen enlargement and genetic diversity of *Plasmodium falciparum* infection in two ethnic groups with different malaria susceptibility in Mali, West Africa. *Transactions of the Royal Society of Tropical Medicine and Hygiene*, *100*(3), 248–257.

Bersaglieri, T., Sabeti, P. C., Patterson, N., Vanderploeg, T., Schaffner, S. F., Drake, J. A., & Hirschhorn, J. N. (2004). Genetic signatures of strong recent positive selection at the lactase gene. *American Journal of Human Genetics*, *74*(6), 1111–1120. Available from https://doi.org/10.1086/421051.

Berstan, R., Dudd, S. N., Copley, M. S., Morgan, E. D., Quye, A., & Evershed, R. P. (2004). Characterisation of 'bog butter' using a combination of molecular and isotopic techniques. *Analyst, 129*(3), 270–275.

Bettinger, R. L., Barton, L., & Morgan, C. (2010). The origins of food production in north China: A different kind of agricultural revolution. *Evolutionary Anthropology: Issues, News, and Reviews, 19*(1), 9–21.

Bininda-Emonds, O. R., Cardillo, M., Jones, K. E., MacPhee, R. D., Beck, R. M., Grenyer, R., & Purvis, A. (2007). The delayed rise of present-day mammals. *Nature, 446*(7135), 507.

Blackburn, D. G., Hayssen, V., & Murphy, C. J. (1989). The origins of lactation and the evolution of milk: A review with new hypotheses. *Mammal Review, 19*(1), 1–26.

Boback, S. M., Cox, C. L., Ott, B. D., Carmody, R., Wrangham, R. W., & Secor, S. M. (2007). Cooking and grinding reduces the cost of meat digestion. *Comparative Biochemistry and Physiology Part A: Molecular & Integrative Physiology, 148*(3), 651–656.

Bocquet-Appel, J.-P. (2011). When the world's population took off: The springboard of the Neolithic Demographic Transition. *Science, 333*(6042), 560–561.

Bocquet-Appel, J.-P., & Bar-Yosef, O. (2008). *The Neolithic Demographic Transition and its consequences*. Springer Science & Business Media.

Bogucki, P. I. (1984). Ceramic sieves of the linear pottery culture and their economic implications. *Oxford Journal of Archaeology, 3*(1), 15–30.

Bollongino, R., Burger, J., Powell, A., Mashkour, M., Vigne, J.-D., & Thomas, M. G. (2012). Modern taurine cattle descended from small number of Near-Eastern founders. *Molecular Biology and Evolution, 29*(9), 2101–2104.

Bose, D., & Welsh, J. (1973). Lactose malabsorption in Oklahoma Indians. *The American Journal of Clinical Nutrition, 26*(12), 1320–1322.

Bradshaw, J. L. (1997). *Human evolution: A neuropsychological perspective*. Psychology Press.

Brand Miller, J., & McVeagh, P. (1999). Human milk oligosaccharides: 130 reasons to breast-feed. *British Journal of Nutrition, 82*(5), 333–335.

Broushaki, F., Thomas, M. G., Link, V., López, S., van Dorp, L., Kirsanow, K., & Díez-del-Molino, D. (2016). Early Neolithic genomes from the eastern Fertile Crescent. *Science, 353*(6298), 499–503.

Brown, K. H., Parry, L., Khatun, M., & Ahmed, G. (1979). Lactose malabsorption in Bangladeshi village children: Relation with age, history of recent diarrhea, nutritional status, and breast feeding. *The American Journal of Clinical Nutrition, 32*(9), 1962–1969.

Brüssow, H. (2013). Nutrition, population growth and disease: A short history of lactose. *Environmental Microbiology, 15*(8), 2154–2161.

Burger, J., Kirchner, M., Bramanti, B., Haak, W., & Thomas, M. G. (2007). Absence of the lactase-persistence-associated allele in early Neolithic Europeans. *Proceedings of the National Academy of Sciences, 104*(10), 3736–3741.

Butterworth, P. J., Warren, F. J., & Ellis, P. R. (2011). Human α-amylase and starch digestion: An interesting marriage. *Starch-Stärke, 63*(7), 395–405.

Carl, G. R., & Robbins, C. T. (1988). The energetic cost of predator avoidance in neonatal ungulates: Hiding versus following. *Canadian Journal of Zoology, 66*(1), 239–246.

Carlson, S. E. (1985). *N*-acetylneuraminic acid concentrations in human milk oligosaccharides and glycoproteins during lactation. *The American Journal of Clinical Nutrition, 41*(4), 720–726.

Carmody, R. N., & Wrangham, R. W. (2009). The energetic significance of cooking. *Journal of Human Evolution*, *57*(4), 379–391.

Carpenter, D., Dhar, S., Mitchell, L. M., Fu, B., Tyson, J., Shwan, N. A., & Armour, J. A. (2015). Obesity, starch digestion and amylase: Association between copy number variants at human salivary (AMY1) and pancreatic (AMY2) amylase genes. *Human Molecular Genetics*, *24*(12), 3472–3480.

Caskey, D., Payne-Bose, D., Welsh, J., Gearhart, H., Nance, M., & Morrison, R. (1977). Effects of age on lactose malabsorption in Oklahoma Native Americans as determined by breath H 2 analysis. *Digestive Diseases and Sciences*, *22*(2), 113–116.

Cederlund, A., Kai-Larsen, Y., Printz, G., Yoshio, H., Alvelius, G., Lagercrantz, H., & Agerberth, B. (2013). Lactose in human breast milk an inducer of innate immunity with implications for a role in intestinal homeostasis. *PLoS One*, *8*(1), e53876.

Chang, M. H., Hsu, H. Y., Chen, C. J., Lee, C. H., & Hsu, J. Y. (1987). Lactose Malabsorption and small-intestinal lactase in normal Chinese children. *Journal of Pediatric Gastroenterology and Nutrition*, *6*(3), 369–372.

Coelho, A. I., Berry, G. T., & Rubio-Gozalbo, M. E. (2015). Galactose metabolism and health. *Current Opinion in Clinical Nutrition & Metabolic Care*, *18*(4), 422–427.

Colledge, S. (2001). *Plant exploitation on Epipalaeolithic and early Neolithic sites in the Levant* (BAR International Series Vol 986). Oxford: British Archaeological Reports.

Cook, G. (1967). Lactase activity in newborn and infant Baganda. *British Medical Journal*, *1*(5539), 527.

Cook, G. (1978). Breath hydrogen concentrations after oral lactose and lactulose in tropical malabsorption and adult hypolactasia. *Transactions of the Royal Society of Tropical Medicine and Hygiene*, *72*(3), 277–281.

Cook, G., & Al-Torki, M. (1975). High intestinal lactase concentrations in adult Arabs in Saudi Arabia. *British Medical Journal*, *3*(5976), 135–136.

Cook, G., & Kajubi, S. (1966). Tribal incidence of lactase deficiency in Uganda. *Lancet*, *1*, 725–730.

Cook, H., Pearson, A., Simmons, N., & Baker, B. (1970). Dall sheep (*Ovis dalli dalli*) milk. I. Effects of stage of lactation on the composition of the milk. *Canadian Journal of Zoology*, *48*(4), 629–633.

Cooke, D., & Gidley, M. J. (1992). Loss of crystalline and molecular order during starch gelatinisation: Origin of the enthalpic transition. *Carbohydrate Research*, *227*, 103–112.

Copley, M., Berstan, R., Dudd, S., Docherty, G., Mukherjee, A., Straker, V., & Evershed, R. (2003). Direct chemical evidence for widespread dairying in prehistoric Britain. *Proceedings of the National Academy of Sciences*, *100*(4), 1524–1529.

Copley, M., Berstan, R., Dudd, S., Aillaud, S., Mukherjee, A., Straker, V., & Evershed, R. (2005). Processing of milk products in pottery vessels through British prehistory. *Antiquity*, *79*(306), 895–908.

Copley, M., Berstan, R., Mukherjee, A., Dudd, S., Straker, V., Payne, S., & Evershed, R. P. (2005). Dairying in antiquity. III. Evidence from absorbed lipid residues dating to the British Neolithic. *Journal of Archaeological Science*, *32*(4), 523–546.

Cordain, L. (1999). *Cereal grains: Humanity s double-edged sword, . Evolutionary Aspects of Nutrition and Health* (Vol. 84, pp. 19–73). Karger Publishers.

Cordain, L., Hickey, M. S., & Kim, K. (2012). *Malaria and rickets represent selective forces for the convergent evolution of adult lactase persistence. Biodiversity in agriculture: Domestication, evolution, and sustainability.* Cambridge University Press.

Craig, O. E., Chapman, J., Heron, C., Willis, L. H., Bartosiewicz, L., Taylor, G., & Collins, M. (2005). Did the first farmers of central and eastern Europe produce dairy foods? *Antiquity*, *79*(306), 882–894.

Cravioto, A., Tello, A., Villafan, H., Ruiz, J., del Vedovo, S., & Neeser, J.-R. (1991). Inhibition of localized adhesion of enteropathogenic *Escherichia coli* to HEp-2 cells by immunoglobulin and oligosaccharide fractions of human colostrum and breast milk. *Journal of Infectious Diseases*, *163*(6), 1247–1255.

Cronin, T., Downey, L., Synnott, C., McSweeney, P., Kelly, E., Cahill, M., & Stanton, C. (2007). Composition of ancient Irish bog butter. *International Dairy Journal*, *17*(9), 1011–1020.

Curtis, E., & Jones, J. (1959). The synthesis of 4-*O*-β-d-galactosyl-D-glucose (lactose). *Canadian Journal of Chemistry*, *37*(2), 358–360.

Day, A. J., Cañada, F. J., Díaz, J. C., Kroon, P. A., Mclauchlan, R., Faulds, C. B., & Williamson, G. (2000). Dietary flavonoid and isoflavone glycosides are hydrolysed by the lactase site of lactase phlorizin hydrolase. *FEBS Letters*, *468*(2–3), 166–170.

Diamond, J. (2002). Evolution, consequences and future of plant and animal domestication. *Nature*, *418*(6898), 700.

Dunbar, R. I. M. (2017). Fire and the biogeography of Palaeo hominins. In J. Cole, J. McNabb, & M. Grove (Eds.), *Landscapes of Human Evolution: Contributions in honour of John Gowlett*. Oxford: ArchaeoPress.

Dunne, J., Evershed, R. P., Salque, M., Cramp, L., Bruni, S., Ryan, K., & di Lernia, S. (2012). First dairying in green Saharan Africa in the fifth millennium BC. *Nature*, *486* (7403), 390.

Eaton, S. B., Eaton, S. B., III, & Cordain, L. (2002). Evolution, diet, and health. In P. S. Ungar, & M. F. Teaford (Eds.), *Human Diet: Its origin and evolution* (pp. 7–18). Westport, CT: Bergen & Garvey.

Edwards, P. C., Meadows, J., Sayej, G., & Westaway, M. (2004). From the PPNA to the PPNB: New views from the Southern Levant after excavations at Zahrat adh-Dhra'2 in Jordan. *Paléorient*, *30*, 21–60.

Enattah, N. S., Jensen, T. G., Nielsen, M., Lewinski, R., Kuokkanen, M., Rasinpera, H., & Khalil, I. F. (2008). Independent introduction of two lactase-persistence alleles into human populations reflects different history of adaptation to milk culture. *The American Journal of Human Genetics*, *82*(1), 57–72.

Evershed, R. P., Payne, S., Sherratt, A. G., Copley, M. S., Coolidge, J., Urem-Kotsu, D., & Nieuwenhuyse, O. (2008). Earliest date for milk use in the Near East and southeastern Europe linked to cattle herding. *Nature*, *455*(7212), 528.

Falk, D., Redmond, J. C., Guyer, J., Conroy, C., Recheis, W., Weber, G. W., & Seidler, H. (2000). Early hominid brain evolution: A new look at old endocasts. *Journal of Human Evolution*, *38*(5), 695–717.

Flatz, G., & Rotthauwe, H. (1973). Lactose nutrition and natural selection. *The Lancet*, *302* (7820), 76–77.

Frayer, D. W. (1980). Sexual dimorphism and cultural evolution in the late Pleistocene and Holocene of Europe. *Journal of Human Evolution*, *9*(5), 399–415.

Fuller, D. Q. (2007). Contrasting patterns in crop domestication and domestication rates: Recent archaeobotanical insights from the Old World. *Annals of Botany*, *100*(5), 903–924.

Fuller, D. Q., Sato, Y.-I., Castillo, C., Qin, L., Weisskopf, A. R., Kingwell-Banham, E. J., & Van Etten, J. (2010). Consilience of genetics and archaeobotany in the entangled history of rice. *Archaeological and Anthropological Sciences*, *2*(2), 115–131.

Fuller, D. Q., Willcox, G., & Allaby, R. G. (2011). Cultivation and domestication had multiple origins: Arguments against the core area hypothesis for the origins of agriculture in the Near East. *World Archaeology*, *43*(4), 628–652.

Gade, D. (1999). *Nature and culture in the Andes*. Madison, WI: University of Wisconsin Press.

Gallagher, E. M., Shennan, S. J., & Thomas, M. G. (2015). Transition to farming more likely for small, conservative groups with property rights, but increased productivity is not essential. *Proceedings of the National Academy of Sciences*, *112*(46), 14218–14223.

Gallego Romero, I., Basu Mallick, C., Liebert, A., Crivellaro, F., Chaubey, G., Itan, Y., & Villems, R. (2011). Herders of Indian and European cattle share their predominant allele for lactase persistence. *Molecular Biology and Evolution*, *29*(1), 249–260.

Gaunitz, C., Fages, A., Hanghøj, K., Albrechtsen, A., Khan, N., Schubert, M., & Bignon-Lau, O. (2018). Ancient genomes revisit the ancestry of domestic and Przewalski's horses. *Science*, *360*(6384), 111–114.

Gerbault, P. (2013). The onset of lactase persistence in Europe. *Human Heredity*, *76*(3–4), 154–161.

Gerbault, P., Gillis, R., Vigne, J. D., Tresset, A., Bréhard, S., & Thomas, M. G. (2016). Statistically robust representation and comparison of mortality profiles in archaeozoology. *Journal of Archaeological Science*, *71*, 24–32.

Gerbault, P., Moret, C., Currat, M., & Sanchez-Mazas, A. (2009). Impact of selection and demography on the diffusion of lactase persistence. *PLoS One*, *4*(7), e6369.

Gerbault, P., Liebert, A., Itan, Y., Powell, A., Currat, M., Burger, J., & Thomas, M. G. (2011). Evolution of lactase persistence: An example of human niche construction. *Philosophical Transactions of the Royal Society of London B: Biological Sciences*, *366* (1566), 863–877.

Gibson, G. R. (1998). Dietary modulation of the human gut microflora using prebiotics. *The British Journal of Nutrition*, *80*(4), S209–S212.

Gibson, G. R., Hutkins, R., Sanders, M. E., Prescott, S. L., Reimer, R. A., Salminen, S. J., & Cani, P. D. (2017). Expert consensus document: The International Scientific Association for Probiotics and Prebiotics (ISAPP) consensus statement on the definition and scope of prebiotics. *Nature Reviews Gastroenterology and Hepatology*, *14*(8), 491.

Gilat, T., Dolizky, F., Gelman-Malachi, E., & Tamir, I. (1974). Lactase in childhood—a non-adaptable enzyme. *Scandinavian Journal of Gastroenterology*, *9*(4), 395.

Griffiths, M., Green, B., Leckie, R. M., Messer, M., & Newgrain, K. (1984). Constituents of platypus and echidna milk, with particular reference to the fatty acid complement of the triglycerides. *Australian Journal of Biological Sciences*, *37*(6), 323–330.

Hardy, K., Brand-Miller, J., Brown, K. D., Thomas, M. G., & Copeland, L. (2015). The importance of dietary carbohydrate in human evolution. *The Quarterly Review of Biology*, *90*(3), 251–268.

Harvey, C., Fox, M., Jeggo, P., Mantei, N., Povey, S., & Swallow, D. (1993). Regional localization of the lactase-phlorizin hydrolase gene, LCT, to chromosome 2q21. *Annals of Human Genetics*, *57*(3), 179–185.

Helmer, D., Gourichon, L., Monchot, H., Peters, J., & Segui, M. S. (2005). Identifying early domestic cattle from Pre-Pottery Neolithic sites on the Middle Euphrates using sexual dimorphism. In J. D. Vigne, D. Helmer, & J. Peters (Eds.), *First steps of animal domestication. New archaeozoological approach*. Oxford: Oxbow Books.

Helmer, D., Gourichon, L., & Vila, E. (2007). The development of the exploitation of products from *Capra* and *Ovis* (meat, milk and fleece) from the PPNB to the Early Bronze in the northern Near East (8700 to 2000 BCcal.). *Anthropozoologica, 42*(2), 41–69.

Heron, C., Craig, O. E., Luquin, A., Steele, V. J., Thompson, A., & Piličiauskas, G. (2015). Cooking fish and drinking milk? Patterns in pottery use in the southeastern Baltic, 3300–2400 cal BC. *Journal of Archaeological Science, 63*, 33–43.

Hofmanová, Z., Kreutzer, S., Hellenthal, G., Sell, C., Diekmann, Y., Díez-del-Molino, D., & Link, V. (2016). Early farmers from across Europe directly descended from Neolithic Aegeans. *Proceedings of the National Academy of Sciences, 113*(25), 6886–6891.

Holden, C., & Mace, R. (1997). Phylogenetic analysis of the evolution of lactose digestion in adults. *Human Biology, 69*(5), 2.

Holden, C., & Mace, R. (2009). Phylogenetic analysis of the evolution of lactose digestion in adults. *Human Biology, 81*(5/6), 597–619.

Huang, S.-S., & Bayless, T. M. (1967). Lactose intolerance in healthy children. *New England Journal of Medicine, 276*(23), 1283–1287.

Hublin, J.-J., Ben-Ncer, A., Bailey, S. E., Freidline, S. E., Neubauer, S., Skinner, M. M., & Harvati, K. (2017). New fossils from Jebel Irhoud, Morocco and the pan-African origin of *Homo sapiens*. *Nature, 546*(7657), 289.

Imtiaz, F., Savilahti, E., Sarnesto, A., Trabzuni, D., Al-Kahtani, K., Kagevi, I., & Järvelä, I. (2007). The T/G − 13915 variant upstream of the lactase gene (LCT) is the founder allele of lactase persistence in an urban Saudi population. *Journal of Medical Genetics, 44*(10), e89.

Inchley, C. E., Larbey, C. D., Shwan, N. A., Pagani, L., Saag, L., Antão, T., & Mitt, M. (2016). Selective sweep on human amylase genes postdates the split with Neanderthals. *Scientific Reports, 6*, 37198.

Ingram, C. J., Elamin, M. F., Mulcare, C. A., Weale, M. E., Tarekegn, A., Raga, T. O., & Bradman, N. (2007). A novel polymorphism associated with lactose tolerance in Africa: Multiple causes for lactase persistence? *Human Genetics, 120*(6), 779–788.

Ingram, C. J., Mulcare, C. A., Itan, Y., Thomas, M. G., & Swallow, D. M. (2009). Lactose digestion and the evolutionary genetics of lactase persistence. *Human Genetics, 124*(6), 579–591.

Ingram, C. J., Raga, T. O., Tarekegn, A., Browning, S. L., Elamin, M. F., Bekele, E., & Swallow, D. M. (2009). Multiple rare variants as a cause of a common phenotype: Several different lactase persistence associated alleles in a single ethnic group. *Journal of Molecular Evolution, 69*(6), 579.

Itan, Y., Jones, B. L., Ingram, C. J., Swallow, D. M., & Thomas, M. G. (2010). A world-wide correlation of lactase persistence phenotype and genotypes. *BMC Evolutionary Biology, 10*(1), 36.

Itan, Y., Powell, A., Beaumont, M. A., Burger, J., & Thomas, M. G. (2009). The origins of lactase persistence in Europe. *PLoS Computational Biology, 5*(8), e1000491.

Iverson, S. J., Bowen, W. D., Boness, D. J., & Oftedal, O. T. (1993). The effect of maternal size and milk energy output on pup growth in grey seals (*Halichoerus grypus*). *Physiological Zoology, 66*(1), 61–88.

Jablonski, N. G., & Chaplin, G. (2010). Human skin pigmentation as an adaptation to UV radiation. *Proceedings of the National Academy of Sciences*, *107*(Suppl. 2), 8962–8968.

Jäger, A. K., & Saaby, L. (2011). Flavonoids and the CNS. *Molecules*, *16*(2), 1471–1485.

Jenness, R. (1974). Biosynthesis and composition of milk. *Journal of Investigative Dermatology*, *63*(1), 109–118.

Jenness, R., Erickson, A. W., & Craighead, J. J. (1972). Some comparative aspects of milk from four species of bears. *Journal of Mammalogy*, *53*(1), 34–47.

Johnson, J., Simoons, F., Hurwitz, R., Grange, A., Mitchell, C., Sinatra, F., & Kretchmer, N. (1977). Lactose malabsorption among the Pima Indians of Arizona. *Gastroenterology*, *73*(6), 1299–1304.

Jones, B. L., Oljira, T., Liebert, A., Zmarz, P., Montalva, N., Tarekeyn, A., & Bradman, N. (2015). Diversity of lactase persistence in African milk drinkers. *Human Genetics*, *134*(8), 917–925.

Jones, B. L., Raga, T. O., Liebert, A., Zmarz, P., Bekele, E., Danielsen, E. T., & Swallow, D. M. (2013). Diversity of lactase persistence alleles in Ethiopia: Signature of a soft selective sweep. *The American Journal of Human Genetics*, *93*(3), 538–544.

Kadwell, M., Fernandez, M., Stanley, H. F., Baldi, R., Wheeler, J. C., Rosadio, R., & Bruford, M. W. (2001). Genetic analysis reveals the wild ancestors of the llama and the alpaca. *Proceedings of the Royal Society of London B: Biological Sciences*, *268*(1485), 2575–2584.

Keusch, G. T., Troncale, F. J., Miller, L. H., Promadhat, V., & Anderson, P. R. (1969). Acquired lactose malabsorption in Thai children. *Pediatrics*, *43*(4), 540–545.

Killian, J. K., Nolan, C. M., Stewart, N., Munday, B. L., Andersen, N. A., Nicol, S., & Jirtle, R. L. (2001). Monotreme IGF2 expression and ancestral origin of genomic imprinting. *Journal of Experimental Zoology Part A: Ecological Genetics and Physiology*, *291*(2), 205–212.

Knol, J., Scholtens, P., Kafka, C., Steenbakkers, J., Gro, S., Helm, K., & Wells, J. (2005). Colon microflora in infants fed formula with galacto-and fructo-oligosaccharides: More like breast-fed infants. *Journal of Pediatric Gastroenterology and Nutrition*, *40*(1), 36–42.

Konuspayeva, G., Faye, B., & Loiseau, G. (2009). The composition of camel milk: A meta-analysis of the literature data. *Journal of Food Composition and Analysis*, *22*(2), 95–101.

Kretchmer, N., Hurwitz, R., Ransome-Kuti, O., Dungy, C., & Alakija, W. (1971). Intestinal absorption of lactose in Nigerian ethnic groups. *The Lancet*, *298*(7721), 392–395.

Kuhn, R., & Low, I. (1949). The occurrence of lactose in the plant kingdom. *Chemische Berichte*, *82*, 477–481.

Kumar, S., Nagarajan, M., Sandhu, J., Kumar, N., Behl, V., & Nishanth, G. (2007). Mitochondrial DNA analyses of Indian water buffalo support a distinct genetic origin of river and swamp buffalo. *Animal Genetics*, *38*(3), 227–232.

Kunz, C., Rudloff, S., Baier, W., Klein, N., & Strobel, S. (2000). Oligosaccharides in human milk: Structural, functional, and metabolic aspects. *Annual Review of Nutrition*, *20*(1), 699–722.

Kunz, T., Oftedal, O., Robson, S., Kretzmann, M., & Kirk, C. (1995). Changes in milk composition during lactation in three species of insectivorous bats. *Journal of Comparative Physiology B: Biochemical, Systemic, and Environmental Physiology*, *164*(7), 543–551.

Labrie, V., Buske, O. J., Oh, E., Jeremian, R., Ptak, C., Gasiūnas, G., & Adamonis, K. (2016). Lactase nonpersistence is directed by DNA-variation-dependent epigenetic aging. *Nature Structural & Molecular Biology*, *23*(6), 566–573.

Ladas, S. D., Katsiyiannaki-Latoufi, E., & Raptis, S. A. (1991). Lactose maldigestion and milk intolerance in healthy Greek schoolchildren. *The American Journal of Clinical Nutrition*, *53*(3), 676–680.

Larson, G., Karlsson, E. K., Perri, A., Webster, M. T., Ho, S. Y., Peters, J., & Fredholm, M. (2012). Rethinking dog domestication by integrating genetics, archeology, and biogeography. *Proceedings of the National Academy of Sciences*, *109*(23), 8878–8883.

Lazaridis, I., Nadel, D., Rollefson, G., Merrett, D. C., Rohland, N., Mallick, S., & Sirak, K. (2016). Genomic insights into the origin of farming in the ancient Near East. *Nature*, *536*(7617), 419–424.

Lefèvre, C. M., Sharp, J. A., & Nicholas, K. R. (2010). Evolution of lactation: Ancient origin and extreme adaptations of the lactation system. *Annual Review of Genomics and Human Genetics*, *11*, 219–238.

Leonardi, M., Gerbault, P., Thomas, M. G., & Burger, J. (2012). The evolution of lactase persistence in Europe. A synthesis of archaeological and genetic evidence. *International Dairy Journal*, *22*(2), 88–97.

Lewinsky, R. H., Jensen, T. G., Møller, J., Stensballe, A., Olsen, J., & Troelsen, J. T. (2005). T-13910 DNA variant associated with lactase persistence interacts with Oct-1 and stimulates lactase promoter activity in vitro. *Human Molecular Genetics*, *14*(24), 3945–3953.

Liebert, A., López, S., Jones, B. L., Montalva, N., Gerbault, P., Lau, W., & Swallow, D. M. (2017). World-wide distributions of lactase persistence alleles and the complex effects of recombination and selection. *Human Genetics*, *136*(11–12), 1445–1453.

Linnaeus, C. (1758). Systema Naturae per Regna Tria Naturae [editio decima, reformata 1 (Regnum animale)]. *Salvii, Stockholm*, *824*.

Lokki, A. I., Järvelä, I., Israelsson, E., Maiga, B., Troye-Blomberg, M., Dolo, A., & Holmberg, V. (2011). Lactase persistence genotypes and malaria susceptibility in Fulani of Mali. *Malaria Journal*, *10*(1), 9.

Lönnerdal, B., Keen, C. L., Glazier, C. E., & Anderson, J. (1984). A longitudinal study of rhesus monkey (*Macaca mulatta*) milk composition: Trace elements, minerals, protein, carbohydrate, and fat. *Pediatric Research*, *18*(9), 911–914.

Luick, J., White, R., Gau, A., & Jenness, R. (1974). Compositional changes in the milk secreted by grazing reindeer I. Gross composition and ash1. *Journal of Dairy Science*, *57*(11), 1325–1333.

Macfarlane, G. T., & Cummings, J. H. (1999). Probiotics and prebiotics: Can regulating the activities of intestinal bacteria benefit health? *British Medical Journal*, *318*(7189), 999.

MacHugh, D. E., Larson, G., & Orlando, L. (2017). Taming the past: Ancient DNA and the study of animal domestication. *Annual Review of Animal Biosciences*, *5*, 329–351.

Macintosh, A. A., Pinhasi, R., & Stock, J. T. (2016). Early Life conditions and physiological stress following the transition to farming in central/Southeast Europe: Skeletal growth impairment and 6000 years of gradual recovery. *PLoS One*, *11*(2), e0148468.

Madsen, O. (2009). *Mammals (mammalia). The Timetree of life* (pp. 459–461). Oxford: Oxford University Press.

Makarewicz, C., & Tuross, N. (2012). Finding fodder and tracking transhumance: Isotopic detection of goat domestication processes in the Near East. *Current Anthropology*, *53* (4), 495–505.

Malmström, H., Linderholm, A., Lidén, K., Storå, J., Molnar, P., Holmlund, G., & Götherström, A. (2010). High frequency of lactose intolerance in a prehistoric hunter-gatherer population in northern Europe. *BMC Evolutionary Biology*, *10*(1), 89.

Mandel, A. L., des Gachons, C. P., Plank, K. L., Alarcon, S., & Breslin, P. A. (2010). Individual differences in AMY1 gene copy number, salivary α-amylase levels, and the perception of oral starch. *PLoS One*, *5*(10), e13352.

Marshall, F., & Hildebrand, E. (2002). Cattle before crops: The beginnings of food production in Africa. *Journal of World Prehistory*, *16*(2), 99–143.

Mathieson, I., Lazaridis, I., Rohland, N., Mallick, S., Patterson, N., Roodenberg, S. A., & Novak, M. (2015). Genome-wide patterns of selection in 230 ancient Eurasians. *Nature*, *528*(7583), 499–503.

McClellan, H. L., Miller, S. J., & Hartmann, P. E. (2008). Evolution of lactation: nutrition v. protection with special reference to five mammalian species. *Nutrition Research Reviews*, *21*(2), 97–116.

McCracken, R. D. (1971). Lactase deficiency: An example of dietary evolution. *Current Anthropology*, *12*(4/5), 479–517.

Meloni, T., Colombo, C., Ruggiu, G., Dessena, M., & Meloni, G. (1998). Primary lactase deficiency and past malarial endemicity in Sardinia. *Italian Journal of Gastroenterology and Hepatology*, *30*(5), 490–493.

Messer, M., & Kerry, K. R. (1973). Milk carbohydrates of the echidna and the platypus. *Science*, *180*(4082), 201–203.

Messer, M., & Mossop, G. (1977). Milk carbohydrates of marsupials I. Partial separation and characterization of neutral milk oligosaccharides of the eastern grey kangaroo. *Australian Journal of Biological Sciences*, *30*(5), 379–388.

Messer, M., & Urashima, T. (2002). Evolution of milk oligosaccharides and lactose. *Trends in Glycoscience and Glycotechnology*, *14*(77), 153–176.

Meurant, G. (1995). *Handbook of milk composition*. Academic Press.

Molleson, T. (1994). The eloquent bones of Abu Hureyra. *Scientific American*, *271*(2), 70–75.

Montalva, N., Adhikari, K., Liebert, A., Mendoza-Revilla, J., Flores, S. V., Mace, R., & Swallow, D. M. (2019). Adaptation to milking agropastoralism in Chilean goat herders and nutritional benefit of lactase persistence. *Annals of Human Genetics*, *83*(1), 11–22.

Mulcare, C. A., Weale, M. E., Jones, A. L., Connell, B., Zeitlyn, D., Tarekegn, A., & Thomas, M. G. (2004). The T allele of a single-nucleotide polymorphism 13.9 kb upstream of the lactase gene (LCT)(C − 13.9 kbT) does not predict or cause the lactase-persistence phenotype in Africans. *The American Journal of Human Genetics*, *74*(6), 1102–1110.

Mummert, A., Esche, E., Robinson, J., & Armelagos, G. J. (2011). Stature and robusticity during the agricultural transition: Evidence from the bioarchaeological record. *Economics & Human Biology*, *9*(3), 284–301.

Myles, S., Bouzekri, N., Haverfield, E., Cherkaoui, M., Dugoujon, J.-M., & Ward, R. (2005). Genetic evidence in support of a shared Eurasian-North African dairying origin. *Human Genetics*, *117*(1), 34–42.

Nei, M. (1986). Stochastic errors in DNA evolution and molecular phylogeny. *Progress in Clinical and Biological Research*, *218*, 133–147.

Németh, K., Plumb, G. W., Jean-Guy, B., Juge, N., Jacob, R., Naim, H. Y., & Kroon, P. A. (2003). Deglycosylation by small intestinal epithelial cell [beta]-glucosidases is a critical step in the absorption and metabolism of dietary flavonoid glycosides in humans. *European Journal of Nutrition*, *42*(1), 29.

Newburg, D. S. (2000). Oligosaccharides in human milk and bacterial colonization. *Journal of Pediatric Gastroenterology and Nutrition*, *30*, S8–S17.

Newburg, D. S., & Neubauer, S. H. (1995). Carbohydrates in milk: Analysis, quantities and significance. In R. G. Jensen (Ed.), *Handbook of milk composition* (Vol. 349, p. 273). New York, NY: Academic Press.

Nielsen, R., Hellmann, I., Hubisz, M., Bustamante, C., & Clark, A. G. (2007). Recent and ongoing selection in the human genome. *Nature Reviews. Genetics*, *8*(11), 857.

Northrop-Clewes, C., Lunn, P., & Downes, R. (1997). Lactose maldigestion in breast-feeding Gambian infants. *Journal of Pediatric Gastroenterology and Nutrition*, *24*(3), 257–263.

O'Brien, M. J., & Bentley, R. A. (2015). The role of food storage in human niche construction: An example from Neolithic Europe. *Environmental Archaeology*, *20*(4), 364–378.

Oftedal, O. T. (1984). Lactation in the dog: Milk composition and intake by puppies. *The Journal of Nutrition*, *114*(5), 803–812.

Oftedal, O. T. (1995). Comparative analysis of nonhuman milks. In R. G. Jensen (Ed.), *Handbook of milk composition* (pp. 749–788). San Diego, CA: Academic Press.

Oftedal, O. T. (2002a). The mammary gland and its origin during synapsid evolution. *Journal of Mammary Gland Biology and Neoplasia*, *7*(3), 225–252.

Oftedal, O. T. (2002b). The origin of lactation as a water source for parchment-shelled eggs. *Journal of Mammary Gland Biology and Neoplasia*, *7*(3), 253–266.

Oftedal, O. T. (2011). Milk of marine mammals. In J. W. Fuquay, P. F. Fox, & P. L. McSweeney (Eds.), *Encyclopedia of dairy sciences*. Academic Press.

Oftedal, O. T., & Jenness, R. (1988). Interspecies variation in milk composition among horses, zebras and asses (Perissodactyla: Equidae). *Journal of Dairy Research*, *55*(1), 57–66.

Oftedal, O. T., Boness, D. J., & Tedman, R. A. (1987). *The behavior, physiology, and anatomy of lactation in the pinnipedia. Current Mammalogy* (pp. 175–245). Springer.

Oftedal, O. T., Boness, D. J., & Bowen, W. D. (1988). The composition of hooded seal (*Cystophora cristata*) milk: An adaptation for postnatal fattening. *Canadian Journal of Zoology*, *66*(2), 318–322.

Ortiz, J., Guilderson, T., Adkins, J., Sarnthein, M., Baker, L., & Yarusinsky, M. (2000). Abrupt onset and termination of the African Humid Period: Rapid climate responses to gradual insolation forcing. *Quaternary Science Reviews*, *19*(1), 347–361.

Outram, A. K., Stear, N. A., Bendrey, R., Olsen, S., Kasparov, A., Zaibert, V., & Evershed, R. P. (2009). The earliest horse harnessing and milking. *Science*, *323*(5919), 1332–1335.

Peter, B. M., Huerta-Sanchez, E., & Nielsen, R. (2012). Distinguishing between selective sweeps from standing variation and from a de novo mutation. *PLoS Genetics*, *8*(10), e1003011.

Plantinga, T. S., Alonso, S., Izagirre, N., Hervella, M., Fregel, R., Van Der Meer, J. W., & De La Rúa, C. (2012). Low prevalence of lactase persistence in Neolithic South-West Europe. *European Journal of Human Genetics*, *20*(7), 778.

Powell, A., Shennan, S., & Thomas, M. G. (2009). Late Pleistocene demography and the appearance of modern human behavior. *Science*, *324*(5932), 1298–1301.

Pribila, B. A., Hertzler, S. R., Martin, B. R., Weaver, C. M., & Savaiano, D. A. (2000). Improved lactose digestion and intolerance among African-American adolescent girls fed a dairy rich-diet. *Journal of the Academy of Nutrition and Dietetics*, *100*(5), 524–528.

Purugganan, M. D., & Fuller, D. Q. (2009). The nature of selection during plant domestication. *Nature*, *457*(7231), 843.

Ranciaro, A., Campbell, M. C., Hirbo, J. B., Ko, W.-Y., Froment, A., Anagnostou, P., & Omar, S. A. (2014). Genetic origins of lactase persistence and the spread of pastoralism in Africa. *The American Journal of Human Genetics*, *94*(4), 496–510.

Rasinperä, H., Savilahti, E., Enattah, N. S., Kuokkanen, M., Tötterman, N., Lindahl, H., & Kolho, K.-L. (2004). A genetic test which can be used to diagnose adult-type hypolactasia in children. *Gut*, *53*(11), 1571–1576.

Reithel, F., & Venkataraman, R. (1956). Lactose in the Sapotaceae. *Science*, *123*(3207), 1083–1084.

Richards, M. P. (2002). A brief review of the archaeological evidence for Palaeolithic and Neolithic subsistence. *European Journal of Clinical Nutrition*, *56*(12).

Richards, M. P., & Hedges, R. E. (1999). Stable isotope evidence for similarities in the types of marine foods used by Late Mesolithic humans at sites along the Atlantic coast of Europe. *Journal of Archaeological Science*, *26*(6), 717–722.

Richards, M. P., Schulting, R. J., & Hedges, R. E. (2003). Archaeology: Sharp shift in diet at onset of Neolithic. *Nature*, *425*(6956), 366.

Rosado, J. L., Gonzalez, C., Valencia, M. E., López, P., Palma, M., López, B., & Báez, M. D. C. (1994). Lactose maldigestion and milk intolerance: A study in rural and urban Mexico using physiological doses of milk. *The Journal of Nutrition*, *124*(7), 1052–1059.

Rossi, M., Maiuri, L., Fusco, M. I., Salvati, V. M., Fuccio, A., Auricchio, S., & Semenza, G. (1997). Lactase persistence versus decline in human adults: Multifactorial events are involved in down-regulation after weaning. *Gastroenterology*, *112*(5), 1506–1514.

Sabeti, P. C., Reich, D. E., Higgins, J. M., & Levine, H. Z. (2002). Detecting recent positive selection in the human genome from haplotype structure. *Nature*, *419*(6909), 832.

Sahi, T. (1994). Genetics and epidemiology of adult-type hypolactasia. *Scandinavian Journal of Gastroenterology*, *29*(Suppl. 202), 7–20.

Salque, M., Bogucki, P. I., Pyzel, J., Sobkowiak-Tabaka, I., Grygiel, R., Szmyt, M., & Evershed, R. P. (2013). Earliest evidence for cheese making in the sixth millennium BC in northern Europe. *Nature*, *493*(7433), 522–525.

Schirru, E., Corona, V., Usai-Satta, P., Scarpa, M., Cucca, F., De Virgiliis, S., & Jores, R.-D. (2007). Decline of lactase activity and c/t-13910 variant in Sardinian childhood. *Journal of Pediatric Gastroenterology and Nutrition*, *45*(4), 503–506.

Schlebusch, C. M., Sjödin, P., Skoglund, P., & Jakobsson, M. (2013). Stronger signal of recent selection for lactase persistence in Maasai than in Europeans. *European Journal of Human Genetics*, *21*(5), 550.

Ségurel, L., & Bon, C. (2017). On the evolution of lactase persistence in humans. *Annual Review of Genomics and Human Genetics*, *18*(1), 297–319.

Shahack-Gross, R., Berna, F., Karkanas, P., Lemorini, C., Gopher, A., & Barkai, R. (2014). Evidence for the repeated use of a central hearth at Middle Pleistocene (300 ky ago) Qesem Cave, Israel. *Journal of Archaeological Science*, *44*, 12–21.

Sherratt, A. (1981). *Plough and pastoralism: Aspects of the secondary products revolution*. Cambridge University Press.

Simoons, F. J. (1969). Primary adult lactose intolerance and the milking habit: A problem in biological and cultural interrelations. *Digestive Diseases and Sciences*, *14*(12), 819–836.

Simoons, F. J. (1970). Primary adult lactose intolerance and the milking habit: A problem in biologic and cultural interrelations. *The American Journal of Digestive Diseases*, *15* (8), 695–710.

Stiner, M. C. (2001). Thirty years on the "Broad Spectrum Revolution" and paleolithic demography. *Proceedings of the National Academy of Sciences*, *98*(13), 6993–6996.

Stiner, M. C., Munro, N. D., Surovell, T. A., Tchernov, E., & Bar-Yosef, O. (1999). Paleolithic population growth pulses evidenced by small animal exploitation. *Science*, *283*(5399), 190–194.

Sverrisdóttir, O. Ó., Timpson, A., Toombs, J., Lecoeur, C., Froguel, P., Carretero, J. M., & Thomas, M. G. (2014). Direct estimates of natural selection in Iberia indicate calcium absorption was not the only driver of lactase persistence in Europe. *Molecular Biology and Evolution*, *31*(4), 975–983.

Swallow, D. M. (2003). Genetic influences on carbohydrate digestion. *Nutrition Research Reviews*, *16*(1), 37–43.

Swallow, D. M., & Hollox, E. J. (2000). The genetic polymorphism of intestinal lactase activity in adult humans. In C. Scriver, A. Beaudet, W. Sly, & D. Valle (Eds.), *Metabolic basis of inherited disease* (pp. 1651–1663). McGraw-Hill.

Swallow, D. M., & Troelsen, J. T. (2016). Escape from epigenetic silencing of lactase expression is triggered by a single-nucleotide change. *Nature Structural and Molecular Biology*, *23*(6), 505–508.

Tilden, C. D., & Oftedal, O. T. (1997). Milk composition reflects pattern of material care in prosimian primates. *American Journal of Primatology*, *41*(3), 195–211.

Timpson, A., Gillis, R. E., Manning, K., & Thomas, M. G. (2018). Modelling caprine age-at-death profiles using the Gamma distribution. *Journal of Archaeological Science*, *99*, 19–26.

Tishkoff, S. A., Reed, F. A., Ranciaro, A., Voight, B. F., Babbitt, C. C., Silverman, J. S., & Osman, M. (2007). Convergent adaptation of human lactase persistence in Africa and Europe. *Nature Genetics*, *39*(1), 31.

Toba, T., Nagashima, S., & Adachi, S. (1991). Is lactose really present in plants? *Journal of the Science of Food and Agriculture*, *54*(2), 305–308.

Urashima, T., Fukuda, K., & Messer, M. (2012). Evolution of milk oligosaccharides and lactose: A hypothesis. *Animal*, *6*(3), 369–374.

Urashima, T., Saito, T., Nakamura, T., & Messer, M. (2001). Oligosaccharides of milk and colostrum in non-human mammals. *Glycoconjugate Journal*, *18*(5), 357–371.

Urashima, T. A. S., & Messer, M. (2007). Milk oligosaccharides. In G. B. JP Kamerling, Y. C. Lee, A. Suzuki, & A. V. N. Taniguchi (Eds.), *Comprehensive glycoscience* (Vol. 4, pp. 695–724). Amsterdam, The Netherlands: Elsevier.

Vigne, J.-D., & Helmer, D. (2007). Was milk a "secondary product" in the Old World Neolithisation process? Its role in the domestication of cattle, sheep and goats. *Anthropozoologica*, *42*(2), 9–40.

Vigne, J.-D., Carrere, I., Briois, F., & Guilaine, J. (2011). The early process of mammal domestication in the Near East: New evidence from the Pre-Neolithic and Pre-Pottery Neolithic in Cyprus. *Current Anthropology*, *52*(S4), S255–S271.

Vorbach, C., Capecchi, M. R., & Penninger, J. M. (2006). Evolution of the mammary gland from the innate immune system? *Bioessays*, *28*(6), 606–616.

Wahlqvist, M. L. (2015). Lactose nutrition in lactase nonpersisters. *Asia Pacific Journal of Clinical Nutrition, 24*(S1), s21–s25.

Wang, Y., Harvey, C., Rousset, M., & Swallow, D. M. (1994). Expression of human intestinal mRNA transcripts during development: Analysis by a semiquantitative RNA polymerase chain reaction method. *Pediatric Research, 36*(4), 514.

Wang, Y., Harvey, C. B., Hollox, E. J., Phillips, A. D., Poulter, M., Clay, P., & Swallow, D. M. (1998). The genetically programmed down-regulation of lactase in children. *Gastroenterology, 114*(6), 1230–1236.

Warinner, C., Hendy, J., Speller, C., Cappellini, E., Fischer, R., Trachsel, C., & Swallow, D. M. (2014). Direct evidence of milk consumption from ancient human dental calculus. *Scientific Reports, 27*(4), 7104.

Watson, A. (2009). *The evolution of international society: A comparative historical analysis reissue with a new introduction by Barry Buzan and Richard Little*. Routledge.

Weiss, E., Kislev, M. E., & Hartmann, A. (2006). Autonomous cultivation before domestication. *Science, 312*(5780), 1608–1610.

Wilde, S., Timpson, A., Kirsanow, K., Kaiser, E., Kayser, M., Unterländer, M., & Thomas, M. G. (2014). Direct evidence for positive selection of skin, hair, and eye pigmentation in Europeans during the last 5,000 y. *Proceedings of the National Academy of Sciences, 111*(13), 4832–4837.

Wollstonecroft, M. M. (2011). Investigating the role of food processing in human evolution: A niche construction approach. *Archaeological and Anthropological Sciences, 3*(1), 141–150.

Wollstonecroft, M. M., Ellis, P. R., Hillman, G. C., Fuller, D. Q., & Butterworth, P. J. (2012). A calorie is not necessarily a calorie: Technical choice, nutrient bioaccessibility, and interspecies differences of edible plants. *Proceedings of the National Academy of Sciences, 109*(17), E991.

Wrangham, R., & Conklin-Brittain, N. (2003). Cooking as a biological trait. *Comparative Biochemistry and Physiology Part A: Molecular & Integrative Physiology, 136*(1), 35–46.

Wrangham, R., & Carmody, R. (2010). Human adaptation to the control of fire. *Evolutionary Anthropology: Issues, News, and Reviews, 19*(5), 187–199.

Wrangham, R. W., Jones, J. H., Laden, G., Pilbeam, D., Conklin-Brittain, N., Brace, C. L., & O'Connell, J. (1999). The raw and the stolen: Cooking and the ecology of human origins. *Current Anthropology, 40*(5), 567–594.

Xie, F., Flanagan, B. M., Li, M., Sangwan, P., Truss, R. W., Halley, P. J., & Dean, K. M. (2014). Characteristics of starch-based films plasticised by glycerol and by the ionic liquid 1-ethyl-3-methylimidazolium acetate: A comparative study. *Carbohydrate Polymers, 111*, 841–848.

Yang, D. Y., Liu, L., Chen, X., & Speller, C. F. (2008). Wild or domesticated: DNA analysis of ancient water buffalo remains from north China. *Journal of Archaeological Science, 35*(10), 2778–2785.

Yang, Y., He, M., Cui, H., Bian, L., & Wang, Z. (2000). The prevalence of lactase deficiency and lactose intolerance in Chinese children of different ages. *Chinese Medical Journal (Beijing), 113*(12), 1129–1132.

Yang, Y., Shevchenko, A., Knaust, A., Abuduresule, I., Li, W., Hu, X., & Shevchenko, A. (2014). Proteomics evidence for kefir dairy in Early Bronze Age China. *Journal of Archaeological Science, 45*, 178–186.

Zeder, M. A. (2008). Domestication and early agriculture in the Mediterranean Basin: Origins, diffusion, and impact. *Proceedings of the National Academy of Sciences, 105* (33), 11597–11604.

Zeder, M. A. (2015). Core questions in domestication research. *Proceedings of the National Academy of Sciences, 112*(11), 3191–3198.

Zeder, M. A., & Hesse, B. (2000). The initial domestication of goats (*Capra hircus*) in the Zagros Mountains 10,000 years ago. *Science, 287*(5461), 2254–2257.

CHAPTER

2 Digestion, absorption, metabolism, and physiological effects of lactose

Andrew Szilagyi

Jewish General Hospital, McGill University School of Medicine, Division of Gastroenterology, Montreal, QC, Canada

2.1 Introduction

The disaccharide lactose composed of glucose and galactose (Fig. 2.1, panel A) represents the primary carbohydrate in mammalian milk. The only mammals which do not produce lactose-based milk are the sea lions and some Pacific marsupials. Nevertheless, milk produced by these mammals is still based on a trisaccharide molecule made up combined of galactose (Coelho, Berry, & Rubio-Gozalbo, 2015). Lactose, consisting of two aldohexoses, chemically classified as O-β-D-galactopyranosyl-(1−4)-β-glucose, is produced only in the mammary gland. There is some debate whether lactose exists in any other taxonomic group of the biological classification system. Initially, some plants, forsythia and tropical shrubs, were reported to contain lactose (Brüssow, 2013; Kuhn and Low, 1949). Similarly, galactose may be present without combining with glucose in some plants. However, some authors failed to detect lactose or galactose (Toba, Nagashima, & Adachi, 1991), while others did report variable contents of galactose (Acosta & Gross, 1995; Gross & Acosta, 1991; Kim, Hartnett, & Scaman, 2007) and trace amounts of lactose in green tea leaves (Shanmugavelan et al., 2013). Therefore while the presence of galactose in small amounts is detectable in several fruits and some vegetables, the finding of lactose remains somewhat more contentious and may be present only in trace amounts.

In the mammary gland transfer of activated uridine diphosphate (UDP)-galactose to an acceptor glucose catalyzed by the enzyme galactosyl transferase forms lactose. During lactation there is also coexpression of α-lactalbumin with galactosyl transferase which leads to the formation of lactase synthase in the right conformation (Adam, Rubio-Texeira, & Polaina, 2004). This process is

Lactose. DOI: https://doi.org/10.1016/B978-0-12-811720-0.00002-7

modulated by prolactin, produced by the pituitary and is coordinated with declining progesterone levels postpartum (Brüssow, 2013). The evolution and teleological developmental functions of maternal milk feeding is reviewed by Power and Schulkin (2013).

In comparison with milk from other species, human milk is considered unique in terms of its high sugar content. It contains about 70 g/L lactose (7%) making up 30%–40% of caloric value in neonates (Schaafsma, 2008). Another 5–8 g/L is added by galactose containing oligosaccharides (galactose-linked by β-glycosidic bond to lactose) in human milk (Bode, 2012), which are not digested by the host but reach the colon intact and are supposed to have prebiotic functions there. The latter is thought to contribute to a predominance of intestinal bifidobacteria (Chichlowski, German, Lebrilla, & Mills, 2011) as discussed in more detail in Chapter 4, Lactose—a conditional prebiotic? In contrast, bovine milk contains 46 g/L (4.6%) of lactose (Wijesinha-Bettoni & Burlingame, 2013) and only a small fraction of bovine milk oligosaccharides, which are much less diverse compared to human milk oligosaccharides (HMO). Milk sugars ensure that galactose levels do not become limiting during postnatal early development, in order to guarantee its crucial structural function, for example, for the brain (Coelho, Berry, & Rubio-Gozalbo, 2015 and references therein, Prado & Dewey, 2014)

A number of questions pertaining to lactose are raised by its unique position in mammalian biology. The disaccharide has a relatively low osmotic load (Coelho, Berry, & Rubio-Gozalbo, 2015 and references therein). On a teleological level it is hypothesized that lactose closely matches with required solubility in both synthesis and secretion (Naim, 2001).

In the food industry lactose is used to improve the quality of taste and food coloring (Lee & Szilagyi, 2012). In the pharmaceutical industry lactose represents an excipient in medications. It is present in both oral medications and inhalants (Kaialy et al., 2011; Mahesh Kumar & Misra, 2006; Nakate et al., 2005). In industry lactose exists in two crystalline forms: α and β lactose. The α and β lactose configurations differ in their solubility and their formation and ratio to each other are temperature dependent (Lee & Szilagyi, 2012). In rat models there may be some differential effects in weight gain with the α form causing greater gain (Baker et al., 1967). The presence of any human-related effects is not established. Lactose in pharmaceutical applications and as a component of dairy foods influenced by the processes in the dairy chain will be discussed further in Chapter 5, Application of lactose in the pharmaceutical industry, and Chapter 6, Lactose in the dairy production chain, respectively.

The digestion of lactose into its monosaccharide components requires the intestinal enzyme lactase (later on just referred to as "lactase"). While glucose directly enters the glycolysis and is largely used for energy (and some structural roles), galactose requires a unique universal metabolic pathway for its catabolism and potential detoxification, but has an important structural function and roles in neurological and immunological processes. Because of the specific functions of galactose it is also surmised that digestion of lactose present in milk ensures adequate supply of this monosaccharide to newborns (Coelho, Berry, & Rubio-Gozalbo, 2015, and references therein).

Under physiological conditions weaning in all mammalian neonates may be triggered by diminishing lactose content in mammary glands of the mother. This event correlates with diminishing intestinal lactase levels in the infant (Brüssow, 2013). The complex physiology of breast-feeding and weaning are reviewed by Lawrence and Lawrence (2016).

A unique aspect in humans with respect to lactose digestion is the evolutionary division into two phenotypes as adults. Those who retain the ability to digest lactose by intestinal lactase are considered as LP while those who lose the ability in adulthood to digest large quantities of lactose at one meal are considered as LNP. The effect of lactose dose is discussed further below and more comprehensively in Chapter 3, Lactose intolerance and other related food sensitivities. The genetic regulation of the lactase gene (LCT) including mutations leading to LP as well as diagnostics, phenotypic characteristics, and consequences of the LP and LNP genotypes are described in more detail in Chapter 1, The evolution of lactose digestion, and Chapter 3, Lactose intolerance and other related food sensitivities, respectively. The division into LP and LNP originates through Mendelian genetic control of the promoter region away from the LCT and results in dominant LP and recessive "wild-type" LNP status. In rodents intestinal lactase may be reinduced in adulthood with lactose feeding establishing lactase as an inducible enzyme in some mammals (Mitra, Sachan, Udupa, Seshadri, & Jayakumar, 2003; Peuhkuri et al., 1997). However, this is not the case in humans (Gilat, Russo, Gelman-Malachi, & Aldor, 1972).

While the evolution of the genetics of lactase is described in Chapter 1, The evolution of lactose digestion, and the clinical relevance of lactose intolerance is discussed in Chapter 3, Lactose intolerance and other related food sensitivities, synthesis, catalytic characteristics, and regulation of the enzyme lactase are outlined here in the frame of lactose digestion. In addition, transport processes involved in the absorption of the digestive products of lactose are described. The metabolic effects of lactose and its digestive products in the human body are reviewed and discussed, giving an overview of the role of lactose in human physiology. The specific importance of lactose in neonatal development as described through the biochemical routes of the lactase enzyme products, glucose, galactose, and liberated ceramides, is emphasized.

Moreover, the importance of intestinal bacterial metabolism of excess lactose not digested by the host in the neonate and human adult is discussed as well. While further details on bacterial lactose fermentation are dealt with in Chapter 4, Lactose—a conditional prebiotic?, the impact of differential handing of lactose in LP and LNP subjects on physiological processes and diseases and the deduced need for applying a Mendelian randomization method to reduce confounding in human studies are addressed in this chapter.

2.2 Digestion of lactose by intestinal lactase in mammalian hosts

Virtually all mammals are born with intestinal lactase and are generally able to digest lactose. The lactase located in the brush border of the upper small intestine, duodenum and jejunum, splits lactose into the monosaccharides, glucose, and galactose. In humans rare cases of congenital lactase deficiency due to mutations of the LCT itself have been described (Robayo-Torres & Nichols, 2007; Uchida et al., 2012). Loss of the enzyme leads to severe symptoms of diarrhea, nephrocalcinosis, acidosis, and hypercalcemia with its related complications (Saarela, Similä, & Koivisto, 1995) upon exposure to dietary lactose.

In mammals LCT expression initiates before birth, remains high during the nursing period and then progressively declines after weaning, resulting in LNP. Lactase activity is first detected around the 12th gestational week and levels remain below those of term neonate through 26–34 weeks of gestation. At 40 weeks' gestation it reaches peak expression (Antonowicz & Lebenthal, 1977; Grand, Watkins, & Torti, 1976; Nilsson, 2016; Rasinpera et al., 2005).

In addition to lactose, the mammary gland also produces an array of bioactive proteins and growth factors with functions in the neonate. One of these is fibroblast growth factor 21 (FGF-21).This growth factor is normally active in liver, white and brown fat and is instrumental in maturation of gut enzymes' expression including lactase. This is achieved by high intestinal expression of the auxiliary coreceptor for FGF-21 found in neonate, but not adult rodents (and hypothesized to be similar to the human state) (Gavaldà-Navarro et al., 2015).

Due to developmental regulation, the concrete mechanism of which still needs to be elucidated in humans' expression of the gene encoding LCT declines with age. This physiological downregulation of lactase expression occurs postweaning without intestinal injury or disease. Moreover, lactase is not affected by dietary changes and lactose intake. The enzyme activity is not induced by lactose feeding and neither reduced by the absence of lactose in the diet (Kogut, Donnell, & Shaw, 1967; Gilat, Russo, Gelman-Malachi, & Aldor, 1972).

In human, due to single genetic alterations in the promoter region situated in a *cis* position, approximately 14,000 kb upstream from lactase, about one-third of the human population, retains the ability to continue to digest lactose into adulthood. In the rest of the populations, intestinal lactase falls to about 10% of original neonatal levels. This Mendelian trait distributes phenotypes in populations along fairly distinctive geographical lines.

As discussed in greater detail in Chapter 1, The evolution of lactose digestion, several polymorphisms in the gene preceding and regulating LCT, which allowed the digestion of lactose after weaning, emerged at similar time periods (7000–10,000 years ago) in different places (Enattah et al., 2002; Gerbault, Moret, Currat, & Sanchez-Mazas, 2009; Tishkoff et al., 2007). In the remaining two-thirds of the world's population, lactase quantity and activity diminishes in a spotty manner along the brush border (Maiuri et al., 1991) at various time periods as the infant ages. Maiuri et al. (1994) also demonstrated that after weaning variability in lactase mRNA alone, protein alone, or activity in some villi appears, while not observed in other villi. These findings suggested that both transcriptional and post-transcriptional factors cause a decline in lactase levels to drive adult lactase hypolactasia (Rossi et al., 1997). As stated above two additional observations play a role in the timing of decline. A polymorphism has been detected which may account for a delay in LCT decline (Baffour-Awuah et al., 2015). The prevalence and ethnic distributions of this polymorphism are not yet clear. Secondly the recent recognition that epigenetic factors participate in control of LCT mRNA may also participate in the decline of LCT with age (Labrie et al., 2016; Oh et al., 2017). As a result persons with reduced levels of intestinal lactase digest suprathreshold quantities of lactose by an alternate route. Intestinal bacteria take over some of the functions of lactase. In addition to adult LNP populations, people suffering from diseases which affect intestinal lactase levels (secondary lactase insufficiency) require metabolism via this alternate route. In the following sections, characteristics, regulation, and functions of intestinal lactase are described.

2.2.1 Nomenclature, structure, and catalytic activity of the intestinal brush border enzyme lactase

Lactase phlorizin hydrolase (LPH, systemic name: lactose galactose hydrolase, referred to as lactase post script) is a glycoside hydrolase. It belongs to the class of β-glycosidases and catalyzes the hydrolysis of the glycosidic bound in the disaccharide lactose (β-D-galactopyranosyl-(1→4)-α-D-glucopyranose) to yield D-galactose and D-glucose (EC 3.2.1.108). The enzyme from intestinal mucosa is isolated as a complex that also catalyzes the reaction of EC 3.2.1.62 glycosylceramidase cf. EC 3.2.1.33 amylo-α-1,6-glucosidase (KEGG database).

The mature human enzyme lactase is an integral glycoprotein that spans the apical membrane of mature enterocytes. It is composed of two identical extracellular 160 kDa polypeptide chains forming a dimer and a short intracytoplasmic part. It is anchored uniformly to the brush border membrane of epithelial cells of all villi in the duodenum and jejunum (Naim, 2001; Skovbjerg, Sjöström, & Norén, 1981). This is accomplished by hydrophobic amino acids at the C-terminus which anchor the enzyme to the lipid hydrophobic membrane. Intraluminal hydrophilic amino acids on the other hand create two active catalytic sites. Lactase has four homologous structural domains, each of which is involved in the intramolecular organization of the enzyme.

2.2.2 Biosynthesis, transcriptional regulation, and maturation of lactase

The biosynthesis of lactase includes tightly differential/tissue regulated transcription and translation into a large precursor, which becomes *N*- and 0-glycosylated during the intracellular transport, and undergoes further posttranslational processing for maturation. It begins with a 55 kilo-base pair DNA segment on chromosome 2q21 (spanning 17 exons), which is transcribed onto mRNA through RNA polymerase (Kruse et al., 1988) employing several transcription factors, enhancers, and activators, such as Oct-1, GATA-6, Cdx2, and HNF1α as established in nonhuman mammals and proposed and partially established for humans (Jensen et al., 2011; Leseva et al., 2018; Lewinsky et al., 2005; and references therein). It is generally accepted that LCT expression is primarily regulated at the transcriptional level. For the nuclear factor, Oct-1, binding to the promoter enhancer region about 14,000 kb upstream of the LCT was shown as well as stimulation of reporter gene expression from lactase promoter constructs in vitro (Lewinsky et al., 2005). In adults with LP phenotype it was reported that the activator Oct-1 is binding more tightly to the mutated T variant of the C/T -13910 polymorphism of the promoter enhancer region thereby leading to the maintenance of lactase activity beyond weaning (also discussed in Chapter 1: The evolution of lactose digestion). The interaction with the wild-type C variant, however, is much weaker supporting the fact that the mutation phenotype (dominant) is causing the LP effect (Lewinsky et al., 2005). Recent studies revealed that the binding affinity of Oct-1 to the TT-variant of the homozygotic mutant genotype is increased due to its lower sensitivity to methylations. In contrast, the wild-type CC-variant is subject to methylations, which interfere with the binding of Oct-1, thereby lowering its binding affinity and in turn activating effect. The heterozygotic C/T variant ranges in between. Significantly higher DNA methylation levels across CpGs in CC versus C/T versus T/T individuals have been demonstrated in samples from pediatric patients between 8 and 23 years of age, which were shown to correlate with their LP/LNP phenotype and lactase enzymatic activity (Labrie et al., 2016; Leseva et al., 2018). Since it is believed that methylation pattern might differ during life span, methylation might be an age-related modulator of gene expression without changing the DNA sequence itself. This epigenetic mechanism might partially explain the developmental regulated LCT expression. Greater binding of Oct-1 at other mutation sites (-14010*C associated with LP in Africa vs ancestor variant -14010*G) has been reported as well with similar effects (Jensen et al., 2011).

Lactase mRNA is translated by membrane-bound ribosomes into a polypeptide of 1927 amino acids forming the so-called pre–pro-lactase with no active lactase function. The preliminary inactive molecule requires further modifications to remove specific molecules which then allow proper enzyme folding and function. Three enzymatic steps are required for maturation: *N*- and *O*-glycosylation, dimerization, and proteolytic cleavage (Amiri, Diekmann, von Köckritz-Blickwede, & Naim, 2015).

The primary translational product, pre–pro-lactase sequence, is transported into the endoplasmic reticulum (Norén & Sjöström, 2001). Here several subunits are cleaved off as the enzyme matures to form pro-lactase. Pro-lactase prevents further cleavage and allows proper folding into the quaternary structure

(Jacob, Bulleid, & Naim, 1995). When pro-lactase blebs off the endoplasmic reticulum, *N*-glycosylation and dimerization take place, which migrate to fuse with the Golgi apparatus. Here it acquires its *O*-glycosylation. As a result enzymatic activity increases fourfold. The enzyme containing vesicle cleaves off and subsequently fuses with the external brush border membrane (Naim et al., 1994). In the brush border enterocyte, pro-lactase is cleaved and forms the quaternary structure of lactase, an 1192 amino acid long, mature LPH (referenced by Naim, Sterchi, & Lentze, 1987; Norén & Sjöström, 2001). Only two remaining amino acids are cleaved off by pancreatic trypsin in the lumen for the final trimming of the enzyme (Diekmann, Behrendt, Amiri, & Naim, 2017; Zecca et al., 1998).

Interestingly, lactase activity is coordinated with sucrase isomaltase and malto glucoamylase 2 α-glucosidases activities, such that when lactase level is decreased the other two intestinal enzymes are increased (Amiri et al., 2015; Sterchi et al., 1990). It is thought that this coordinated interplay of enzymes is regulated by posttranscriptional mechanisms (Van Beers et al., 1995). The inverse relationship helps to distinguish between primary adult onset hypolactasia and secondary hypolactasia caused by general diseases involving the brush border of the intestine (differences and diagnostics are discussed in Chapter 3: Lactose intolerance and other related food sensitivities).

2.2.3 Enzyme substrates and catalytic sites of lactase

Substrates catalyzed by lactase include lactose, cellobiose, flavonoids and isoflavone glycosides, aryl-β-glycosides, for example phlorizin, *o*- and *m*-nitrophenyl-β-glycosides and glycosyl-β-ceramides (Day et al., 2000; Zecca et al., 1998). Interestingly, contrary to lactose, many of these substrates are of plant origin. Phlorizin, 1-[2-(β-D-glucopyranosyloxy)-4,6-dihydroxyphenyl]-3-(4-hydroxyphenyl)-1-propanone, consisting of the flavonoid phloretin and glucose, for example, is mainly derived from the bark of apple trees, but can also be chemically synthesized for therapeutic approaches, for example in diabetes management. Origin and function of glycoceramides as important glycolipids in mammals are discussed in Section 2.5.5. For the chemical structures of these substrates, see Fig. 2.1.

According to sequence homology, mature lactase contains two different regions (III and IV), each of them homologous to family 1 glycosidases. Indeed, the functional enzyme has two active catalytic sites, both hydrophilic and luminally located (Day et al., 2000; Sterchi et al., 1990) but with distinct substrate specificities. Domain IV harbors the lactase β-D-galactoside galactohydrolase activity, mainly catalyzing the hydrolysis of lactose into glucose and galactose, and domain III the phlorizin-glycosyl-*N*-acylsphingosine glucohydrolase activity, mainly catalyzing the hydrolysis of phlorizin into phloretin and glucose. These two main catalytic activities are reflected in the term "lactase phlorizin hydrolase," the earlier mentioned complete name of the enzyme.

An early in vitro experiment performed by Büller et al. (1989) using homogenates of total small intestines from rats of different ages, in which total and

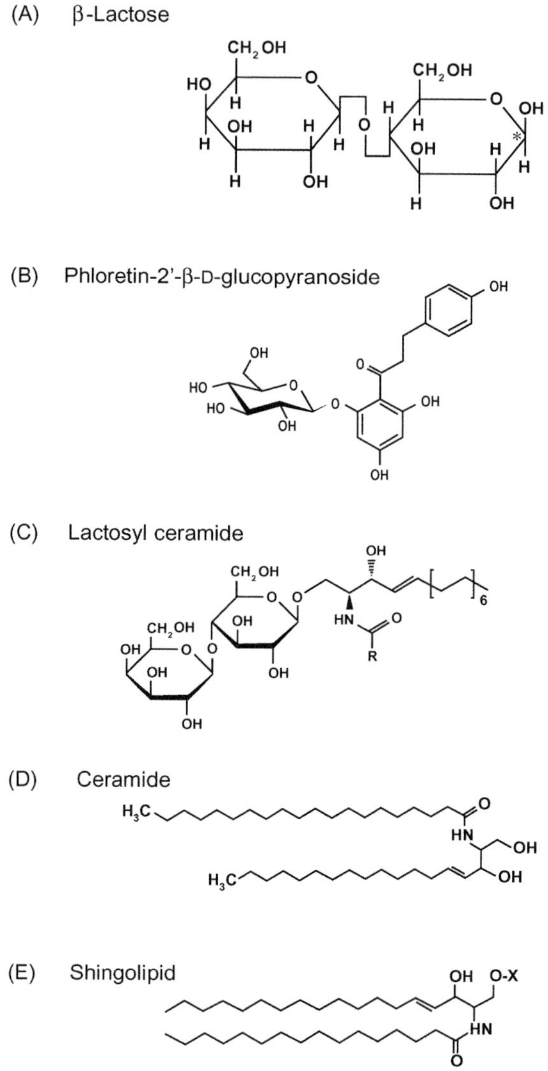

FIGURE 2.1

Chemical structures of various substrates of lactase and some important structural compounds related to early neonatal development in which lactase has an initial role are displayed. (A) Lactose, the disaccharide consisting of glucose and galactose. Lactose exists in two stereoisomers, β and α form, differing in the position of the hydroxyl group at the anomeric carbon (*) of the glucose moiety. β-D-Lactose is depicted here. (B) Phloretin-2′-β-D-glucopyranoside, the metabolic product of phlorizin, the flavonoid, which is the other major substrate for lactase. (C) Lactosylceramide, an example of a glycosylated

(Continued)

specific enzyme activities of LPH in comparison to other disaccharases were determined, revealed the developmental pattern of the enzyme and remaining lactase activity in the adult rat. Lactose hydrolysis was inhibited by glycoceramide (galactosyl- and lactosylceramide) by 38% in a noncompetitive enzymatic fashion. Therefore the authors concluded that the two substrates are likely catalyzed on separate sites. In addition, it was found that the enzyme showed 40- to 55-fold higher affinity for the glycolipids than for lactose. The authors suggested that lactase may still remain an important enzyme in adulthood in the rats playing an important role in the digestion of glycolipids, since lactose is not part of the adult rat diet (Büller et al., 1989).

Lactase is the primary intraluminal enzyme for the hydrolysis of flavonoids. Although phlorizin is considered to be a flavonoid, the digestion of the majority of other flavonoids and isoflavone glycosides actually occurs at the same active site as digestion of lactose (Day et al., 2000). Of note is that phlorizin and glycosylated ceramides (such as lactosylceramide) are currently the only known substrates which are digested at this second catalytic site.

2.2.4 Hormonal regulation of lactase function

In addition, genetic control of lactase expression developmental and functional hormones can affect lactase activity in mammals. However, whether such hormonal control is also present in humans is not well established. Most studies are carried out in small animal models like the rat or in vitro cell cultures, some of which are derived from human sources.

In rat models thyroxine (thyroid gland hormone) administered for 4 days decreased intestinal lactase activity (Chaudhry, Mahmood, & Mahmood, 2008; Yeh, Yeh, & Holt, 1991). In line with that, decreasing the amount of thyroid hormones led to a reversible increase in rat intestinal lactase activity (Freund et al., 1991). Administration of triiodothyronine (active molecule of thyroid hormone) to rats also decreased lactase mRNA expression (implying control at the transcription level) (Hodin, Chamberlain, & Upton, 1992). These studies were consistent. However, a more recent study by Suzuki, Mochizuki, and Goda (2014) using an in vitro model of IEC-6 (rat intestinal cells) reported that thyroxine, triiodothyronine, or glucocorticoids all enhanced the expression of lactase. Cortisone was shown to increase intestinal lactase activity in two other postnatal rat studies

◀ ceramide, a lipid—in this case—attached to lactose. Lactase has a major function in cleaving the carbohydrate portion of this molecule. (D) represents the unglycosylated ceramide structure, composed of sphingosine and a fatty acid. (E) Sphingolipid, any lipid containing the organic aliphatic amino alcohol sphingosine or a substance structurally similar to it. Ceramide is the most simple sphingolipid, where X is a hydrogen. Ceramide and other sphingolipids play significant roles in cell–cell communications involved with immunity surface structure and neurological development as described in Section 2.5.5.

(Chaudhry, Mahmood, & Mahmood, 2008; Yeh, Yeh, & Holt, 1991). In contrast, insulin did not affect lactase activity (Chaudhry, Mahmood, & Mahmood, 2008). As well, no effect of estrogen or progesterone on lactase was observed in an in vitro model of human intestinal cells (Salomon et al., 1996).

Currently the direct relevance of these animal or in vitro models to effects on human lactase is not clear, since hormonal effects in humans are less clearly demonstrated. Effects of thyroid gland dysfunctions on lactase are variable and unpredictable (Ebert, 2010; Heckl et al., 2016). Hormonal effects of thyroid and estrogens or progesterone on digestion or symptoms related to lactose may also be governed by hormonal effects on gut motility. These concepts are further discussed in Chapter 3: Lactose intolerance and other related food sensitivities.

2.3 Absorption and fate of lactose digestion products

In general most of the ingested lactose is broken down into its monosaccharide products, which are subsequently absorbed in the small intestine. However, in adult males Pimentel et al. found small amounts of intact blood lactose after ingestion of 0.8 L of milk. Levels were low at a median of 0.02 mmol/L and the level was inversely related to serum galactose (approximately 0.2 mmol/L) independent of lactase status (Pimentel et al., 2017) indicating that small amounts of lactose can be absorbed intact.

The bulk of ingested lactose, however, will be absorbed in the form of its D-monosaccharide products, glucose, and galactose, resulting from digestion. The sugars are absorbed from the upper intestinal lumen (duodenum and jejunum) into the bloodstream by essentially the same pathway. Both glucose and galactose are taken up into the enterocyte by cotransport with sodium using the same energy-dependent transporter (active transport) and are released into the circulation by facilitated diffusion using another transport protein—again the same for both monosaccharides. Both glucose and galactose are used for energy utilization in the body. In the newborn infant, lactose may provide as much as 40% of the caloric intake but only 3%–4% in the adult due to proportionally lower milk intake (and in addition lower lactose content in bovine vs human milk) (Kishnani & Chen, 2013). However, both have also structural functions, and galactose, in particular, is involved in multiple structural and cellular communication functions as well. The fate of both monosaccharides is described separately below.

2.3.1 Glucose

2.3.1.1 Glucose absorption

Glucose, one of the products from lactose hydrolysis, is an aldohexose and like most sugars in nature, normally existing as enantiomer D-glucose. It is absorbed through the enterocyte by several transporters. The primary transporters for glucose are the high affinity polar compounds, sodium–glucose cotransporters (also called symporters), SGLT1 and SGLT2. They carry a molecule of glucose together with two molecules of sodium across the brush border membrane from the intestinal lumen into the enterocyte against a gradient (uphill analogy). Cotransport is driven by active sodium extrusion by the basolateral sodium/potassium-ATPase, which requires energy provided by ATP hydrolyses (Ehrenkranz et al., 2005). While SGLT1 can transport either a molecule of glucose or galactose, SGLT2 is specific for glucose, but not for galactose transport (Harada & Inagaki, 2012). From the enterocyte glucose (as well as galactose and fructose) are transported across the basolateral membrane into the circulation mainly by GLUT2 through facilitated diffusion (downhill analogy) without utilization of energy. Other GLUT transporters can also come into play under certain nutritional conditions and in different tissues. These transporters are related to the solute carrier family, SCL2A2 class specifically (reviewed by Mueckler & Thorens, 2013).

Remarkably, phlorizin, one of the substrates for the enzyme lactase, is found to inhibit the energy-dependent glucose transporters, SGLT1 and SGLT2, but does not inhibit the energy-independent carrier GLUT. Phlorizin is converted to glucose and phloretin by lactase. The latter conversion of which then blocks GLUT, transport of chloride, bicarbonate, and lithium. However, phloretin facilitates the transport of urea and glycerol. It does not block SGLT (Ehrenkranz et al., 2005).

2.3.1.2 Glucose metabolism

From the circulation, glucose enters the liver, also via GLUT transporters, where it is metabolized further. Glucose catabolism starts with the first phosphorylation by hexokinase at the sixth position. The resulting glucose-6-P either can enter the glycolytic pathway (via conversion to fructose-6-phosphate) and subsequent Krebs cycle to generate ATP, and thus serve energy utilization, or can also enter a number of different pathways. One of those is the pentose phosphate pathway, which serves anabolic functions and is important for providing reducing equivalents (NADPH) and precursors for nucleotide synthesis. Another pathway of glucose-6-phosphate metabolism starts also with the formation of fructose-6-phosphate, which in turn serves as starting point of the hexosamine pathway producing UDP-*N*-acetylglucosamine (used to make glycosaminoglycans, proteoglycans, and glycolipids). Pyruvate generated in the glycolysis can be converted to acetyl-CoA, which serves as a precursor, for example, fatty acids (Adeva-Andany et al., 2016). Isomerization of glucose-6-phosphate to glucose-1-phosphate by phosphoglucomutase can be followed by transformation into UDP-glucose. This molecule is a precursor to glycogen formation (storage form of glucose) used for multiple functions, but mostly energy. Glycogen can be formed from the direct conversion of glucose or galactose. Alternatively, an indirect pathway is used starting from the formation of three carbon molecules, such as pyruvate (Barosa et al., 2012). The other metabolites produced from UDP-glucose, UDP-gluconate, and UDP galactose (detoxified form of galactose) are used for glycosylation of lipids and proteins with multiple functions (Adeva-Andany et al., 2016) as described in Section 2.5.5 on galactose functions. Fig. 2.2 outlines the metabolic pathway of glucose and the interconnections with the metabolism of galactose.

2.3.2 Galactose

Galactose has the same chemical formula as glucose, that is, $C_6H_{12}O_6$. It is similar to glucose in its structure, differing only in the position of one hydroxyl group. This difference, however, gives galactose different chemical and biochemical properties to glucose. Galactose, like glucose, is an ubiquitous aldohexose, which also exists as a D-isomer in nature. It is found in most living cells, including yeasts and bacteria. In addition to its presence as part of the milk sugar lactose in dairy foods, it is also a nutrient in some fruits and vegetables (Acosta & Gross, 1995; Coelho, Berry, & Rubio-Gozalbo, 2015). It can exist in free form or as bound galactose in which it forms complex carbohydrates (oligosaccharides, polysaccharides, and glycoproteins) (Coelho, Berry, & Rubio-Gozalbo, 2015, and other references in this chapter). The biological importance of galactose, however, goes beyond its importance as a nutrient and a metabolite. Galactose appears to have been selected by evolutionary pressure to also exert a crucial structural role. Indeed, despite the fact that it differs from glucose in the configuration of the hydroxyl group at the carbon-4 position only, galactose has a myriad of specific functional and structural roles in living organisms that cannot be exerted by glucose. Since galactose is considered to be toxic in its free form, virtually all living cells from unicellular to multicellular organisms have the metabolic process to detoxify galactose. This system is called the Leloir pathway, discovered by Federico Leloir, 1970, who was awarded with the Nobel Prize for his work. Apart from its importance in detoxifying galactose and energy production from this sugar, the Leloir pathway is crucial for the glycosylation of complex molecules, such as myelin. Galactocerebroside is the predominant glycolipid in this—therefore galactose was initially also called cerebrose—due to its functional significance for the brain (Coelho, Berry, & Rubio-Gozalbo, 2015).

2.3.2.1 Galactose absorption

The absorption of galactose from the intestinal lumen into the blood circulation follows the same pathway as described above for glucose, mainly employing the transporters SGLT1 and GLUT2.

2.3.2.2 Galactose metabolism

Absorbed β-D-galactose enters the portal vein and is transported into the liver. The majority of galactose load is retained in the liver, but small fractions are delivered directly to the brain and also in suckling mammals to the mammary

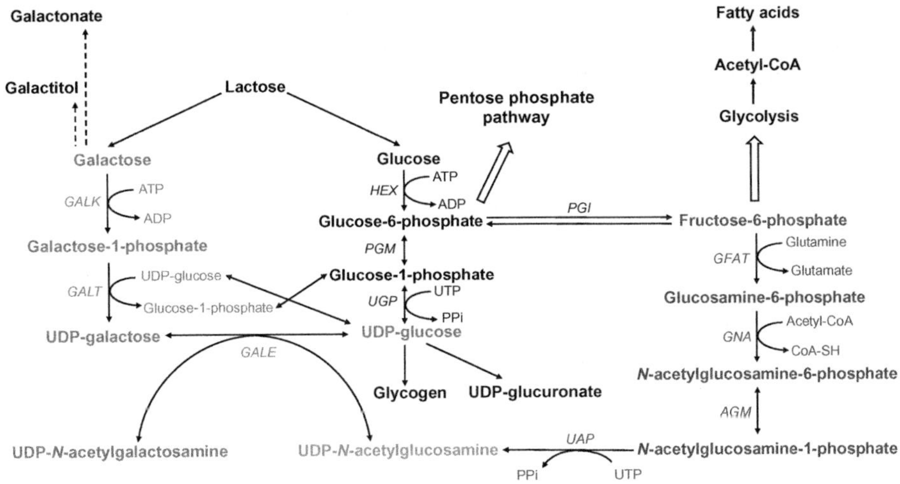

FIGURE 2.2

Biochemical pathways linking the metabolism of glucose, galactose, fructose, and *N*-acetylglucosamine/*N*-acetyl galactosamine in the liver. Galactose is metabolized by the universal Leloir pathway. Four enzymes are involved: the first one, galactose mutarotase, is not shown here but converts L-galactose to D-galactose, galactokinase (GALK) phosphorylates galactose, galactose-1-phosphate uridyltransferase (UDP) (GALT) converts galactose-1-phosphate to uridyldiphosphate (UDP) galactose, and UDP-galactose-4-epimerase (GALE) catalyzes the interconversion of UDP-galactose and UDP-glucose and their *N*-acetylated molecules. Glucose is phosphorylated by hexokinase (HEX) to glucose-6-phoshate. This compound can either be further converted via glucose-1-phosphate to UDP-glucose using the enzymes, phosphoglucomutase (PGM) and UTP—glucose-1-phosphate uridyltransferase—enter the pentose phosphate pathway or be converted to fructose-6-phosphate by glucose-6-phosphate isomerase (PGI). This compound initiates the hexosamine and glycolytic pathways. The enzymes involved in the hexosamine pathway are glucosamine-6-phosphate synthetase (GFAT), glucosamine-6-phosphate acetyltransferase (GNA), *N*-acetylglucosamine-phosphate mutase (AGM), and UDP-*N*-acetylglucosamine pyrophosphorylase (UAP). The other abbreviations are ATP, adenosine triphosphate, ADP, adenosine diphosphate, UTP, uridine triphosphate, UDP, uridine diphosphate, PPi, diphosphate, CoA, coenzyme A.

The figure was drawn based on reference book knowledge and information from the references: Adeva-Andany, M. M., Pérez-Felpete, N., Fernández-Fernández, C., Donapetry-García, C., & Pazos-García, C. (2016). Liver glucose metabolism in humans. Bioscience Reports, 36*(6). pii: e00416. Coelho, A. I., Berry, G. T., & Rubio-Gozalbo, M. E. (2015). Galactose metabolism and health.* Current Opinion in Clinical and Nutrition & Metabolic Care, 18, *422–427. Sanders, R. D., Sefton, J. M., Mobe, K. H., & Fridovich-Keil, J. L. (2010). UDP-galactose 4' epimerase (GALE) is essential for development of* Drosophila melanogaster. Disease Models & Mechanism, 3,*628–638.*

glands to reform lactose. Galactose is epimerized to its α-D form by galactose mutarotase, the first enzyme in the Leloir pathway (see Fig. 2.2). Galactokinase (GALK) phosphorylates α-galactose to galactose-1-phosphate, which requires energy from ATP hydrolysis. GALT removes an uridine monophosphate from UDP-glucose to form UDP-galactose and glucose-1-phosphate which reenters the glucose cycle. UDP galactose can act as a galactosyl donor for formation of glycoproteins (e.g., galectins; immune functions) and glycolipids (e.g., ceramides; structural and immune functions), as discussed in Section 2.5.5. In addition, UDP-galactose can be converted back to UDP-glucose by the fourth major enzyme of the Leloir pathway, UDP-GALE (Lai, Elsas, & Wierenga, 2009). The epimerase can also convert UDP-glucose to UDP-galactose to produce endogenous galactose, when nutritional galactose is limited to safeguard the availability of the molecule for its critical role in brain development.

There are also minor pathways of galactose metabolism. One pathway by aldose reductase leads to galactitol which is an end product and is excreted in urine. Oxidation of galactose by galactose dehydrogenase leads to the formation of galactonate which can be excreted or further metabolized through the pentose phosphate pathway (Coelho, Berry, & Rubio-Gozalbo, 2015). These pathways are important to process excess accumulations of galactose.

In its detoxified UDP form, galactose serves a number of essential functions (discussed in Section 2.5.5) in the fetus and neonate which cannot be replaced by glucose. In addition to lactose, endogenous galactose can be formed as well. It was hypothesized by Gitzelmann (1995) that a metabolic reaction of UDP-galactose pyrophosphorylase was the basis of endogenous galactose production. This source may contribute to toxicity in galactosemia as discussed in Section 2.3.3. However, as suggested by Coelho et al. the more likely source of additional galactose may be the turnover of galactosylated proteins and lipids in the Golgi apparatus (Berry & Walter, 2012). Endogenous synthesis of galactose is inversely related to age dependent with greater proportional quantities produced in the neonate than older children or adults (Schadewaldt, Kamalanathan, Hammen, & Wendel, 2004).

2.3.3 Toxic effects of unmodified galactose: galactosemia

Despite its important role for structural development in bound form (see Section 2.5.5) galactose is considered to be toxic in its free form and increases senescence (Cui et al., 2006). The toxic effects of incompletely utilized galactose are exemplified by a group of neonatal inborn errors of metabolic diseases which involve mutations in the key enzymes of the Leloir pathway. The diseases are rare. The incidence of the most common type (classical galactosemia) has been calculated to be about 1 in 47,000 based on genetic frequency in Caucasian neonates (Suzuki, West, & Beutler, 2001). The incidence in other ethnic groups may be similar but different dedicated mutations can be detected (Petry & Reichardt, 1998). These metabolic conditions start interfering with normal function in the neonate from the time they start to consume lactose. Mutations in three of the Leloir pathway enzymes give rise to galactosemia with variable toxic effects in neonates and may continue into adulthood.

2.3.3.1 Forms of galactosemia

To date no disease has been linked to galacto mutarotase. Deficiency of the second enzyme GALK leads primarily to neonatal cataracts (clouding of the lens of the eye leading to poor vision) (galactosemia II). This form is devoid of other organ problems that include the liver and ovaries and therefore more mild than the other forms of galactosemias (i.e., I and III).

Deficiency of the third enzyme GALT leads to classical galactosemia, the most common form (galactosemia I). Mutations of the fourth enzyme, UDP-GALE (galactosemia III), are also rare, and can lead to similar deficits as in galactosemia I. These range from mild to severe (galactosemia III) (Holden, Rayment, & Thoden, 2003).

2.3.3.2 Treatment of galactosemia

The primary treatment of all forms of galactosemia is restriction of galactose in the diet (Welling et al., 2017). Successful treatment requires early detection of metabolic abnormalities. Early detection reverses and improves clinical outcome in neonates. However, it has been observed that even patients on galactose restricted diet may develop long-term effects. This could be partly due to failure to restrict all galactose containing foods (Gross & Acosta, 1991). In a very small case series, however, Berry reported that in adult galactosemic patients, despite accounting for dietary intake of galactose from fruits and vegetables, metabolic measurements of urinary galactitol was consistently higher than expected based on intake. The authors concluded that the contribution of endogenous production of galactose can account for a failure to completely achieve dietary restrictions (Berry et al., 1993). Interestingly, since the beginning of the new millennium (2000) it has been observed in a number of case reports that despite liberalization of galactose intake in adults with genetic mutation of galactosemia I, there does not appear to be further deterioration of organ damage. One hypothesis for this is maturation and/or expansion of UDP-glucose pyrophosphorylase, apparently the only other enzyme that can convert galactose-1-phosphate to UDP-galactose. As a result of these observations the lifelong need for galactose-free diets has been questioned. The age and frequency of stable post neonatal metabolic development is still not well defined for now (reviewed by Van Calcar et al., 2014).

2.3.3.3 Physiological effects and mechanism of galactosemia

The main organs affected by the condition include the brain, eyes, liver, and ovaries reflecting the main functional targets linked to galactose. In the nervous system, abnormal glycosylation of glycoproteins, glycolipids, and mucopolysaccharides possibly during the prenatal period are thought to play a role in disease pathogenesis (discussed by Van Calcar et al., 2014). In the long term the concept of "self-intoxication" was proposed, based on the findings that large quantities of galactose are produced likely from production and degradation of different glycoconjugates (references in Van Calcar et al., 2014). The specific toxic effects are hypothesized to be due to either accumulation of galactose-1-phosphate or the deficiency of UDP-galactose. The first hypothesis is that the metabolic block due to GALT deficiency leads to the accumulation of galactose-1-phosphate. UDP galactose is the primary galactosyl donor for formation of glycolipids and glycoproteins. The metabolic accumulation of galactose-1-phosphate is associated with over-glucosylation and reduced galactosylation in brain tissue. Contribution of galactitol or galactonate from minor metabolic pathways adds to the toxic effects (reviewed by Ridel, Leslie, & Gilbert, 2005). However, as discussed by Van Calcar et al., the pathogenesis is likely to be more complex than simple accumulation of galactose-1-phosphate.

The second hypothesis is that toxicity occurs due to deficiency of UDP-galactose. UDP-galactose also serves as a substrate for the enzyme UDP-galactose: ceramide galactosyl transferase (CGT). It was postulated that consistent reduction of UDP-galactose at a threshold dose affects CGT as well, leading to further aggravation of the metabolic disturbance induced by GALT or GALE deficiency (Lebea & Pretorius, 2005). In galactosemia I intellectual retardation, neurological deficits (ataxia, tremor, dysarthria), cataract formation (intraocular hemorrhage), immunological deficits, liver and ovarian dysfunction are manifest into adulthood (Ridel, Leslie, & Gilbert, 2005). However, tolerance for galactose with aging improves and as stated the production of endogenous galactose diminishes compared with the neonatal and early childhood period.

Chronic exposure to D-galactose by subcutaneous injections rather than a dietary source is a standard method in rodent models to produce Alzheimer disease amyloid plaques. These are made up of fragments of degraded protein between neurons. In addition, the acute phase reactant protein serum amyloid Aβ may participate as well (Zhu et al., 2014; also reviewed by Van Dam & De Deyn, 2011). These models are then used to test various protective mechanisms. Abnormal D-galactose metabolism leads to reactive oxygen damage which induces aging and inflammation (Cannizzo et al., 2011).

While neonatal cataract and eye involvement are part of the clinical spectrum of galactosesmia I and II, there is some controversy whether ordinary senile cataracts (cataracts in elderly) are related to milk drinking and LP status (Birlouez-Aragon et al., 1993; Chacon-Camacho et al., 2014; Karas et al., 2003). The presence of LP is associated with increased capacity to consume dairy. However, the relation to increased dairy consumption is not clearly established since galactosemia may affect adults as well. This observation may lead to confusion in assigning etiological significance. Possible defects in handling of galactose in normal aging adults need to be further studied but have not been established.

Similarly ovarian dysfunction is part of galactosemia I and can develop in adulthood despite the elimination of galactose from diet. The role of GALT in normal ovarian function has been hypothesized and abnormally formed glycoprotein with follicle stimulating hormone in women with classical galactosemia has been reported (Petry & Reichardt, 1998). The role of galactose in non-galactosemic adults is unproven; however, the toxicity of galactose was put forth by Cramer as a hypothesis to explain some cases of ovarian cancer (Cramer, 1989). This is discussed in some more detail in Section 2.5.9.

2.4 Intestinal bacterial fermentation of lactose

Lactose escaping digestion by intestinal lactase (due to either a temporary high lactose ingestion in young children or low or absent lactase activity in premature infants or older LNP children and adults) can be transferred to the lower intestine. The absence of lactose digestion leaves the disaccharide intact and due to osmotic forces exerted by the intact molecule propels lactose into the lower intestine. With exception of the condition of bacterial overgrowth (Krajicek & Hansel, 2016), in which lactose is prematurely fermented in the upper gastrointestinal tract by invading bacteria, residual lactose is metabolized by lower intestinal bacteria instead. Premature infants have reduced lactase activity and this may occur in some neonates as well. In addition, diseases which compromise the integrity of the small intestinal villi could lead to reduced lactose digestion (Vesa, Marteau, & Korpela, 2000). Congenital lactase deficiency is somewhat different in that there is an absolute loss or mutation of intestinal lactase and is followed by severe symptoms in neonates soon after lactose is introduced in the diet (Robayo-Torres & Nichols, 2007; Saarela, Similä, & Koivisto, 1995; Uchida et al., 2012) as described in the section about digestion in Section 2.2. However, from a population perspective, adult LNP persons with lactase less than 10% of original levels (Enattah et al., 2007), who consume larger single doses of dairy foods and milk, would most likely depend on bacterial metabolism of lactose.

Variables which affect the likelihood of bacterial mediated lactose metabolism as well as magnitude of symptoms in LNP populations include the dose(s) ingested in a single bolus or during a limited period of time as well as a number of physiological conditions. Most of these modulate intestinal transit and the rate of delivery of substrate to the colon per unit time (these concepts are further described in Chapter 3: Lactose intolerance and other related food sensitivities).

2.4.1 End products of intestinal microbial metabolism in general

The nutrition of the intestinal microbiota is dependent on the diet of the host. Nutrients that bypass host digestion and metabolism traverse the gut and are metabolized by bacteria. The end products are derived from protein and mainly carbohydrate fermentation. End products of protein fermentation are branched chain fatty acids, such as valeric, isovaleric acid, 2-methlbutyrate derived from branch chain amino acids, as well as ammonia, thiols, amines, phenols, indoles, hydrogen sulfide. Those molecules may adversely affect the host. End products of carbohydrate fermentation are generally regarded as beneficial and include lactate, the short chain fatty acids (SCFAs), acetate, propionate, and butyrate, and less commonly, formate and methanolic acid as well as carbon dioxide, methane, and hydrogen gas (Macfarlane & Macfarlane, 2012). The latter gases might contribute to intestinal discomfort, when produced in large amounts. Acetate and propionate enter the host's portal system and contribute to the host's nutrition, for example, by the well-known hepatic conversion of propionate to glycogen and the role of acetate in de novo lipogenesis. Butyrate has multiple antiinflammatory and antineoplastic functions, and is also the main nutrient for colonocytes (Koh, De Vadder, Kovatcheva-Datchary, & Bäckhed, 2016).

2.4.2 Metabolic capacity of intestinal microbiota

The human microbiota consists of bacteria that have multiple abilities to metabolize different substrates, the generalists, and those that are more restricted in substrate specificity, the specialists. The predominant phyla of the adult gut microbiota (90%) consist of the Gram-negative generalists, Bacteroidetes, and the Gram-positive, largely specialists, Firmicutes (Cockburn & Koropatkin, 2016). Nutrition among the bacterial community largely depends on synergy or cross feeding. In this paradigm, specific species depend on the metabolic potential of other species that can degrade more complex polysaccharides (Rivière et al., 2015). The resulting smaller metabolites, for example acetic acid, lactic acid, or oligosaccharides derived from polysaccharides are then metabolized and stimulate the growth of others, for example, butyrate producing bacteria, such as *Eubacterium hallii*, *Faecalibacterium prausnitzii*, some *Roseburia* strains, and others (Duncan, Louis, & Flint, 2004; Louis & Flint, 2017; Rios-Covan, Gueimonde, Duncan, Flint, & de los Reyes-Gavilan, 2015). This principle also explains how bacteria-like lactobacilli and bifidobacteria, which do not themselves produce butyric acid, could lead to induction of butyric acid formation. The reason why specific carbohydrates (e.g., prebiotics) lead to expansion of these lactic acid producing bacteria rests on the finding that they compete more efficiently for these substrates (Falony, Calmeyn, Leroy, & de Vuyst, 2009; Falony, Lazidou et al., 2009).

2.4.3 β-Galactosidase in intestinal bacteria

The fermentation of lactose depends on bacteria that have enzymes with β-galactosidase activity able to hydrolyze the glyosidic bond in the disaccharide. β-Galactosidase activity has been found in the majority of fecal bacteria isolated and cultured from Chinese subjects (He, Priebe, Vonk, & Welling, 2005). More information on β-galactosidase expressing gut bacteria can be found in Chapter 4, Lactose—a conditional prebiotic? The resulting monosaccharides, glucose, and galactose enter biochemical pathways in the bacteria that are similar to the one of the host. As mentioned earlier the Leloir pathway handles galactose in most living organisms.

2.4.4 End products of intestinal lactose fermentation and their effects on the host

Symptoms produced by excess lactose entering the lower intestine of humans are attributed by He et al. (2008) to products of bacterial metabolism, such as SCFAs and gases like hydrogen, carbon dioxide, and methane. This concept that metabolic products from bacteria contribute to symptoms of lactose intolerance is reproduced in Fig. 2.3.

While SCFAs are typical metabolic end products of the adult gut microbiota in all humans, methane production has been observed only in 41% in men and 49% in women of the general population (Pitt, de Bruijn, Beeching, Goldberg, & Blendis, 1980). Methane production has also been hypothesized to contribute to gastrointestinal symptoms but the role is not settled, except the role methane plays in increasing accuracy of the lactose breath test. Hydrogen breath tests are used to diagnose lactose maldigestion as described in more detail and discussed further in Chapter 3: Lactose intolerance and other related food sensitivities.

However, continued consumption of suprathreshold concentrations of lactose by LNP can result in reduction of symptoms (Habte, Sterby, & Jijalmassen, 1973; Sadre & Karbasi, 1979). This observable symptomatic adaptation to lactose may have several causes, including intestinal microbial adaptation, but is not

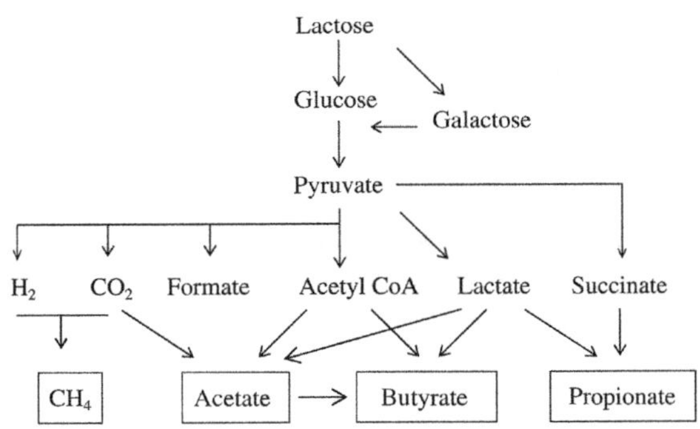

FIGURE 2.3

Metabolism of lactose by intestinal bacteria. The enzyme β-galactosidase splits lactose. The Leloir pathway converts galactose to glucose, which is further converted to pyruvate. From pyruvate SCFAs are formed via different routes and intermediates. Butyrate is also produced by second tier stimulation of butyrogenic bacteria. Each SCFA is in turn used by the host for a variety of purposes. In addition, SCFA gases are formed and used by bacteria for production of methane.

He, T., Venema, K., Priebe, M. G., Welling, G. W., Brummer, R. J., & Vonk, R. J. (2008). The role of colonic metabolism in lactose intolerance. European Journal of Clinical Investigation, 38, *541–547.*

associated with the availability of bacterial β-galactosidase. Using feces from 28 Chinese LNP persons, who had undergone two lactose challenges, He et al. found that 80 % of the cultured fecal bacteria possessed beta-galactosidase activity, independently if the feces donors have reported lactose intolerance symptoms or not (He, Priebe, Vonk, & Welling, 2005).

It was reported that expansion of lactic acid producing bacteria (e.g., lactobacilli and bifidobacteria) was correlated with increased lactose metabolism (Jiang & Savaiano, 1997a). The authors hypothesized that since these bacteria produce less hydrogen, expansion of lactic acid bacteria in humans may account for reduced hydrogen in breath after a lactose test. Lactic acid bacteria produce less hydrogen compared with *Bacteroides* and *Clostridia* and symptoms of lactose intolerance can be correlated with hydrogen output during breath tests (Ladas, Papanikos & Arapakis, 1982; Szilagyi et al., 2005). Symptoms in general attributed to lactose may be due to several other factors. As will be noted in Chapter 3, Lactose intolerance and other related food sensitivities, there is some blurring of distinct causes in both LP and LNP persons. As a result the focus of association with symptoms has shifted to metabolic products of the intestinal microbiota. In the case of lactose intolerance the metabolome has been investigated by Windey et al. in LP and LNP persons with or without lactose intolerance in vitro. By analyzing fecal samples from these persons incubated in vitro with lactose, the authors noted different metabolomic signatures among the three groups. 5-Methyl-2-furancarboxaldehyde levels were increased in the LNP group irrespective of tolerance pattern compared with the LP group. In addition, an unidentified aldehyde was also significantly increased in the LNP group, however, uniquely in the intolerant subgroup (Windey, Houben, Deroover, & Verbeke, 2015). Continued consumption of lactose over a few weeks results in changes in the microbiome which are measurable by the hydrogen breath test after repeated lactose challenges (Hertzler & Savaiano, 1996).

2.5 Physiological effects of lactose and its metabolites and impact on human health

Lactose exerts its metabolic effects either upon upper intestinal digestion by host lactase into its monosaccharide products, glucose, and galactose, and subsequent absorption, or by lower intestinal fermentation to lactate and SCFA by gut bacteria. In order to exert direct effects as an intact disaccharide, lactose needs to be exposed undigested to effective cells or tissues. Although lactose is digested slower than other disaccharides, that is, 50% slower than sucrose (He et al., 2008), the process is still quite rapid and is not the rate limiting factor in lactose utilization (Stellaard et al., 2000). This rate is about 134 μmol/min (Beyreiss, Hoepffner, Scheerschmidt, & Müller, 1985). In LNP people, stable lactose remains until the portion not digested by residual lactase reaches the lower intestine where it is also rapidly hydrolyzed by bacteria. Direct effects of lactose before metabolism would be exerted only during small bowel to colon transit. Whether the effects of excess lactose, which decrease intestinal transit time (i.e., faster transit), are adequate to perform the various immunological and metabolic outcomes that have been reported in in vitro models remain poorly defined. It would seem reasonable to consider that in vivo the primary metabolic effects of lactose depend on the effects on the microbiome. In LNP persons continued consumption of suprathreshold doses of lactose leads to emergence of lactic acid bacteria as first-line metabolizing species (Jiang & Savaiano, 1997b). This unequal process of lactose metabolism can be construed to modulate the microbiota which could manifest primarily in regular lactose consuming LNP persons (Szilagyi, 2004). In addition, the capacity of lactose to shift fecal microbiota toward bacteria generally regarded as beneficial has been determined in vitro and some prebiotic potential noted (Sanz, Gibson, & Rastall, 2005). However, the method does not take into account the physiological benefit of those shifts for the host required to call a substrate a prebiotic by definition. The concept of prebiotics is further discussed in Chapter 4, Lactose—a conditional prebiotic?

Information on the impact of lactose intake on health status, apart from the huge amount of reports on symptoms of lactose maldigestion and lactose intolerance (see Chapter 3: Lactose intolerance and other related food sensitivities), is rather scarce and scattered and probably compromised by the lack of proper distinction and stratification into different types regarding the LP/LNP genotype.

In the following, current literature on the physiological effects of lactose is discussed.

2.5.1 Effect of lactose on taste and the reward system

The first contact upon intake of a lactose-containing food item, such as dairy products, human milk, or infant formula, occurs in the mouth. Sweet receptors (Hoon et al., 1999; Nelson et al., 2001; Zhao et al., 2003) signal sugar molecules, for example, lactose, and the accompanied sweet taste is perceived. For the perceived sweetness of lactose, values between 16% and 40% are given (depending on concentration, pH, temperature, and other food components present). An average of 30% is generally accepted, although a recent review mentioned approximately 20% sweetness compared to that of sucrose (Clemens et al., 2016). The perception of sweet taste is mediated by the G-protein-coupled heterodimeric receptor TAS1R2/TAS1R3 present in taste bud cells. The temporal sweetness intensity profile of lactose is similar to that of sucrose. Because of its lower sweetness, lactose is unable to induce the release of mediators such as endorphins or dopamine and therefore is free of reward effects, in contrast with other sugars (Delaveau, 2002). Studies (Blass & Smith, 1992) testing the potency of different sugars as calming agents in human infant showed interesting differences. Small amounts (0.1 mL aliquots) of 0.51 M sucrose, fructose, glucose, or lactose were ingested orally by 1- to 2-day-old infants who were crying spontaneously. Fructose and sucrose showed similar calming effects, glucose slightly less, but lactose was not at all effective, similar to water. The oral administration (but not intragastric) of sucrose doubles the pain threshold in rats (Blass & Fitzgerald, 1988; Shide & Blass, 1989) and causes a 50% reduction in crying of human infants undergoing painful procedures such as blood collection through a heel prick (Blass & Hoffmeyer, 1991)

These studies show a distinctive different physiological response to lactose compared to other sugars.

2.5.2 Cariogenic effects of lactose compared to other sugars

It has been repeatedly shown that sweets, particularly, sucrose, leads to dental caries through altered pH and demineralization of teeth (Mundorff-Shrestha et al., 1994; Touger-Decker & van Loveren, 2003). However, sugars incorporated into cellular structures (e.g., part of other molecules) of foods are not known to have similar adverse effects (Sheiham, 2001). The anticariogenic effects of milk and dairy foods have been recognized in the last eight decades (Schweigert, Shaw, Zepplin, & Elvenhjem, 1946). Protective effects have been attributed to protein and calcium. Lactose, however, is listed as having a cariogenic potential because of its fermentability (Touger-Decker & van Loveren, 2003). Nevertheless, according to Moynihan (2002), lactose is considered as the least cariogenic of fermentable sugars and is not thought to have significant adverse dental effects (Sheiham, 2001).

Some dairy foods, such as cheese and yogurt without added sugar, are even protective against damage to teeth (Kashket & DePaola, 2002). In these products most of the lactose has been fermented to lactate during the production process (see also Chapter 6: Lactose in the dairy production chain describing the fate of lactose in the dairy chain). The format of dairy food may also be relevant. Whether pure lactose independent of the dairy matrix has adverse effects is less clear. Galactose monosaccharide is not considered to be cariogenic. Further work on comparison of cariogenicity may be required to ascertain the lower effects of lactose compared with sucrose.

2.5.3 Glycemic index of lactose and implications for metabolic health

Depending on their gastrointestinal handling, carbohydrates can be classified as "glycemic carbohydrates" (carbohydrates that are digested and absorbed in the small intestine and cause a rise in blood glucose, for example, free sugars, maltodextrins, and starch) and as "nonglycemic" (carbohydrates that are not absorbed in the small intestine and move down to become fermented in the colon). Examples of the latter are resistant starch, nonstarch polysaccharides, and sugar alcohols (Englyst & Englyst, 2005).

The glycemic index (GI) is a value associated with the carbohydrates in food that indicates the effect of the carbohydrate on a person's blood glucose level—simply said—a measure of how quickly a food causes blood sugar levels to rise. A value of 100 represents the standard, an equivalent amount of pure glucose or white bread (Wolever, 1990). The three GI ratings are low (55 or less), medium (56–69), and high (70 or more). In addition to being less sweet than sucrose (equivalent to about one-third of sucrose; Clemens et al., 2016), lactose has a lower GI (Hughes, Atchison, Hazelrig, & Boshell, 1989). The listed GI of lactose is 46 compared with glucose listed with 100 (https://www.gisymbol.com/gi-sugar/) and therefore falls into the low-GI category. In comparison, sucrose has a GI of 65, maltose of 105, maltodextrins of 85–105, and different starches have a GI ranging between 40 and 110 depending on the food matrix (http://www.nutrients-review.com/carbs/polysaccharides-starch.html). In a Swedish study with nine healthy volunteers the glycemic response following milk/milk protein–based test meals, containing 25 g intrinsic or added lactose, was significantly lower in comparison with a white wheat bread meal providing 25 g starch consisting of only glucose molecules (Gunnerud, Holst, Östman, & Björck, 2012). Other studies have shown that low-GI diets may result in weight loss, reduce blood sugar levels, and lower the risks of heart disease and type 2 diabetes. Potential mechanisms for reduction of type 2 diabetes include evidence that low-GI diets improve insulin sensitivity and β-cell function in people with type 2 diabetes and those at risk for type 2 diabetes (Augustin et al., 2015; Philippou & Al Dhaheri, 2017).

As a result lactose is considered to be more acceptable as a carbohydrate in diabetics. The glucose and insulinogenic response to galactose 25 g, glucose 25 g, or lactose 50 g are similarly low and thought to be due to a gut hormonal effect (incretin) (Ercan, Nuttall, Gannon, Redmon, & Sheridan, 1993).

However, more recently, in a clinical Mendelian randomization study from Denmark, consumption of dairy foods and milk was not found to reduce the risk of diabetes (Bergholdt, Nordestgaard, & Ellervik, 2015). Still further studies will be needed to confirm or refute clinical benefit of lactose in diabetes.

Stahel et al. hypothesized that galactose would result in greater insulin sensitivity than glucose and fructose due to its lower lipogenicity and propensity to be stored as glycogen. Indeed, in a recent study with rats, they could show that galactose at 15% of daily intake for a period of 9 weeks improved hepatic insulin sensitivity compared to glucose and fructose as determined by hyperinsulinemic-euglycemic clamp technique. Moreover, galactose caused an increase in fed-state hepatic glycogen content. Interestingly it also caused a shift in gut microbial populations—a decreased fecal Firmicutes:Bacteroidetes ratio generally regarded as favorable (Stahel, Kim, Xiao, & Cant, 2017).

In a prospective cohort study to investigate the relations between dietary GI, glycemic load, and other carbohydrate measures (added sugars, total sugars, glucose, sucrose, lactose, fructose, starch, carbohydrate) and depression in postmenopausal women, higher consumption of lactose was found to be significantly associated with lower odds of incident depression (Gangwisch et al., 2015).

2.5.4 Role of lactose in energy supply and endurance

Animal studies have shown that the switch in nutrition from parenteral (via placenta) in the fetus to enteral via colostrum, milk, and lactose in the newborn requires a multihormonal response to allow for gluconeogenesis (Hammon, Steinhoff-Wagner, Schönhusen, Metges, & Blum, 2012). Early neonatal fasting studies in guinea pigs suggest that there is early rapid glycogenolysis which maintains fasting glucose for 24 hours (Duee et al., 1983). However, in normal suckling pigs, gluconeogenesis is detected. Glycogen synthesis is derived from a direct pathway utilizing conversion of glucose-6-phosphate exchanged with fructose-6-phosphate and an indirect pathway of converting three carbon molecules (pyruvate) (Barosa et al., 2012). Galactose enters the pathway through conversion of galactose-1-phosphate to UDP galactose. These interconversions are displayed in Fig. 2.3. Using differential carbohydrate and heavy water labeling (which correlate with complete body water and fructose-6-phosphate conversion to glucose-6-p conversion), Barosa et al. determined that in healthy adults galactose contributed about 19% to glycogen synthesis in the liver, after a single breakfast meal.

The effects of galactose on energy utilization have been evaluated prior to exercise. Jentjens and Jeukendrup evaluated the glycemic and insulin response in trained athletes by providing drinks with either glucose, galactose, or trehalose. As expected glucose was the most glycemic with both galactose and trehalose being significantly less. Glucose also raised plasma insulin twofold higher than galactose or trehalose (Jentjens and Jeukendrup, 2003).

In high-intensity and prolonged exercise, supply of glucose is essential. Although muscle glycogen is the predominant source, hepatic glycogenolysis contributes significantly. In endurance exercise and athletes liver glycogen is reported to be lower. It is thus suggested that replenishment of carbohydrates of any type (>1.5 g/min) be used to prevent glycogen depletion (Gonzalez et al., 2016). In this scheme O'Hara et al. compared glucose or galactose oxidation during exercise but given 30 min before commencement of vigorous (cycling) 2 hours' exercise. Placebo was compared with 75 g glucose or galactose tracers in nine healthy athletes in random order. The main findings were that while peak oxidation of glucose was higher, the mean oxidation rates were not significantly different. Also while glucose was the source of oxidation in the first hour, galactose oxidation was delayed to the second hour. The authors concluded that both glucose and galactose supply exogenous carbohydrate for oxidation, but galactose delays utilization and as such reduces reliance on liver derived glycogen (O'Hara et al., 2012).

In adults energy utilization was compared with different carbohydrates to increase postexercise liver glycogen synthesis. Decombaz et al. carried out a controlled trial in 10 participants of postexercise glycogen synthesis. Using a mixture of either maltodextrin with fructose, or glucose or galactose, these authors found that the drinks with fructose or galactose were twice as effective as glucose in increasing glycogen synthesis (Décombaz et al., 2011).

In addition to the specific effects of lactose the consumption of milk has been evaluated as a potential replacement for hydration following exercise. The rationale was that milk is not associated with increased urine output when rehydration is carried out with a 150% (of sweat loss) volume repletion (Shirreffs, Watson, & Maughan, 2007).

The role of galactose utilization in LNP persons has not been specifically studied; however, as discussed in Chapter 3, Lactose intolerance and other related food sensitivities, most LNP persons can easily tolerate and assimilate at least 10 g of lactose at one meal. Above the approximate threshold of 15 g some lactose will enter the lower intestine. At that level both glucose and galactose are fermented by intestinal bacteria. The caloric salvage is then derived from SCFAs, gases, and other volatile metabolites. There is a body of literature which supports the adaptive metabolic and nutritional role of the microbiome in endurance athletes (reviewed by Mach & Fuster-Botella, 2017). As such the microbiome can contribute to prolonged energy response and further work may shed light on its value in LNP persons.

These studies suggest that lactose can supply a significant portion of carbohydrates for energy utilization and contribute to a balanced energy supply during endurance exercise.

2.5.5 Functions involving galactose

As described in the previous section, galactose is taken up by the liver where, at least in adults, it is converted to glucose and preferentially incorporated into hepatic glycogen (Barosa et al., 2012).

But also the brain (similarly to liver cells) and to a lesser extent the muscles have the capacity to take up and metabolize galactose (Roser et al., 2009). In a rat model, Roser et al. demonstrated that galactose was taken up by brain tissue and converted to bioactive amino acids, glutamine, and glutamate, and the important neuronal transmission inhibitor gamma amino butyric acid (GABA) (Roser et al., 2009). All three amino acids—glutamate, glutamine, and GABA—are essential neurotransmitters, with glutamine mainly serving as precursor for the other amino acid neurotransmitters, glutamate and aspartate (excitatory), and GABA (inhibitory) (Albrecht, Sidoryk-Węgrzynowicz, Zielińska, & Aschner, 2010; Struzyńska & Sulkowski, 2004), and were increased upon galactose feeding (Roser et al., 2009). Remarkably, the increase in the amino acid concentrations after galactose administration persists for longer periods of time than after glucose administration (Roser et al., 2009).

The functions served by galactose also include participation in protein and lipid glycosylation via UDP-galactose. This is a pivotal compound in the Leloir pathway, which requires the metabolism of galactose (or, alternatively, the conversion from UDP-glucose). The products resulting from glycosylations using UDP-galactose are necessary for fetal and neonatal cellular structural epithelial formation and immune cellular communications and effectors as well as neurological development (Coelho, Berry, & Rubio-Gozalbo, 2015). Galactose as part of the N-linked glycan moiety of glycosylated proteins, for example, is involved in the correct folding of the glycoprotein in the endoplasmic reticulum. Glycosylation of lipids and proteins also plays a role in cell-to-cell signaling and adhesion (a mechanism employed by cells of the immune system) via sugar-binding proteins called lectins (Varki et al., 2009). These carbohydrate-binding proteins are highly specific for sugar moieties.

Galactose is an important ligand of lectins. Lectins-binding galactose are called galectins. Upon galactose binding they form a characteristic β-sandwich molecular conformation characterized by a double connected layer of molecules (two opposing antiparallel β-sheets). To date 15 different galectins have been identified, of which 10 are active in humans.

They have multiple roles in recognition on the cellular and molecular level acting from within nuclei to cell surface and extracellular matrix (Kaltner et al., 2017). Their functions include immunological control (Kucuksezer & Ozdemir, 2016) and epithelial organization (Delacour et al., 2006; Viguier, Advedissian, Delacour, Poirier, & Deshayes, 2014) including neural outgrowth (Kaltner et al., 2017). Galectins are associated with the patho-etiology of a number of diseases as well. Some of the conditions which are published include obesity and type 2 diabetes (Menini, Iacobini, Blasetti Fantauzzi, Pesce, & Pugliese, 2016), chronic kidney disease (Chen & Kuo, 2016), metastatic cancer potentiation (Cousin & Cloninger, 2016), and cardiovascular disease (van der Hoeven et al., 2016). Although not completely evaluated, the functions of galectins continue to be important targets of research (Kaltner et al., 2017).

Many of the neuromodulatory functions of galactose are linked with ceramide (with which galactose—as part of the glycosylation moieties—forms sphingolipids and gangliosides). The important neuromodulatory and other functions of ceramides are discussed in the following section.

Via the formation of building blocks, such as precursors for myelination, galactose contributes to the maturation and regeneration of the nervous system.

In addition, human milk contains about 200 different oligosaccharides, called HMO (Bode, 2012), with galactose at the reducing end attached in repeating units to *N*-acetylglucosamine linked to lactose (Mills, Ross, Hill, Fitzgerald, & Stanton, 2011). In the neonate these oligosaccharides serve a prebiotic function which promote *Bifidobacterium* as the dominant genus in the infant intestinal microbiota. More information about HMOs and lactose, their potential prebiotic effects and synthesis can be found in Chapter 4, Lactose—a conditional prebiotic? and Chapter 6: Lactose in the dairy production chain and their immunological functions are reviewed by Triantis, Bode, and Van Neerven (2018).

2.5.5.1 *Glycosphingolipids containing galactose and ceramide (N-acylsphingosine)*

Galactose, including glucose, *N*-acetylgalactosamine, and *N*-acetylneuraminic acid, can be part of the carbohydrate moiety (glycan) of a very important group of molecules, the glycosylated sphingolipids (Fig. 2.4). Glycosphingolipids, or

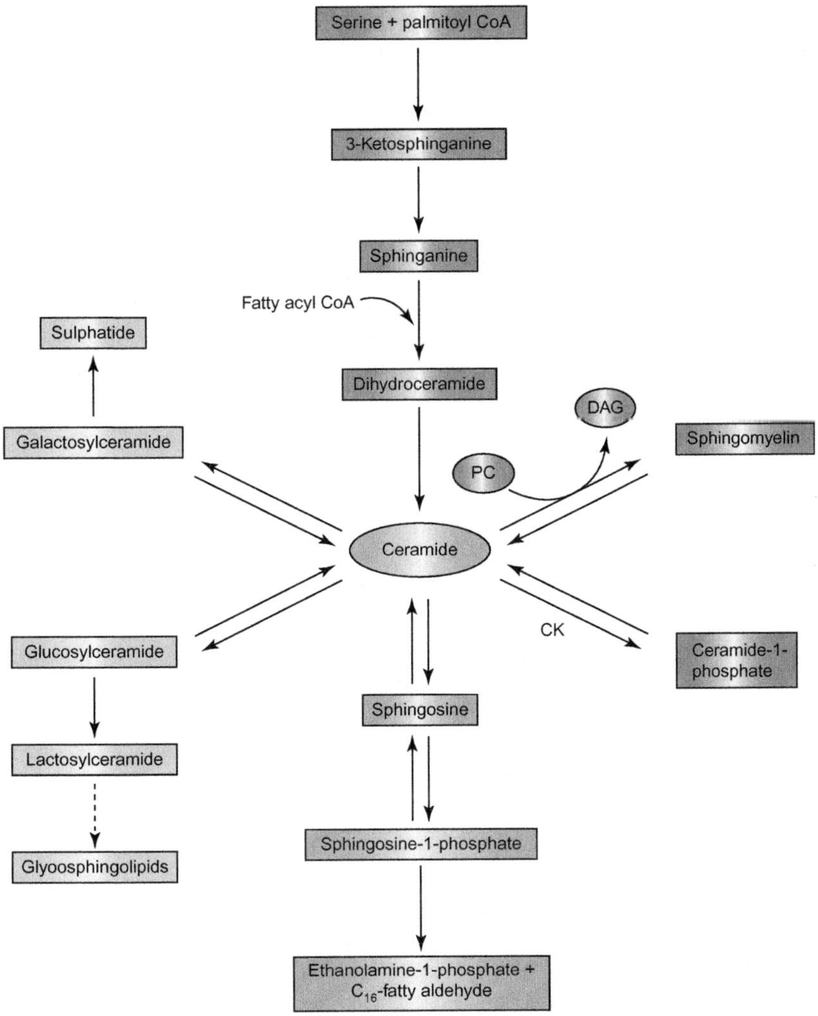

FIGURE 2.4

Overview of molecules involved in the biosynthesis of glycolipids. The central role of ceramide in the biosynthesis of sphingolipids is depicted. Sphingosine, sphingosine-1-phosphate, ceramide-1-phosphate, sphingomyelin, and the family of glycosphingolipids can all be derived from ceramide.

Figure adapted with permission from Ogretmen, B., & Hannun, Y. A. (2004). Biologically active sphingolipids in cancer pathogenesis and treatment. Nature Reviews Cancer, 4*(8), 604–616.*

short glycolipids, are glycoconjugates that are predominantly found on the extracellular surface of cells as part of cell membranes, where they fulfill multiple important functions in cell-to-cell communication, that is, signal transmission and cell recognition. According to Kopitz (2017), most of these functions are based on two principles: (1) glycosphingolipids are spontaneously contributing to the formation of lipid rafts in biological membranes, thereby forming functional complexes (the so-called glycosynapses) with receptor proteins and ion channels and (2) their glycan moieties are bound by receptors recognizing carbohydrate structures, such as galectins (protein−glycan recognition) or cognate glycans (glycan−glycan recognition; Kopitz, 2017).

Glycosphingolipids are composed of a ceramide backbone—a molecule composed of a fatty acid and sphingosine, an 18-carbon amino alcohol with an unsaturated hydrocarbon chain (Fig. 2.5) with a wide variety of carbohydrate groups (mono- or oligosaccharides) attached to carbon 1 of sphingosine. While the glycan part is involved in the various biological activities, the ceramides can be seen as carrier for those glycans. The four principal classes of glycosphingolipids are the cerebrosides, sulfatides, globosides, and gangliosides. All of them can contain galactose.

Cerebrosides, or monoglycoceramides, have a single sugar group linked to ceramide. The sugar residue can be either glucose or galactose, with galactose being the most common sugar in these structures (galactocerebrosides) and glucose to a minor level (glucocerebrosides). Galactocerebrosides are synthesized from ceramide and UDP-galactose and are found predominantly in neuronal cell membranes. By contrast, glucocerebrosides are found in membranes of other tissues, but only as intermediates of the synthesis of complex gangliosides, and thus at low levels. Elevated levels of glucocerebrosides in cells such as monocytes and macrophages are indicative of a metabolic defect (https://themedicalbiochemistry-page.org/sphingolipids.php cite updated June 2018; Gault, Obeid, & Hannun, 2010).

Sulfatides are sulfuric acid esters of galactocerebrosides. Globosides represent cerebrosides that contain additional carbohydrates, predominantly galactose, glucose, or *N*-acetylgalactosamine. Lactosylceramide is a central globoside, which plays pivotal roles as a precursor in the biosynthesis of complex glycosphingolipids. Lactosylceramide has also been shown to be a bioactive lipid involved in various cell signaling cascades (Bieberich, 2012; Chatterjee & Pandey, 2007; Nakamura et al., 2013, 2017).

Cerebrosides can also be further glycosylated to form complex polysaccharide-containing gangliosides. Gangliosides are very similar to globosides, but distinct from them by the fact that they also contain sialic acid (*N*-acetylneuraminic acid) in varying amounts.

All groups of gangliosides contain galactose and ceramide (Ryan, Rice, & Mitchell, 2013).

2.5.5.2 The role of glycosphingolipids and lactase in neonatal neural development and homeostasis

The milk fat globule membrane, a component in mammalian milks, carries sources of polar lipids which have interactive and interdependent functions (Fig. 2.1 outlines some of these important molecules, which are also part of the metabolic pathways shown in Fig. 2.4). These include the sphingolipids: sphingomyelin, glycosphingolipids, and gangliosides. The breakdown products of sphingomyelin result in bioactive products of ceramide, sphingosine, and sphingosine-1-phosphate. These metabolic products have important structural and immune functions in both the gut and the brain. The intermediate metabolites, ceramides and membrane G-coupled receptors such as sphingosine-1-phosphate, provide intracellular signaling functions. These include regulation of cell growth, cell differentiation, apoptosis, and immune cell migration. The importance of these structures in human physiology is also reflected by the fact that sphingosine-1-phosphate receptors 1 and 5 have become therapeutic targets for small molecular agonists in multiple sclerosis (Gusman & Shoemake, 2017; Vállez García, Doorduin, de Paula Faria, Dierckx, & de Vries, 2017) and inflammatory bowel diseases (Karuppuchamy et al., 2017; Sandborn et al., 2016). Sphingosine-1-phosphate receptor modulators regulate the process of remyelination in multiple sclerosis and also regulate lymphocyte trafficking in both multiple sclerosis and ulcerative colitis. Whether these molecules also benefit Crohn's disease is not yet known but trials are underway.

Ceramide (*N*-acylsphingosine) represents a key backbone which has a function of anchoring sphingolipids in the lipid bilayer, but also has signaling functions not all of which have been completely elucidated. Lactase splits ceramide from gluco-, galacto-, and lactosylceramide present in milk which is then absorbed and metabolized. There are several sites in the intestine ceramide which can be handled [By which ceramide, direct digestion at the level of small intestine, liberation by lactase, less likely but possible intestinal transport via the bile acid bacterial pathway: farnesoid X receptor (FXR) discussed later]. Sphingomyelin from the milk fat globule is sequentially hydrolyzed at the brush border by the bile acid released nucleotide phosphodiesterase pyrophosphatase 7. The brush border enzyme neutral ceramidase splits ceramide from its acyl residue and produces sphingosine. This molecule is rapidly absorbed through the membrane, converted to palmitic acid in the mucosa and transported into the cell. Palmitoyl-CoA then attaches to serine and through a series of steps is reassembled into ceramide (Nilsson, 2016).

Ceramide is essential for neural development contributing to ganglioside synthesis in utero. These represent about 6% of all phospholipids in the nervous system (Ryan, Rice, & Mitchell, 2013). These hydrophobic molecules combine in the Golgi apparatus, attaching to carbohydrates and the resulting length of molecules have different specific functions. During the development of the nervous system the ganglioside profile changes from simple species early in embryogenesis to the more complex gangliosides later in embryogenesis. The first in the metabolic series, already present in early embryogenesis and biochemically simplest ganglioside, GM3, is made by galactosylation of glucoceramide and catalyzed by lactosylceramide synthase (Prokazova, Samovilova, Gracheva, & Golovanova, 2009). GM3 is actually the principal ganglioside of extra neural tissue, while at least five major groups of gangliosides take part in neuronal and brain development, as well as neural repair. After birth gangliosides in mother's milk continue to improve cognitive functions. An early meta-analysis of the effect of breast-feeding on cognitive function showed a 3.16 (95% CI: 2.35, 3.98) score benefit over formula-fed infants. The impact on premature neonates was even larger (Anderson, Johnstone, & Remley, 1999). Other functions of gangliosides include Ca^{++} homeostasis, cell—cell interactions, and immunity.

It is not clear to what extent residual lactase contributes to ceramide digestion in adults and information is limited on the possible role of residual lactase activity regarding glycosylated ceramide digestion. This question is especially relevant to adult LNP populations. This issue was addressed in an in vitro rat model by Büller et al., which evaluated lactase activity for lactose and glycosylated ceramides in adult rats. The study was discussed in Section 2.2.3. Because of different responses along the rat intestine to lactose and glycolipids the authors concluded that two different catalytic sites were present on lactase. Since the adult rat diet does not include lactose, residual lactase in the adult may serve to supply ceramide to the adult rat. This second site is the phlorizin catalytic site (Büller et al., 1989). However, whether it applies in humans as well is not clear.

In adult mice, the intestinal nuclear FXR regulates genes involved in bile acid synthesis and ceramide synthesis in the small intestine. This is mediated via altering the composition and concentration of the FXR, influenced by gut bacteria and bile acids. In obese mice, this results in increased levels of agonists (e.g., murocholic acid in rodents and chenodeoxycholic acid in humans) and stimulated ceramide synthesis, while antagonists of FXR (increased by altered levels of specific gut bacteria) reduce ceramide serum levels via repression of FXR signaling (Gonzalez, Jiang, Xie, & Patterson, 2017; Trauner, Claudel, Fickert, Moustafa, &Wagner, 2010). The increased levels of ceramide are correlated in humans with features of the metabolic syndrome (nonalcoholic fatty liver, obesity, insulin resistance, and type 2 diabetes mellitus) (Han, 2018; Promrat, Longato, Wands, & de la Monte, 2011). The role of the FXR in neonates is less clear. In newborn rodent models these receptors may be less mature to take part in bile acid synthesis regulation (Halpern et al., 2010) and supposedly ceramide synthesis regulation as well. As such, ceramide formation may also be inadequate in human neonates compared with adults. In the neonate lactase they may have an essential function to provide ceramide for neurological development.

To augment ceramide availability, it is hypothesized that the maternal microbiome in late phase pergnancy could contribute. In the third trimester there is some evidence that maternal microbiome becomes similar to that found in obesity. This alteration is teleologically designed to increase the role of bacteria in processing undigested nutrients from the host which then contributes to increased nutrients in the mother. In this schema maternal FXR which regulates bile acid, lipid, and glucose metabolism as well as provides ceramide source may be available to the fetus via amniotic fluid. Such a possibility has been recognized in Down's syndrome (Charkiewicz, Blachnio-Zabielska, Zbucka-Kretowska, Wolczynski, & Laudanski, 2015). Further work may support this hypothesis.

These studies raise the question of an expanding role of lactase in the neonate. In addition to providing glucose and galactose for calories and as building blocks in structural formations, the enzyme may have an important role in unlocking of ceramide from various sources, which is seen as part of the structural control of a number of essential developmental functions.

2.5.6 Effects on immune system

Breast milk contains many biologically active compounds including lactose and polymers of lactose, mostly galacto-oligosaccharides (Martin, Ling, & Blackburn, 2016), which in part are further modified by glycosidic functional groups, sialic acid, or fucose to form different HMOs, as described earlier. Biologically active molecules modulate postnatal intestinal functions, development of neonatal immunity, brain development, and modulate the intestinal microbiota (Triantis, Bode, & Van Neerven, 2018). Interestingly acidic galacto-oligosaccharides from cow's milk do not apparently exert immunomodulating effects in humans, suggesting a possible species-specific activity (Triantis, Bode, & Van Neerven, 2018).

Lactose itself has immunomodulatory functions which are evident in experimental conditions either in vitro or small animal models (mice). The effects of lactose on immunity are partly mediated through interactions with galectins—the group of lectins specifically combined with galactose (as β-galactoside)—as already described earlier. The family of galectins expressed along the gastrointestinal tract has different functions as structural sorting molecules and immune modulators (Ashraf et al., 2017; Kaltner et al., 2017). Galectin-9 particularly has been found to be a versatile immune modulator and eosinophil attractant. In animal models galectin-9 attenuated asthma attacks in Guinea pigs and inhibited cutaneous anaphylaxis in mice. In vitro it was shown that galectin-9 prevented degranulation of mast cells by binding IgE. However, the addition of lactose to the medium abrogated the inhibitory effect of galectin-9 (Niki et al., 2009). The inhibitory effect of lactose is thought to be due to prevention of the binding of galectin-9 to cellular glycans (Niki et al., 2009; Triantis, Bode, & Van Neerven, 2018). The finding that lactose modified the effect of galectin-9 suggested a possible immune modulating role for this disaccharide and related human milk saccharides.

The wide function of galectin-9 in immunomodulation was also shown to include regulation of T cell response. Impairment of helper T cells, Th1 and Th17, is putatively involved in autoimmune diseases (e.g., psoriasis, inflammatory bowel disease). T helper cells are regulated by a class of lymphocytes designated as Treg. Galectin-9 is able to downregulate the secretion of proinflammatory cytokines mediated by Treg. In an in vitro study using cocultures of human peripheral monocytes, lactose was reported to inhibit this downregulation of cytokine secretion by galectin-9. Levels of cytokines IL-17 and interferon-γ did not change in the presence of lactose (Paasela, Kolho, Vaarala, & Honkanen, 2014).

These studies suggest a proinflammatory effect of lactose under the given experimental conditions.

However, lactose has additional effects which demonstrate antibacterial and antiinflammatory effects also. For example, the expression of an antimicrobial peptide, cathelicidin, was stimulated by a molecule in breast milk that was identified by NMR as lactose. Lactose was subsequently used to identify the gene encoding this antimicrobial peptide in human cell cultures. In addition, in vitro lactose acted in a synergistic manner with butyrate and phenyl butyrate (bacterial metabolites of carbohydrate fermentation) to induce the antimicrobial peptide (Cederlund et al., 2013).

Lactose exhibited antiinflammatory effects in a mouse model of intraperitoneal injected caerulein induced acute pancreatitis (inflammation of the pancreas with virtually no known pharmaceutical therapy). In this model intraperitoneal injected lactose 100 mg/kg prevented and improved acute pancreatitis. Lactose decreased the production of tumor necrosis factor α (TNF-α) and macrophages, which increased in response to TNF-α and to a lesser extent decreased monocyte chemotactic protein 1. These effects decreased inflammation and attenuated pancreatitis. The outcome produced by lactose was also reproduced by an antagonist to galectin-3, *N*-acetyl-D-lactosamine, suggesting that this galectin modulated proinflammatory effects of caerulein on the pancreas. In this case lactose exhibited antiinflammatory effects (Pan et al., 2018).

However, it is not clear if these observations also pertain to adults with lactose spillover (LNP) or the role if any of the small quantity of lactose absorbed intact (Pimentel et al., 2017). As such these observations raise the question whether lactose-containing foods may aggravate certain diseases, such as bowel carcinogenesis and inflammatory bowel diseases (Crohn's disease and idiopathic ulcerative colitis). However, on epidemiological levels, dairy foods have protective impacts on both colorectal cancer (CRC) (Huncharek, Muscat, & Kupelnick, 2009; Szilagyi, Nathwani, Vinokuroff, Correa, & Shrier, 2006) and inflammatory bowel diseases (Octoratou, Merikas, Malgarinos, Stanciu, & Triantafillidis, 2012; Opstelten et al., 2016). However, the relationship of the protective effect of milk and dairy foods on these diseases is complex and might be due to the aggregate action of many components and not (solely) to lactose. Such in vitro versus in vivo discrepancies in outcome may reflect a generalized matrix effect of milk and dairy foods as has been proposed for cardiovascular diseases and metabolic syndrome (Thorning et al., 2017).

2.5.7 Effects of lactose on mineral absorption

Milk and dairy foods have impact on health. One of the important functions is supply of minerals, such as calcium, phosphorus, magnesium, potassium, and zinc, which modulates the risk factors for diseases. These include factors related to bone health (osteoporosis and fractures), hypertension, colorectal carcinogenesis, and inflammation in the gut (Suchy et al., 2010). Although dairy foods are an important source of daily calcium requirements (due to both high content and bioavailability), which are difficult to achieve without dairy foods intake (Rozenberg et al., 2016), it is unsettled whether lactose normally present in the complex dairy matrix also independently enhances mineral absorption and retention. There are a number of animal and human studies addressing this question, which are reviewed later. Note that early studies rather focused on the effects of lactose as isolated ingredients, whereas the concept of effects by the dairy matrix as a whole is from more recent date.

The mechanism by which lactose might enhance absorption of calcium is thought to be independent of vitamin D and proceeds via the paracellular route (as opposed to transcellular energy and vitamin D–dependent absorption). The effect is expected to occur in the colon due to a lower pH resulting from the lactose-induced higher production of fermentation end products (SCFA) by intestinal bacteria. Although only about 10% calcium is normally absorbed in the colon, this mechanism is believed to contribute to additional calcium bioavailability and thus in turn calcium status (reviewed by Areco et al., 2015).

2.5.7.1 Animal studies

Multiple studies mostly in weanling rats appear in the literature which examined the possible impact of lactose on mineral absorption. The outcomes were conflicting, with some studies showing that lactose increased calcium absorption (Buchowski & Miller, 1991; Bushnell & DeLuca, 1981; Ghishan, Stroop, & Meneely, 1982; Schaafsma & Visser, 1980), while others showed a failure to increase calcium absorption with lactose (Ghishan, Stroop, & Meneely, 1982; Heijnen, Brink, Lemmens, & Beynen, 1993). The outcome of the results varied according to whether animals were raised in a germ-free or conventional environment (Andrieux & Sacquet, 1983; Andrieux, Guéguen, & Sacquet, 1980; Andrieux, Sacquet, & Gueguen, 1982). When a 10% lactose solution was incorporated in feeds of germ-free rats the tendency overall was to increase calcium absorption in the cecum, which was not seen with conventional raised rats. Feeding lactose to cats with increased or decreased dose of lactose failed to increase mineral absorption (Beynen & Yu, 2003). In adolescent, but not suckling, rats lactose tended to increased absorption of minerals including calcium (Ghishan, Stroop, & Meneely, 1982).

2.5.7.2 Human studies

Early studies on humans did report enhanced calcium and other mineral absorption by the consumption of lactose and this is still considered relevant (Michaelsen, Nielsen, Roos, Friis, & Mølgaard, 2011). This was shown in both adults—although with some conflicting information—and more consistently in neonates. All the studies however included only small numbers of subjects. In a well-conducted study by Ziegler and Fomon (1983) on six neonates in crossover design with washout periods, using lactose-free formulae with or without added lactose, a small but statistically significant increased absorption and retention of calcium, magnesium, and manganese was found with added lactose. Using measurements of fecal excretion of calcium in healthy term neonates, Moya et al. (1992) showed that lactose added to formula led to more calcium retention and lower excretion than the addition of the placebo polycose suggesting a specific effect of lactose compared to the glucose polymer. In a later study, however, Moya et al. suggested in a metabolic study that a lactose-free formula (which may be of benefit in premature neonates with reduced intestinal lactase) may be as good as a lactose-replete formula. The authors showed that the lactose-free formula was associated with magnesium and calcium retention despite the absence of lactose and, therefore, useful for nutrition in neonates with reduced intestinal lactase. It was noted, however, that the infants that fed the lactose-free formula had ingested more calcium as well (Moya, Lifschitz, Ameen, & Euler, 1999).

In a more recent study by Abrams et al., 18 full-term infants were evaluated for calcium and zinc absorption using a double isotope method. Formula supplemented with or without lactose was used and fed to these children in a blinded crossover study for 2 weeks. The authors found a 10.3% significant increase in calcium absorption with added lactose compared with the lactose-free formula; however, this quantity (60 mg/day) was not clinically significant. There was no significant increase of absorption of zinc (Abrams, Griffin, & Davila, 2002).

It should also be noted that many cases of congenital lactase deficiency that have been reported are associated with hypercalcemia and nephrocalcinosis (Fazeli, Kaczmarek, Kirschstein, & Santer, 2015; Saarela, Similä, & Koivisto, 1995). These cases are accompanied with acidosis. Acidosis or excess lactose in the colon (due to failure to digest lactose) will alter colonic pH as well, and could lead to increased calcium absorption resulting in elevated calcium status.

The mechanism of enhanced mineral absorption, thus, likely depends on the need for lactose to reach the colon. In support of this notion a study by Zittermann et al. showed no benefit of a 35 g lactose load on calcium bioavailability in 11 healthy lactose "tolerant" participants. Digestion status was studied with a standard lactose tolerance/rise in glucose test (Zittermann et al., 2000). It would be highly interesting to see in comparison if and how the result would be different in LNP subjects.

In conclusion the available data on lactose and enhanced absorption of calcium are promising but are not yet sufficient to establish a cause-and-effect relationship as required by the European Food Safety Authority, who has not recognized the health claim that lactose improves calcium absorption (EFSA Panel on Dietetic Products, Nutrition and Allergies (NDA), 2011). However, improved methodologies such as the double isotope method described earlier and further work may provide further proof of enhanced mineral absorption with lactose.

When viewed from a clinical perspective, dairy foods and bone disease (mainly osteoporosis and fractures) are associated with controversies, as reviewed by Rozenberg et al. (2016). From a practical perspective LNP is associated with lower intakes of dairy (Almon, Sjöström, & Nilsson, 2013; Shrier, Szilagyi, & Correa, 2008). However, these attributes may not predict a higher likelihood of osteoporosis (Enattah et al., 2004; Rozenberg et al., 2016) and intake of calcium may be a better predictor (Balk et al., 2017; Laaksonen et al., 2009). Intake of milk and calcium (as well as other important nutrients in dairy: proteins, potassium, magnesium, and phosphorous) is also associated with improved bone density and structure, especially when intake starts in childhood and adolescence. However, the relationship with fracture risk may be more controversial (reviewed by Caroli, Poli, Ricotta, Banfi, & Cocchi, 2011).

2.5.8 Effects of lactose in neonates and malnourished children

Lactose is the major source of carbohydrate in most mammals and might serve several functions. One important aspect is osmolarity. While most carbohydrates transported through mammalian fluids are monosaccharides – most prominently glucose, the milk of humans and many other mammals contain disaccharides and oligosaccharides. Remarkably, the osmolarity of disaccharides is only half as high as the osmolarity of the two equivalent monosaccharides. Besides its regulatory role in milk production (see Chapter 6, Lactose in the Dairy Chain), this would less likely cause postprandial osmotic stress in infants, when receiving larger amounts of carbohydrates with their milk to match their high caloric requirements. The immediate absorption of the released monosaccharides after lactose digestion warrants the maintenance of low osmolarity in the lumen of the intestine.

As stated earlier, lactose in milk together with the other macronutrients, milk fat and protein, meets the energy density requirements for neonatal nourishment. It provides about 40% of daily nutritional requirements and thus, in addition to milk fat and proteins, a major contribution to the energy requirements of the neonate. In addition to a source of energy, its digestion products, glucose and galactose, also fulfill many structural, communication, and developmental functions as already described earlier.

Although after absorption galactose can be converted to glucose and used for energy, and the galactose required for the formation of galactolipids can be provided by the enzymatic epimerization from glucose, it is speculated that preformed galactose can be of benefit for infants. The presence of galactose in mammalian milk either bound in lactose or oligosaccharides (see Chapter 1, The Evolution of Lactose Digestion), suggests some common unique requirement for young, growing mammals. Most young mammals and in particular human infants undergo a period of rapid brain development and myelination during the nursing period, which requires large amounts of galactoceramides as described in Section 2.5.5. In addition, enzymatic capacity of the liver might not be as high as in adults. Therefore, galactose (lactose) in the milk may ensure that galactose levels do not become limited for galactosylceramide synthesis, myelination and brain development in the young (Jensen, 1995).

An other important feature of lactose is also its low sweetness compared with a number of other saccharides including sucrose (Clemens et al., 2016), which enhances taste and palatability of foods without incurring increased dental and periodontal ill effects. In cases of inadequate intestinal lactase levels, whether developmental or secondary to disease conditions, undigested quantities of lactose are delivered to the lower intestine. As described earlier, bacterial metabolism of lactose leads to caloric salvage through the production of SCFA, which can either enter the host's portal system to be metabolized in the liver or fuel enterocytes directly. In addition, altered pH in the colon due to bacterial lactose metabolism may also have benefits for calcium and mineral absorption. The reduction of the osmotic effects of undigested lactose by the microbiome could lead to reduction of diarrhea in cases of neonates and children with moderately severe malnutrition (a condition which may be associated with reduced intestinal villus height or bacterial overgrowth). Therefore addition of lactose to formulae could be of benefit. However, in severe cases of malnutrition, the addition of lactose may aggravate diarrhea (also compare the thoughts on the putative dual effect of undigested lactose reaching the colon described as "two sides of the same coin" concept in Chapter 4: Lactose—a conditional prebiotic?). As a result, Grenov et al. (2016) suggest to evaluate the addition of lactose to replacement formulas after assessing the risk and benefits on an individual basis.

The notion that lactose may exert potentially beneficial effects is also supported by a study in neonates with allergy to cow's milk (CMA) protein. Treatment of such patients requires withdrawal of milk. There is, however, some evidence that adding lactose to formulae may benefit such infants. Francavilla et al. carried out a crossover study in full-term infants to examine the effects of adding 3.8% lactose to formula fed to infants with cow's milk protein allergy. They compared the fecal microbiome between healthy controls without dietary restrictions to each phase of the study on CMA patients for 2 months at a time. They reported that the counts of bifidobacteria and lactic acid bacteria increased significantly while *Bacteroides* and *Clostridia* decreased in CMA patients when they were given added lactose. In healthy controls the bacterial profiles showed no significant differences. In infants with CMA, however, lactose addition to the diet increased median concentrations of SCFAs (especially acetic and butyric acids) and other volatile organic metabolites with supposed beneficial effects for the host. While intestinal enzymes were not evaluated the authors speculated that in the healthy neonate, lactose is incompletely hydrolyzed allowing some to enter the lower intestine (Francavilla et al., 2012). If this would be the case lactose could contribute to the shaping of the microbiota in infants.

There is some controversy as to when brush border lactase fully develops in the term neonate. Lifschitz et al. in a small clinical study found that 29% of 4–5 weeks' old, breast-fed infants produced measured breath hydrogen (indicating failure of intestinal lactase in these individuals). However, their stools were of normal pH and contained no glucose suggesting full colonic metabolism of host undigested lactose (Lifschitz, Smith, & Garza, 1983). Contrary to this, studies which evaluated neonatal lactase levels suggested that lactase matures quickly after birth. A report by Weaver et al. evaluated 40 healthy neonates for lactose assimilation and found that by day 5 after delivery, lactose hydrolysis achieved 98% (Weaver, Laker, & Nelson, 1986). Also a more recent Cochrane review evaluated that the possible benefit of adding oral lactase to formulae for preterm infants in a single randomized controlled trial, however, did not find any benefits (Tan-Dy & Ohlsson, 2013).

The conclusion from these studies is that neonatal digestion of lactose may not be a problem for supplementing children with CMA, underpinning the fact that CMA and lactose intolerance are distinct conditions, but the mechanisms of how lactose can be tolerated in the neonate, intestinal lactase maturation, or microbial fermentation remain unclear. Certainly the findings on the microbiome by Francavilla suggest that a microbiota modulating effect did occur in their study, which was most likely due to lactose entering the colon. These outcomes should be confirmed in future studies.

2.5.9 Noncommunicable diseases and possible impact of the microbiome/metabolome

Other metabolic effects of lactose which may be relevant to human conditions relate to the changes in the metabolome and microbiome that could affect pathogenesis of different diseases. The bacterial consortium of the lower intestine is particularly malleable by dietary intake (Wu, 2016). Arumugam et al. (2011) described three different dominant types of bacterial populations, enterotypes, which can be identified in different populations, independent of race. In this paradigm, the genera from the phylum Bacteroidetes characterize the predominant bacteria in western societies and are the associates of "western" lifestyle diseases. These diseases are largely considered to be promoted by industrialization and driven partly by western style of diets high in fat, sugar, and processed food. Agrarian lifestyles are dominated by the genus *Prevotella* and are associated with lower rates of western-type diseases.

Pathogenesis of many diseases is associated with decreased diversity and richness of the bacteria composition in the microbiome as well as the so far less well-studied other members of the microbiome (including archaea, viruses, fungi, and other unicellular organisms) (Seksik & Landman, 2015). The term dysbiosis (reduced richness, diversity, loss of beneficial bacteria, or emergence of potential pathogens or pathobionts) is applied to the alterations of the microbiome and can precede or accompany active disease. Some gastrointestinal diseases thought to involve dysbiosis include inflammatory bowel disease (Kostic, Xavier, & Gevers, 2014), irritable bowel syndrome (Rangel et al., 2015), CRC (Gao, Gao, Huang, & Qin, 2017), and hepatic encephalopathy (Bajaj et al., 2017). However, other diseases not previously thought to be involved with the gastrointestinal tract are now also being evaluated for possible relationship with the microbiome. These include multiple sclerosis (Mielcarz & Kasper, 2015), Parkinson's disease (Obata & Pachnis, 2016), autism (Vuong & Hsiao, 2017), obesity and metabolic syndrome

(Musso, Gambino, & Cassader, 2011), and both types of diabetes (Sanz, Olivares, Moya-Pérez, & Agostoni, 2015) as well as other diseases.

The disturbed microbiome has been targeted for possible correction by the use of probiotics (live organisms which survive transport through the gastrointestinal tract and exert benefits to the host) and prebiotics (generally carbohydrates which promote the growth of beneficial bacteria which exert beneficial effects to the host). The impact of pre- and probiotics on different diseases tends to be modest and difficult to demonstrate. Moreover, long-term studies on the use of these agents to combat diseases are not readily available. A review of those is beyond the scope of this chapter. Within this concept lactose in LNP populations might exert prebiotic "like" effects through the alteration of the microbiome and these may be of benefit. Despite this suggestion, little epidemiological studies exist.

2.5.9.1 Cirrhosis and hepatic encephalopathy

A very specific and convincing example of a prebiotic-like effect by lactose is provided by some older studies on hepatic encephalopathy, a metabolic-induced neuromuscular disturbance accompanied by mental status changes, which can be a complication of cirrhosis. This condition might be accompanied with high blood ammonia levels. Lactulose (a synthetic disaccharide consisting of D-galactose and fructose) is used as a standard therapy—together with or without antibiotics—of altered mental status in early mild encephalopathy (Nardone et al., 2016). The ameliorating effect is accomplished by a growth stimulation of lactic acid bacteria, which do not produce ammonia as protein fermenting bacteria do, but have saccharolytic fermentation of lactulose instead, which leads to the production of SCFA and in turn to a decrease of the pH. The lowered colonic pH prevents conversion of NH_4 to NH_3 which is then trapped in the colon. Over 30 years ago it was reported by Uribe et al. that in LNP Mexican cirrhotic patients, that lactose consumption (50 g twice a day) could substitute for lactulose in improving mental status in these patients. Uribe et al. (1980, 1987) also demonstrated that lactose can reduce ammonia levels.

2.5.9.2 Colonic cancer

An example where LP/LNP mechanisms may have an impact mediated by the activity of the microbiome are the epidemiological studies investigating the possible protection of dairy foods against CRC (Aune et al., 2012; Cho et al., 2004; Park, Leitzmann, Subar, Hollenbeck, & Schatzkin, 2009). In a meta-analysis of 68 different studies from different regions of the world (43 high LP/low LNP, 17 intermediate LP/LNP, and 8 high LNP/low LP countries) which examined the impact of dairy intake on the risk of CRC, outcomes varied depending on national LP/LNP distributions (Szilagyi, Nathwani, Vinokuroff, Correa, & Shrier, 2006). Countries in which studies were carried out were grouped according to Mendelian distributions of LP or LNP homozygous populations (defined as countries with <20% LNP or <20% LP respectively). Heterozygous populations ranged between 21% and 79% LNP with a mean of 50%. In both homozygous populations the highest total dairy food intake compared to lowest showed a statistically significant 15%–18% protective effect against CRC. However, in studies with heterozygous populations, protection was a nonsignificant 5% only. These observations are shown in Fig. 2.5.

This study raises the question whether lactose undigested by the host in LNP populations participates in protection against colon cancer via alterations in the intestinal microbiota. Protection against colon cancer was seen in both LP and LNP populations suggesting that in this disease there may be at least two or more different mechanisms responsible for the effect.

In dairy consuming populations the development of colon cancer was postulated to arise from aberrant crypt foci which are cellular irregularities arising from the lower parts of colonic pits. Irregular cellular multiplications were hypothesized to be inhibited by calcium (Lipkin & Newmark, 1995; Wargovich et al., 2000). More recently it was proposed that other nutrients in dairy can have antineoplastic effects too. These are outlined in Table 2.1.

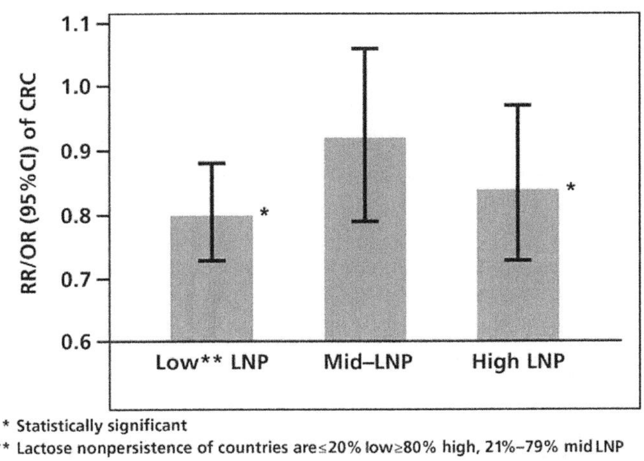

FIGURE 2.5

Summary relative risks of the effects of dairy food consumption on CRC are shown with 95% confidence intervals presented as bar graphs form from three geographical areas of the world divided into groups of countries with high (≥80%), low (≤20%), and mid [average 50% (21%–79%)] LNP proportions of the population. Modest reduction of CRC rates by high dairy intake are found in both high and low LNP countries. In studies emanating from intermediate range of LNP countries the effects of dairy foods on CRC are diminished and are not statistically significant. The graph is based on meta-analyses of relevant studies from 1975 to 2005.

Modified figure from Szilagyi, A., Nathwani, U., Vinokuroff, C., Correa, J. A., & Shrier, I. (2006). The effect of lactose maldigestion on the relationship between dairy food intake and colorectal cancer: A systematic review. Nutrition and Cancer, 55, *141–150.*

Table 2.1 Nutrients in milk and dairy foods with theoretical anticolorectal cancer activity.

Nutrient	Reference
Calcium	Zhang et al. (2016)
Vitamin D	Aggarwal et al. (2016)
Milk fat globule	Zanabria, Tellez, Griffiths, and Corredig (2013)
α-Lactalbumin	Yamaguchi, Takai, Hosono, and Seki (2014)
Medium chain fatty acids	Narayanan, Baskaran, Amalaradjou, & Venkitanarayanan (2015)
Whey protein	Xiao, Badger, and Simme (2005)
Conjugated linoleic acid	Mohammadzadeh, Faramarzi, Mahdavi, Nasirimotlagh, and Asghari Jafarabadi (2013)
Lactose (LNP)[a]	Szilagyi, Nathwani, Vinokuroff, Correa, and Shrier (2006)

[a]*Hypothetical, unproven, may enhance effects in LNP population consuming lower amounts* or *having lower intakes of milk and dairy foods.*

Aberrant crypt foci could lead to benign growths called polyps or adenomas and some of these lead to cancer (Leslie, Carey, Pratt, & Steele, 2002). The initial protective effect indeed was attributed to calcium and several studies on elemental calcium ingestion suggested that recurrent adenomas were prevented after removal through endoscopy (Baron et al., 1999; Keum, Lee, Greenwood, Zhang, & Giovannucci, 2015; Veettil et al., 2017). The dose recommended in both studies using elemental calcium and dairy foods was more than 1.2 g/day. Such levels are rarely achieved from dairy foods in Asia since consumption is less in these areas (Fang, Li, Shi, He, & Li, 2016; Seol, Cho, Kim, & Lee, 2015). In addition, dairy foods may not reduce polyp growth but prevent cancer progression (Szilagyi, Nathwani, Vinokuroff, Correa, & Shrier, 2006; Um, Fedirko, Flanders, Judd, & Bostick, 2017). Finally several studies appeared which contradicted the role of high calcium reducing polyps (Baron et al., 2015; Pommergaard, Burcharth, Rosenberg, & Raskov, 2016).

The second mechanism hypothesized is a potential microbiome effect of lactose in LNP populations which supplants the effect of calcium leading to protective effects at lower quantities of dairy foods (Szilagyi, Nathwani, Vinokuroff, Correa, & Shrier, 2006). Pre- and probiotics have been suggested to be protective against CRC (Gayer, Butler, & Howarth, 2006; Rivera-Huerta et al., 2017; Sharma & Shukla, 2016). This mechanism for lactose, however, is unlikely to be relevant in LP populations since large quantities can be digested, thus impeding lactose from reaching the colon and bacterial fermentation.

The possible impact of lactose is also supported by the failure to demonstrate a significant impact of dairy intake on colon cancer, in mixed LP/LNP populations. In studies from these areas a diluting effect is noted, which was not explained by the type of studies performed (case–control compared to cohort studies showed similar outcomes on analysis) (Szilagyi, Nathwani, Vinokuroff, Correa, & Shrier, 2006).

Another possibility in the light of other anticarcinogens in dairy is that the dairy matrix is the major beneficial agent (Thorning et al., 2017) in anticarcinogenesis. While this is indeed possible further studies would be required. However, a matrix effect also does not explain well the diminished effect noted in heterozygous populations.

2.5.9.3 Other diseases

Other diseases and cancers related to perturbations in the microbiome may be affected unequally between LP and LNP upon dairy food consumption due to possible improvement in dysbiosis. Whether the notion that lactose contributes to benefits in some diseases through the intestinal microbiota can be supported only through further studies employing proper LP/LNP characterization of the subjects, for example, a Mendelian randomization method as described further later.

A second interaction with diseases may stem from the fact that LP populations consume more dairy foods than LNP. This difference also stems from cultural practices. There are several diseases which may be affected. These include prostate and testicular cancers. The risk of prostate and testicular cancers may be increased with increasing amount of dairy intake (Davies, Palmer, & Lipscombe, 1998; Stang, Ahrens, & Baumgrdt-Elms, 2006; Torfadottir et al., 2012; Travis et al., 2012). In the case of prostate, high calcium-rich diet is suspected to promote carcinogenesis (Capiod et al., 2018). In addition, epidemiological studies suggest that LNP phenotype is deficient in patients with prostate cancer (Agarwal et al., 2008). The well described reduced dairy intake in LNP populations (Lember et al., 2006; Morales et al., 2011; Shrier, Szilagyi, & Correa, 2008), therefore, seems to impact on these cancers. However, the role of lactose in these cancers is not clear. The possible effects of dairy foods and the different mechanisms involved are reviewed by Szilagyi (2015). In particular ovarian cancer is discussed further because lactose consumption and specifically galactose toxicity were initially hypothesized to be pathogenetic.

2.5.9.4 Ovarian cancer

In the later part of the 20th century, ovarian epithelial cancer was postulated to be increased with increased milk consumption. Cramer (1989) first reported an epidemiological observation that ovarian cancer was correlated with high dairy food consumption and LP status, while rates were lower in LNP populations. He postulated that possibly an associated abnormal metabolism of galactose (as also seen in galactosemia—see Section 2.3.3) in adults without obvious inborn error of the Leloir pathway led to ovarian toxicity. Also reported was that conversion of galactose-1-phosphate to UDP-galactose was found to be reduced in premenopausal women with first-degree family history of ovarian cancer (Cramer et al., 1994). However, despite some reports of increased risk (Faber et al., 2012), in more recent studies no clear effect of galactose or dairy food consumption was noted in clinical trials and meta-analyses (Kuokkanen et al., 2005; Liu et al., 2015; Merritt, Cramer, Vitonis, Titus, & Terry, 2013; Merritt, Poole, Hankinson, Willett, & Tworoger, 2014; Qin et al., 2005).

2.5.9.5 The role of Mendelian randomization to reduce genetic confounding

In general it was suggested that in genetically segregated populations a Mendelian randomization method should be used to study the effects of milk and dairy foods on outcome in order to address the dichotomy of populations with respect to lactose digestion (Minelli, Thompson, & Abrams, 2004). There have been several studies employing this methodology; however, some studies have been published dealing with a variety of conditions and disorders including vitamin D levels (Alharbi & El-Sohemy, 2017), obesity, diabetes (Bergholdt, Nordestgaard, & Ellervik, 2015), cardiovascular disease (Bergholdt, Nordestgaard, Varbo, & Ellervik, 2015; Smith et al., 2016), hypertension (Ding et al., 2017), height and weight (Gugatschka et al., 2005), and bitter taste receptor distribution associated with LP (Sacerdote et al., 2007). These physiological features including diseases were evaluated for effects of dairy foods on outcome. To date most of these studies did not show significant impact on outcome but individual diseases need to be assessed. A drawback of those studies is also that the largely only included the European C/T 13910 polymorphism, while the other four lactase polymorphisms have not been similarly evaluated. The use of the Mendelian randomization method seems to be a legitimate way to evaluate the possible impacts of dairy foods on many diseases with some limitations if the genetic trait studied influences the primary outcome (this would constitute a horizontal pleiotropy bias). The use of this methodology is discussed in a review by Timpson, Greenwood, Soranzo, Lawson, and Richards (2018).

2.6 Summary and conclusion

If lactose is the primary carbohydrate in mammalian milk, intestinal lactase is essential for neonatal development. However, there is a spectrum on its necessity as can be seen in congenital lactase deficiency where substitute carbohydrates are lifesaving. In addition, the role of lactose as an additive and the mechanism(s) by which this disaccharide can contribute to improved nutrition in neonates deserve further study. Other substrates are also important for fetal and neonatal development, and absence of lactase may require circuitous routes for production of essential glycosylated fats and proteins. The metabolism of galactose also seems to be essential. As seen, galactose is an important component of phospholipids which serve to control neurological, gut epithelial, and immunological development. Substitution of glucose without the ability to convert to galactose can lead to toxic effects on various organs in the host. There is limited knowledge in

Table 2.2 Remaining questions on the role of lactose and lactase functions.

- Why did the disaccharide lactose become the major milk sugar in the evolution of mammals?
- What specific function(s) does lactose play in human adult nutrition?
- Is there a differential function of lactose between lactase persistent or nonpersistent populations?
- Is there a functional impact on the microbiome in lactase nonpersistent populations?
- What role does lactose play, if any, in disease modification (that involve the microbiome)?
- Is there an impact on disease rates?
- Are there any consequences to the reduction of intestinal lactase activity in two-thirds of the world's adult population which is independent of lactose digestion?
- Are there any important residual lactase effects in adult LP/LNP populations which involve other substrates of lactase, for example, ceramide, sphingolipid synthesis, or metabolism of phlorizin?

congenital lactase deficiency and alternate routes for galactose handling. However, the universality of the Leloir pathway suggests that it may serve the host as an energy saving process. In general, consumption of milk and dairy foods provide benefits on multiple levels (Thorning et al., 2016), which of those might be attributed to lactose is not clearly established yet.

An interesting—yet still largely unexplored—phenomenon is the decline of lactase activity with progressing age. The elucidation of the mechanisms behind this developmentally regulated process requires profound future studies. Geographically varying programming might add to the complexity of this subject and needs attention, too.

In adults distinct geographical patterns of LP/LNP occurred based on early evolutionary pressures. These patterns of distribution vary in the modern world due to easy migrations. The LNP populations generally consume less milk and dairy foods but have salvage metabolic mechanism through the microbiome. In these populations, long-term lactose consumption could lead to alteration of the gut microbiome composition and metabolome. These alterations may duplicate parts of the effects of milk and dairy foods in LP populations, possibly, however, at different quantities of intake.

There is limited information on long-term effects of milk and dairy food consumption in LNP populations. It is still not clear whether disease outcomes are different or similar between LP and LNP and studies addressing these questions are highly desired. Introduction of Mendelian randomization techniques for assessing possible differential effects is a promising tool, particularly in studies of mixed LP/LNP populations.

In addition to incomplete knowledge about lactose there is even less information on whether other substrates associated with lactase have any impact in LP/LNP population divide. Table 2.2 outlines some remaining questions regarding lactase and lactose.

References

Abrams, S. A., Griffin, I. J., & Davila, P. M. (2002). Calcium and zinc absorption from lactose-containing and lactose-free infant formulas. *American Journal of Clinical Nutrition*, *76*(2), 442–446.

Acosta, P. B., & Gross, K. C. (1995). Hidden sources of galactose in the environment. *European Journal of Pediatrics*, *154*(7 Suppl. 2), S87–S92.

Adam, A. C., Rubio-Texeira, M., & Polaina, J. (2004). Lactose: The milk sugar from a biotechnological perspective. *Critical Reviews in Food Science and Nutrition*, *44*(7–8), 553–557.

Adeva-Andany, M. M., Pérez-Felpete, N., Fernández-Fernández, C., Donapetry-García, C., & Pazos-García, C. (2016). Liver glucose metabolism in humans. *Bioscience Reports*, *36*(6), pii: e00416.

Aggarwal, A., Höbaus, J., Tennakoon, S., Prinz-Wohlgenannt, M., Graça, J., Price, S. A., et al. (2016). Active vitamin D potentiates the anti-neoplastic effects of calcium in the colon: A cross talk through the calcium-sensing receptor. *The Journal of Steroid Biochemistry and Molecular Biology*, *155*(Pt B), 231–238.

Agarwal, M. M., Rana, S. V., Mandal, A. K., Malhotra, S., Khandelwal, N., Kumar, S., et al. (2008). Lactose intolerance in prostate cancer patients: Incidence and associated factors. *Scandinavian Journal of Gastroenterology*, *43*(3), 270–276.

Alharbi, O., & El-Sohemy, A. (2017). Lactose intolerance (LCT-13910C > T) genotype is associated with plasma 25-hydroxyvitamin D concentrations in Caucasians: A Mendelian randomization study. *Journal of Nutrition*, *147*(6), 1063–1069. Available from https://doi.org/10.3945/jn.116.246108.

Albrecht, J., Sidoryk-Węgrzynowicz, M., Zielińska, M., & Aschner, M.' (2010). Roles of glutamine in neurotransmission. *Neuron Glia Biology*, *6*(4), 263–276. Available from https://doi.org/10.1017/S1740925X11000093.

Almon, R., Sjöström, M., & Nilsson, T. K. (2013). Lactase non-persistence as a determinant of milk avoidance and calcium intake in children and adolescents. *Journal of Nutrition Science*, *2*, e26. Available from https://doi.org/10.1017/jns.2013.11.

Amiri, M., Diekmann, L., von Köckritz-Blickwede, M., & Naim, H. Y. (2015). The diverse forms of lactose intolerance and the putative linkage to several cancers. *Nutrients*, *7*, 7209–7230.

Anderson, J. W., Johnstone, B. M., & Remley, D. T. (1999). Breast-feeding and cognitive development: A meta-analysis. *American Journal of Clinical Nutrition*, *70*, 525–535.

Andrieux, C., Guéguen, L., & Sacquet, E. (1980). Effects of lactose and mode of sterilization of a lactose diet on mineral metabolism in germ-free and conventional rats. *Reproduction, Nutrition, Development*, *20*, 119–138.

Andrieux, C., & Sacquet, E. (1983). Effect of microflora and lactose on the absorption of calcium, phosphorus and magnesium in the hindgut of the rat. *Reproduction, Nutrition, Development*, *23*, 259–271.

Andrieux, C., Sacquet, E., & Gueguen, L. (1982). Microbial flora in the digestive tract and action of lactose on mineral metabolism. *Reproduction, Nutrition, Development*, *22*, 387–394.

Antonowicz, I., & Lebenthal, E. (1977). Developmental pattern of small intestinal enterokinase and disaccharidase activities in the human fetus. *Gastroenterology*, *72*, 1299–1303.

Areco, V., Rivoira, M. A., Rodriguez, V., Marchionatti, A. M., Carpentieri, A., & Tolosa de Talamoni, N. (2015). Dietary and pharmacological compounds altering intestinal calcium absorption in humans and animals. *Nutrition Research Reviews*, *28*, 83–99.

Arumugam, M., Raes, J., Pelletier, E., Le Paslier, D., Yamada, T., Mende, D. R., et al. (2011). *Nature*, *473*(7346), 174–180.

Ashraf, G. M., Perveen, A., Zaidi, S. K., Ahmad, A., Shakil, S., Firoz, C. K., et al. (2017). Galectins—a potential target for cardiovascular therapy. *Current Vascular Pharmacology*, *15*(4), 296–312.

Augustin, L. S. A., Kendall, C. W. C., Jenkins, D. J. A., Willett, W. C., Astrup, A., Barclay, A. W., et al. (2015). Glycemic index, glycemic load and glycemic response: An International Scientific Consensus Summit from the International Carbohydrate Quality Consortium (ICQC). *Nutrition, Metabolism and Cardiovascular Diseases*, *25* (9), 795–815.

Aune, D., Lau, R., Chan, D. S., Vieira, R., Greenwood, D. C., Kampman, E., & Norat, T. (2012). Dairy products and colorectal cancer risk: A systematic review and meta analysis of cohort studies. *Annals of Oncology*, *23*, 37–45.

Baffour-Awuah, N. Y., Fleet, S., Montgomery, R. K., Baker, S. S., Butler, J. L., Campbell, C., et al. (2015). Functional significance of single nucleotide polymorphisms in the lactase gene in diverse US patients and evidence for a novel lactase persistence allele at -13909 in those of European ancestry. *Journal of Pediatric Gastroenterology and Nutrition*, *60*, 182–191.

Bajaj, J. S., Kassam, Z., Fagan, A., Gavis, E. A., Liu, E., Cox, I. J., et al. (2017). Fecal microbiota transplant from a rational stool donor improves hepatic encephalopathy: A randomized clinical trial. *Hepatology*. Available from https://doi.org/10.1002/hep.29306.

Baker, D. H., Becker, D. E., Jensen, A. H., & Harmon, B. G. (1967). Response of the weanling rat to alpha- or beta-lactose with or without an excess of dietary phosphorus. *Journal of Dairy Science*, *50*(8), 1314–1318.

Balk, E. M., Adam, G. P., Langberg, V. N., Earley, A., Clark, P., Ebeling, P. R., et al. (2017). Global dietary calcium intake among adults: A systematic review. *Osteoporosis International*, *28*(12), 3315–3324. Available from https://doi.org/10.1007/s00198-017-4230-x.

Baron, J. A., Barry, E. L., Mott, L. A., Rees, J. R., Sandler, R. S., Snover, D. C., et al. (2015). A trial of calcium and vitamin D for the prevention of colorectal adenomas. *New England Journal of Medicine*, *373*, 1519–1530.

Baron, J. A., Beach, M., Mandel, J. S., van Stolk, R. U., Haile, R. W., Sandler, R. S., et al. (1999). Calcium supplements for the prevention of colorectal adenomas. Calcium Polyp Prevention Study Group. *New England Journal of Medicine*, *340*, 101–107.

Barosa, C., Silva, C., Fagulha, A., Barros, L., Caldeira, M. M., Carvalheiro, M., & Jones, J. G. (2012). Sources of hepatic glycogen synthesis following a milk-containing breakfast meal in healthy subjects. *Metabolism*, *61*(2), 250–254. Available from https://doi.org/10.1016/j.metabol.2011.06.022.

Bergholdt, H. K., Nordestgaard, B. G., & Ellervik, C. (2015). Milk intake is not associated with low risk of diabetes or overweight-obesity: A Mendelian randomization study in 97,811 Danish individuals. *American Journal of Clinical Nutrition*, *102*, 487–496.

Bergholdt, H. K., Nordestgaard, B. G., Varbo, A., & Ellervik, C. (2015). Milk intake is not associated with ischaemic heart disease in observational or Mendelian randomization analyses in 98,529 Danish adults. *International Journal of Epidemiology*, *44*(2), 587–603.

Berry, G. T., Palmieri, M., Gross, K. C., Acosta, P. B., Henstenburg, J. A., Mazur, A., et al. (1993). The effect of dietary fruits and vegetables on urinary galactitol excretion in galactose-1-phosphate uridyltransferase deficiency. *Journal of Inherited Metabolic Disease*, *16*, 91–100.

Berry, G. T., & Walter, J. H. (2012). Disorders of galactose metabolism. In J. M. Saudubray, G. van den Berghe, & J. H. Walter (Eds.), *Inborn metabolic diseases: Diagnosis and treatment*. Germany: Springer.

Beynen, A. C., & Yu, S. (2003). Magnesium balance in adult cats fed a dry food rich in lactose. *Journal of Animal Physiology and Animal Nutrition (Berlin)*, *87*, 245–250.

Beyreiss, K., Hoepffner, W., Scheerschmidt, G., & Müller, F. (1985). Digestion and absorption rates of lactose, glucose, galactose, and fructose in three infants with congenital glucose-galactose malabsorption: Perfusion studies. *Journal of Pediatric Gastroenterology and Nutrition*, *4*, 887–892.

Bieberich, E. (2012). It's a lipid's world: Bioactive lipid metabolism and signaling in neural stem cell differentiation. *Neurochemical Research*, *37*(6), 1208–1229. Available from https://doi.org/10.1007/s11064-011-0698-5.

Birlouez-Aragon, I., Ravelontseheno, L., Villate-Cathelineau, B., Cathelineau, G., & Abitbol, G. (1993). Disturbed galactose metabolism in elderly and diabetic humans is associated with cataract formation. *Journal of Nutrition.*, *123*, 1370–1376.

Blass, E. M., & Fitzgerald, E. (1988). Milk-induced analgesia and comforting in 10-day-old rats: Opioid mediation. *Pharmacology, Biochemistry and Behavior*, *29*, 9–13.

Blass, E. M., & Hoffmeyer, L. B. (1991). Sucrose as an analgesic in newborn humans. *Pediatrics*, *87*, 215–218.

Blass, E. M., & Smith, B. A. (1992). Differential effects of sucrose, fructose, glucose, and lactose on crying in 1- to 3-day-old human infants: Qualitative and quantitative considerations. *Developmental Psychology*, *5*, 804–810.

Bode, L. (2012). Human milk oligosaccharides: Every baby needs a sugar mama (review). *Glycobiology*, *22*(9), 1147–1162.

Brüssow, H. (2013). Nutrition, population growth and disease: A short history of lactose. *Environmental Microbiology*, *15*, 2154–2161.

Buchowski, M. S., & Miller, D. D. (1991). Lactose, calcium source and age affect calcium bioavailability in rats. *Journal of Nutrition*, *121*, 1746–1754.

Büller, H. A., Van Wassenaer, A. G., Raghavan, S., Montgomery, R. K., Sybicki, M. A., & Grand, R. J. (1989). New insights into lactase and glycosylceramidase activities of rat lactase-phlorizin hydrolase. *American Journal of Physiology*, *257*, G616–G623.

Bushnell, P. J., & DeLuca, H. F. (1981). Lactose facilitates the intestinal absorption of lead in weanling rats. *Science*, *211*(4477), 61–63.

Cannizzo, E. S., Clement, C. C., Sahu, R., Follo, C., & Santambrogio, L. (2011). Oxidative stress, inflammaging and immunosenescence. *Journal of Proteomics*, *74*, 2313–2323.

Capiod, T., Barry Delongchamps, N., Pigat, N., Souberbielle, J. C., & Goffin, V. (2018). Do dietary calcium and vitamin D matter in men with prostate cancer? *Nature Reviews Urology*, *15*. Available from https://doi.org/10.1038/s41585-018-0015-z.

Caroli, A., Poli, A., Ricotta, D., Banfi, G., & Cocchi, D. (2011). Invited review: Dairy intake and bone health: A viewpoint from the state of the art. *Journal of Dairy Science*, *94*(11), 5249–5262. Available from https://doi.org/10.3168/jds.2011-4578.

Cederlund, A., Kai-Larsen, Y., Printz, G., Yoshio, H., Alvelius, G., Lagercrantz, H., et al. (2013). Lactose in human breast milk an inducer of innate immunity with implications for a role in intestinal homeostasis. *PLoS One*, *8*, e53876. Available from https://doi.org/10.1371/journal.pone.0053876.

Chacon-Camacho, O. F., Buentello-Volante, B., Velázquez-Montoya, R., Ayala-Ramirez, R., & Zenteno, J. C. (2014). Homozygosity mapping identifies a GALK1 mutation as the cause of autosomal recessive congenital cataracts in 4 adult siblings. *Gene.*, *534*, 218–221.

Charkiewicz, K., Blachnio-Zabielska, A., Zbucka-Kretowska, M., Wolczynski, S., & Laudanski, P. (2015). Maternal plasma and amniotic fluid sphingolipids profiling in fetal down syndrome. *PLoS One*, *10*(5), e0127732. Available from https://doi.org/10.1371/journal.pone.0127732.

Chatterjee, S., & Pandey, A. (2007). The Yin and Yang of lactosylceramide metabolism: Implications in cell function. *Biochimica et Biophysica Acta*, *1780*(3), 370–382.

Chaudhry, K. K., Mahmood, S., & Mahmood, A. (2008). Hormone induced expression of brush border lactase in suckling rat intestine. *Molecular and Cellular Biochemistry*, *312*, 11–16. Available from https://doi.org/10.1007/s11010-008-9715-7.

Chen, S. C., & Kuo, P. L. (2016). The role of galectin-3 in the kidneys. *International Journal of Molecular Science.*, *17*, 565. Available from https://doi.org/10.3390/ijms17040565.

Chichlowski, M., German, J. B., Lebrilla, C. B., & Mills, D. ,A. (2011). The influence of milk oligosaccharides on microbiota of infants: Opportunities for formulas. *Annual Review of Food Science and Technology*, *2*, 331–351.

Cho, E., Smith-Warner, S. A., Spiegelman, D., Beeson, W. L., van den Brandt, P. A., Colditz, G. A., et al. (2004). Dairy foods, calcium, and colorectal cancer: A pooled analysis of 10 cohort studies. *Journal of National Cancer Institute*, *96*, 1015–1022.

Clemens, R. A., Jones, J. M., Kern, M., Lee, S.-Y., Mayhew, E. J., Joanne, L., & Zivanovic, S. (2016). Functionality of Sugars in Foods and Health Institute of Food Technologists. *Reviews in Food Science and Food Safety*, *15*, 433–470.

Cockburn, D. W., & Koropatkin, N. M. (2016). Polysaccharide degradation by the intestinal microbiota and its influence on human health and disease. *Journal of Molecular Biology*, *428*, 3230–3252.

Coelho, A. I., Berry, G. T., & Rubio-Gozalbo, M. E. (2015). Galactose metabolism and health. *Current Opinion in Clinical and Nutrition & Metabolic Care*, *18*, 422–427.

Cousin, J. M., & Cloninger, M. J. (2016). The role of galectin-1 in cancer progression, and synthetic multivalent systems for the study of galectin-1. *International Journal of Molecular Science*, *17*. Available from https://doi.org/10.3390/ijms17091566, pii: E1566.

Cramer, D. W. (1989). Lactase persistence and milk consumption as determinants of ovarian cancer risk. *American Journal of Epidemiology*, *130*, 904–910.

Cramer, D. W., Muto, M. G., Reichardt, J. K., Xu, H., Welch, W. R., Valles, B., & Ng, W. G. (1994). Characteristics of women with a family history of ovarian cancer. I. Galactose consumption and metabolism. *Cancer*, *74*, 1309–1317.

Cui, X., Zuo, P., Zhang, Q., Li, X., Hu, Y., Long, J., et al. (2006). Chronic systemic D-galactose exposure induces memory loss, neurodegeneration, and oxidative damage in mice: Protective effects of R-alpha-lipoic acid. *Journal of Neuroscience Research*, *84*(3), 647654.

Davies, T. W., Palmer, C. R., & Lipscombe, J. M. (1998). Adolescent milk, dairy product and fruit consumption and testicular cancer. *British Journal of Cancer*, *74*, 657–660.

Day, A. J., Cañada, F. J., Díaz, J. C., Kroon, P. A., Mclauchlan, R., Faulds, C. B., et al. (2000). Dietary flavonoid and isoflavone glycosides are hydrolysed by the lactase site of lactase phlorizin hydrolase. *FEBS Letters*, *468*, 166–170.

Delacour, D., Cramm-Behrens, C. I., Dobecq, H., Le Bivic, A., Naim, H. Y., & Jacob, R. (2006). Requirement for galectin-3 in apical protein sorting. *Current Biology*, *16*, 408–414.

Delaveau, P. (2002). Le lactose dans le lait; hypothese sur son importance biologique. *Annales Pharmaceutiques Françaises*, *61*, 340–342.

Diekmann, L., Behrendt, M., Amiri, M., & Naim, H. Y. (2017). Structural determinants for transport of lactase phlorizin-hydrolase in the early secretory pathway as a multi-domain membrane glycoprotein. *Biochimica et Biophysica Acta*, *1861*, 3119–3128.

Ding, M., Huang, T., Bergholdt, H. K., Nordestgaard, B. G., Ellervik, C., & Qi, L. (2017). Dairy consumption, systolic blood pressure, and risk of hypertension: Mendelian randomization study; CHARGE Consortium. *British Medical Journal*. Available from https://doi.org/10.1136/bmj.j1000, 356, j1000.

Décombaz, J., Jentjens, R., Ith, M., Scheurer, E., Buehler, T., Jeukendrup, A., et al. (2011). Fructose and galactose enhance postexercise human liver glycogen synthesis. *Medicine & Science in Sports & Exercise*, *43*(10), 1964–1971. Available from https://doi.org/10.1249/MSS.0b013e318218ca5a.

Duee, P. H., Pégorier, J. P., Bois-Joyeux, B., & Girard, J. (1983). Fuel metabolism and energy stores in fasting or suckling newborn guinea pigs. *Journal of Developmental Physiology*, *5*(6), 383–393.

Duncan, S. H., Louis, P., & Flint, H. J. (2004). Lactate—utilizing bacteria, isolated from human feces, that produce butyrate as a major fermentation product. *Applied and Environmental Microbiology*, *70*, 5810–5817.

Ebert, E. C. (2010). The thyroid and the gut. *Journal of Clinical Gastroenterology*, *44*, 402–406. Available from https://doi.org/10.1097/MCG.0b013e3181d6bc3e.

EFSA Panel on Dietetic Products, Nutrition and Allergies (NDA). (2011). Scientific opinion on the substantiation of health claims related to lactose and increase in calcium absorption leading to an increase in calcium retention (ID 668) pursuant to article 13 (1) of regulation (EC) No 1924/2006. *EFSA Journal*, *9*, 1–13, 2234.

Ehrenkranz, J. R., Lewis, N. G., Kahn, C. R., & Roth, J. (2005). Phlorizin: A review. *Diabetes/Metabolism and Research Reviews*, *21*, 31–38.

Enattah, N., Välimäki, V. V., Välimäki, M. J., Löyttyniemi, E., Sahi, T., & Järvelä, I. (2004). Molecularly defined lactose malabsorption, peak bone mass and bone turnover rate in young Finnish men. *Calcified Tissue International*, *75*(6), 488–493.

Enattah, N. S., Kuokkanen, M., Forsblom, C., Natah, S., Oksanen, A., Jarvela, I., et al. (2007). Correlation of intestinal disaccharidase activities with the C/T-13910 variant and age. *World Journal of Gastroenterology*, *13*, 3508–3512.

Enattah, N. S., Sahi, T., Savilahti, E., Terwilliger, J. D., Peltonen, L., & Järvelä, I. (2002). Identification of a variant associated with adult-type hypolactasia. *Nature Genetics*, *30*, 233–237.

Englyst, K. N., & Englyst, H. N. (2005). Carbohydrate bioavailability. *British Journal of Nutrition*, *94*(1), 1–11.

Ercan, N., Nuttall, F. Q., Gannon, M. C., Redmon, J. B., & Sheridan, K. J. (1993). Effects of glucose, galactose, and lactose ingestion on the plasma glucose and insulin response in persons with non-insulin-dependent diabetes mellitus. *Metabolism*, *42*(12), 1560–1567.

Faber, M. T., Jensen, A., Søgaard, M., Høgdall, E., Høgdall, C., Blaakaer, J., et al. (2012). Use of dairy products, lactose, and calcium and risk of ovarian cancer—results from a Danish case-control study. *Acta Oncology*, *51*, 454–464.

Falony, G., Calmeyn, T., Leroy, F., & de Vuyst, L. (2009). Coculture fermentations of *Bifidobacterium* species and *Bacteroides thetaiotaomicron* reveal a mechanistic insight into the prebiotic effect of inulin-type fructans. *Applied and Environmental Microbiology*, *75*, 2312–2319.

Falony, G., Lazidou, K., Verschaeren, A., Weckx, S., Maes, D., & de Vuyst, L. (2009). In vitro analysis of fermentation of prebiotic inulin-type fructans by *Bifidobacterium* species reveals four different phenotypes. *Applied and Environmental Microbiology*, *75*, 454–461.

Fang, A. P., Li, K. J., Shi, H. Y., He, J. J., & Li, H. (2016). Habitual dietary calcium intakes and calcium metabolism in healthy adults Chinese: A systematic review and meta-analysis. *Asia Pacific Journal of Clinical Nutrition*, *25*(4), 776–784. Available from https://doi.org/10.6133/apjcn.092015.30.

Fazeli, W., Kaczmarek, S., Kirschstein, M., & Santer, R. (2015). A novel mutation within the lactase gene (LCT): The first report of congenital lactase deficiency diagnosed in Central Europe. *BMC Gastroenterology*, *15*, 90. Available from https://doi.org/10.1186/s12876-015-0316-0.

Francavilla, R., Calasso, M., Calace, L., Siragusa, S., Ndagijimana, M., Vernocchi, P., et al. (2012). Effect of lactose on gut microbiota and metabolome of infants with cow's milk allergy. *Pediatric Allergy and Immunology*, *23*(5), 420–427.

Freund, J. N., Foltzer-Jourdainne, C., Duluc, I., Galluser, M., Gosse, F., & Raul, F. R. (1991). Rat lactase activity and mRNA expression in relation to the thyroid and corticoid status. *Cell Molecular Biology*, *37*, 463–466.

Gangwisch, J. E., Hale, L., Garcia, L., Malaspina, D., Opler, M. G., Payne, M. E., et al. (2015). High glycemic index diet as a risk factor for depression: Analyses from the Women's Health Initiative. *American Journal of Clinical Nutrition*, *102*(2), 454–463.

Gao, R., Gao, Z., Huang, L., & Qin, H. (2017). Gut microbiota and colorectal cancer. *European Journal of Clinical Microbiology & Infectious Diseases*. Available from https://doi.org/10.1007/s10096-016-2881-8.

Gault, C. R., Obeid, L. M., & Hannun, Y. A. (2010). An overview of sphingolipid metabolism: From synthesis to breakdown. *Advances in Experimental Medicine and Biology*, *688*, 1–23.

Gavaldà-Navarro, A., Hondares, E., Giralt, M., Mampel, T., Iglesias, R., & Villarroya, F. (2015). Fibroblast growth factor 21 in breast milk controls neonatal intestine function. *Science Reports*, *5*, 13717.

Gayer, M. S., Butler, R. N., & Howarth, G. S. (2006). Probiotics, prebiotics and synbiotics: A role in chemoprevention for colorectal cancer? *Cancer Biology & Therapy*, *5*, 1265–1269.

Gerbault, P., Moret, C., Currat, M., & Sanchez-Mazas, A. (2009). Impact of selection and demography on the diffusion of lactase persistence. *PLoS One*, *4*(7), e6369. Available from https://doi.org/10.1371/journal.pone.0006369.

Ghishan, F. K., Stroop, S., & Meneely, R. (1982). The effect of lactose on the intestinal absorption of calcium and zinc in the rat during maturation. *Pediatric Research*, *16*, 566–568.

Gilat, T., Russo, S., Gelman-Malachi, E., & Aldor, T. A. (1972). Lactase in man: A nonadaptable enzyme. *Gastroenterology*, *62*, 1125–1127.

Gitzelmann, R. (1995). Galactose-1-phosphate in the pathophysiology of galactosemia. *European Journal of Pediatrics*, *154*(Suppl. 2), S45–S49.

Gonzalez, J. T., Fuchs, C. J., Betts, J. A., & van Loon, L. J. (2016). Liver glycogen metabolism during and after prolonged endurance-type exercise. *American Journal of Physiology–Endocrinology and Metabolism*, *311*(3), E543–E553. Available from https://doi.org/10.1152/ajpendo.00232.2016.

Gonzalez, F. J., Jiang, C., Xie, C., & Patterson, A. D. (2017). Intestinal farnesoid X receptor signaling modulates metabolic disease. *Digestive Diseases*, *35*, 178–184.

Grand, R. J., Watkins, J. B., & Torti, F. M. (1976). Development of the human gastrointestinal tract. A review. *Gastroenterology*, *70*, 790–810.

Grenov, B., Briend, A., Sangild, P. T., Thymann, T., Rytter, M. H., Hother, A. L., et al. (2016). Undernourished children and milk lactose. *Food and Nutrition Bulletin*, *37*, 85–99.

Gross, K. C., & Acosta, P. B. (1991). Fruits and vegetables are a source of galactose: Implications in planning the diets of patients with galactosaemia. *Journal of Inherited Metabolic Disease*, *14*, 253–258.

Gugatschka, M., Dobnig, H., Fahrleitner-Pammer, A., Pietschmann, P., Kudlacek, S., Strele, A., et al. (2005). Molecularly-defined lactose malabsorption, milk consumption and anthropometric differences in adult males. *Quarterly Journal of Medicine*, *98*, 857–863.

Gunnerud, U., Holst, J. J., Östman, E., & Björck, I. (2012). The glycemic, insulinemic and plasma amino acid responses to equi-carbohydrate milk meals, a pilot-study of bovine and human milk. *Nutrition Journal*, *11*(1), 83.

Gusman, D. H., & Shoemake, C. (2017). Evaluation and optimization of designed sphingosine-1-phosphate (S1P) receptor subtype 1 modulators for the management of multiple sclerosis. *Yale Journal of Biology and Medicine*, *90*, 15–23, [eCollection].

Habte, D., Sterby, G., & Jijalmassen, B. (1973). Lactose malabsorption in Ethiopian children. *Acta Paediatrica Scandinavica*, *62*, 649–654.

Halpern, M. D., Weitkamp, J. H., Mount Patrick, S. K., Dobrenen, H. J., Khailova, L., Correa, H., et al. (2010). Apical sodium-dependent bile acid transporter upregulation is associated with necrotizing enterocolitis. *American Journal of Physiology: Gastrointestinal and Liver Physiology*, *299*, G623–G631.

Hammon, H. M., Steinhoff-Wagner, J., Schönhusen, U., Metges, C. C., & Blum, J. W. (2012). Energy metabolism in the newborn farm animal with emphasis on the calf: Endocrine changes and responses to milk-born and systemic hormones. *Domestic Animal Endocrinology*, *43*(2), 171–185. Available from https://doi.org/10.1016/j.domaniend.2012.02.005.

Han, C. Y. (2018). Update on FXR biology: Promising therapeutic target? *International Journal of Molecular Science*, *19*(7). Available from https://doi.org/10.3390/ijms19072069, pii: E2069.

Harada, N., & Inagaki, N. (2012). Role of sodium-glucose transporters in glucose uptake of the intestine and kidney. *Journal of Diabetes Investigation*, *3*(4), 352–353. Available from https://doi.org/10.1111/j.2040-1124.2012.00227.x.

He, T., Priebe, M. G., Vonk, R. J., & Welling, G. W. (2005). Identification of bacteria with beta-galactosidase activity in faeces from lactase non-persistent subjects. *FEMS Microbiology Ecology*, *54*, 463–469.

He, T., Venema, K., Priebe, M. G., Welling, G. W., Brummer, R. J., & Vonk, R. J. (2008). The role of colonic metabolism in lactose intolerance. *European Journal of Clinical Investigation*, *38*, 541–547.

Heckl, S., Reiners, C., Buck, A. K., Schäfer, A., Dick, A., & Scheurlen, M. (2016). Evidence of impaired carbohydrate assimilation in euthyroid patients with Hashimoto's thyroiditis. *European Journal of Clinical Nutrition*, *70*, 222–228. Available from https://doi.org/10.1038/ejcn.2015.167.

Heijnen, A. M., Brink, E. J., Lemmens, A. G., & Beynen, A. C. (1993). Ileal pH and apparent absorption of magnesium in rats fed on diets containing either lactose or lactulose. *British Journal of Nutrition*, *70*, 747–756.

Hertzler, S. R., & Savaiano, D. A. (1996). Colonic adaptation to daily lactose feeding in lactose maldigesters reduces lactose intolerance. *American Journal of Clinical Nutrition*, *64*, 232–236.

Hodin, R. A., Chamberlain, S. M., & Upton, M. P. (1992). Thyroid hormone differentially regulates rat intestinal brush border enzyme gene expression. *Gastroenterology*, *103*, 1529–1536.

Holden, H. M., Rayment, I., & Thoden, J. B. (2003). Structure and function of enzymes of the Leloir pathway for galactose metabolism. *Journal of Biological Chemistry*, *278*, 43885–43888.

Hoon, M. A., Adler, E., Lindemeier, J., Battey, J. F., Ryba, N. J. P., & Zuker, C. S. (1999). Putative mammalian taste receptors: A class of taste-specific GPCRs with distinct topographic selectivity. *Cell*, *96*, 541–551.

Hughes, T. A., Atchison, J., Hazelrig, J. B., & Boshell, B. R. (1989). Glycemic responses in insulin-dependent diabetic patients: Effect of food composition. *American Journal of Clinical Nutrition*, *49*, 658–666.

Huncharek, M., Muscat, J., & Kupelnick, B. (2009). Colorectal cancer risk and dietary intake of calcium, vitamin D, and dairy products: A meta-analysis of 26,335 cases from 60 observational studies. *Nutrition and Cancer*, *61*, 47–69.

Jacob, R., Bulleid, N. J., & Naim, H. Y. (1995). Folding of human intestinal lactase-phlorizin hydrolase. *Journal of Biological Chemistry*, *270*(31), 18678–18684.

Jensen, R. G. (1995). *Handbook of Milk Composition*. San Diego: Academic Press.

Jensen, T. G., Liebert, A., Lewinsky, R., Swallow, D. M., Olsen, J., & Troelsen, J. T. (2011). The -14010*C variant associated with lactase persistence is located between an Oct-1 and HNF1α binding site and increases lactase promoter activity. *Human Genetics*, *130*, 483–493.

Jentjens, R. L., & Jeukendrup, A. E. (2003). Effects of pre-exercise ingestion of trehalose, galactose and glucose on subsequent metabolism and cycling performance. *European Journal of Applied Physiology*, *88*(4-5), 459–465.

Jiang, T., & Savaiano, D. A. (1997a). Modification of colonic fermentation by bifidobacteria and pH in vitro. Impact on lactose metabolism, short-chain fatty acid, and lactate production. *Digestive Diseases and Sciences*, *42*, 2370–2377.

Jiang, T., & Savaiano, D. A. (1997b). In vitro lactose fermentation by human colonic bacteria is modified by *Lactobacillus acidophilus* supplementation. *Journal of Nutrition*, *1237*, 1489–1495.

Kaialy, W., Martin, G. P., Ticehurst, M. D., Royall, P., Mohammad, M. A., Murphy, J., et al. (2011). Characterisation and deposition studies of recrystallised lactose from binary mixtures of ethanol/butanol for improved drug delivery from dry powder inhalers. *AAPS Journal*, *13*, 30–43.

Kaltner, H., Toegel, S., Caballero, G. G., Manning, J. C., Ledeen, R. W., & Gabius, H. J. (2017). Galectins: Their network and roles in immunity/tumor growth control.

Histochemistry and Cell Biology, *147*, 239–256. Available from https://doi.org/10.1007/s00418-016-1522-8.

Karas, N., Gobec, L., Pfeifer, V., Mlinar, B., Battelino, T., & Lukac-Bajalo, J. (2003). Mutations in galactose-1-phosphate uridyltransferase gene in patients with idiopathic presenile cataract. *Journal of Inherited Metabolic Disease*, *26*, 699–704.

Karuppuchamy, T., Behrens, E. H., González-Cabrera, P., Sarkisyan, G., Gima, L., Boyer, J. D., et al. (2017). Sphingosine-1-phosphate receptor-1 (S1P1) is expressed by lymphocytes, dendritic cells, and endothelium and modulated during inflammatory bowel disease. *Mucosal Immunology*, *10*, 162–171.

Kashket, S., & DePaola, D. P. (2002). Cheese consumption and the development and progression of dental caries*Nutrition Reviews*, *60*(4), 97–103, [KEGG database] . Available from https://www.genome.jp/dbget-bin/www_bget?ec:3.2.1.108.

Keum, N., Lee, D. H., Greenwood, D. C., Zhang, X., & Giovannucci, E. L. (2015). Calcium intake and colorectal adenoma risk: Dose-response meta-analysis of prospective observational studies. *International Journal of Cancer*, *136*, 1680–1687.

Kim, H. O., Hartnett, C., & Scaman, C. H. (2007). Free galactose content in selected fresh fruits and vegetables and soy beverages. *Journal of Agricultural and Food Chemistry*, *55*, 8133–8137.

Kishnani, P. S., & Chen, Y.-T. (2013). Disorders of carbohydrate metabolism. In R. R. Howell (Ed.), *Emery and Rimoin's principles and practice of medical genetics* (6th ed.). Philadelphia, PA: Elsevier Saunders.

Kogut, M. D., Donnell, G. N., & Shaw, K. N. (1967). Studies of lactose absorption in patients with galactosemia. *Journal of Pediatrics*, *71*, 75–81.

Koh, A., De Vadder, F., Kovatcheva-Datchary, P., & Bäckhed, F. (2016). From dietary fiber to host physiology: Short-chain fatty acids as key bacterial metabolites. *Cell*, *165*, 1332–1345.

Kopitz, J. (2017). Lipid glycosylation: A primer for histochemists and cell biologists. *Histochemistry and Cell Biology*, *147*(2), 175–198. Available from https://doi.org/10.1007/s00418-016-1518-4.

Kostic, A. D., Xavier, R. J., & Gevers, D. (2014). The microbiome in inflammatory bowel disease: Current status and the future ahead. *Gastroenterology*, *146*, 1489–1499.

Krajicek, E. J., & Hansel, S. L. (2016). Small intestinal bacterial overgrowth: A primary care review. *Mayo Clinic Proceedings*, *91*, 1828–1833.

Kruse, T. A., Bolund, L., Grzeschik, K. H., Ropers, H. H., Sjostrom, H., Noren, O., et al. (1988). The human lactase-phlorizin gene is located on chromosome 2. *FEBS Letters*, *240*, 123–126.

Kuhn, R., & Low, I. (1949). The occurrence of lactose in the plant kingdom. *Chemische Berichte*, *82*, 479–481.

Kucuksezer, U. C., & Ozdemir, C. (2016). Role of Galectins in Allergic Disorders. *Recent Pat Inflamm Allergy Drug Discov*, *10*, 2–12.

Kuokkanen, M., Butzow, R., Rasinperä, H., Medrek, K., Nilbert, M., Malander, S., et al. (2005). Lactase persistence and ovarian carcinoma risk in Finland, Poland and Sweden. *International Journal of Cancer*, *117*, 90–94.

Laaksonen, M. M., Impivaara, O., Sievänen, H., Viikari, J. S., Lehtimäki, T. J., Lamberg-Allardt, C. J., et al. (2009). Associations of genetic lactase non-persistence and sex with bone loss in young adulthood. *Bone*, *44*(5), 1003–1009. Available from https://doi.org/10.1016/j.bone.2008.12.019.

Labrie, V., Buske, O. J., Oh, E., Jeremian, R., Ptak, C., Gasiūnas, G., et al. (2016). Lactase non persistence is directed by DNA-variation-dependent epigenetic aging. *Nature Structural & Molecular Biology*, *23*, 566–573.

Ladas, S., Papanikos, J., & Arapakis, G. (1982). Lactose malabsorption in Greek adults: Correlation of small bowel transit time with the severity of lactose intolerance. *Gut*, *23*, 968–9673.

Lai, K., Elsas, L. J., & Wierenga, K. J. (2009). Galactose toxicity in animals. *IUBMB Life*, *61*, 1063–1074.

Breastfeeding: A guide for the medical profession In R. A. Lawrence, & R. M. Lawrence (Eds.), *Physiology of lactation*. (8th ed.). Philadelphia, PA: Elsevier, [Chapter 3].

Lebea, P. J., & Pretorius, P. J. (2005). The molecular relationship between deficient UDP-galactose uridyl transferase (GALT) and ceramide galactosyltransferase (CGT) enzyme function: A possible cause for poor long-term prognosis in classic galactosemia. *Medical Hypotheses*, *65*(6), 1051–1057.

Lee, B., & Szilagyi, A. (2012). Lactose uses in industry and biomedical importance; lactose a functional disaccharide. In V. R. Preedy (Ed.), *Food and nutritional components, in focus no. 3. Dietary sugars: Chemistry, analysis, function and effects* (pp. 806–820). The Royal Society of Chemistry, [Chapter 46].

Lember, M., Torniainen, S., Kull, M., Kallikorm, R., Saadla, P., Rajasalu, T., et al. (2006). Lactase non-persistence and milk consumption in Estonia. *World Journal of Gastroenterology*, *12*, 7329–7331.

Leseva, M. N., Grand, R. J., Klett, H., Boerries, M., Busch, H., Binder, A. M., & Michels, K. B. (2018). Differences in DNA methylation and functional expression in lactase persistent and non-persistent individuals. *Scientific Reports*, *8*, 5649.

Leslie, A., Carey, F. A., Pratt, N. R., & Steele, R. J. (2002). The colorectal adenoma-carcinoma sequence. *British Journal of Surgery*, *89*, 845–860.

Lewinsky, R. H., Jensen, T. G., Møller, J., Stensballe, A., Olsen, J., & Troelsen, J. T. (2005). T-13910 DNA variant associated with lactase persistence interacts with Oct-1 and stimulates lactase promoter activity in vitro. *Human Molecular Genetics*, *14*(24), 3945–3953.

Lifschitz, C. H., Smith, E. O., & Garza, C. (1983). Delayed complete functional lactase sufficiency in breast-fed infants. *Journal of Pediatric Gastroenterology and Nutrition*, *2*(3), 478–482.

Lipkin, M., & Newmark, H. (1995). Calcium and the prevention of colon cancer. *Journal of Cellular Biochemistry*, *22*(Suppl.), 65–73.

Liu, J., Tang, W., Sang, L., Dai, X., Wei, D., Luo, Y., et al. (2015). Milk, yogurt, and lactose intake and ovarian cancer risk: A meta-analysis. *Nutrition and Cancer*, *67*(1), 68–72.

Louis, P., & Flint, H. J. (2017). Formation of propionate and butyrate by the human colonic microbiota. *Environmental Microbiology*, *19*(1), 29–41.

Macfarlane, G. T., & Macfarlane, S. (2012). Bacteria, colonic fermentation, and gastrointestinal health. *Journal of AOAC International*, *95*, 50–60.

Mach, N., & Fuster-Botella, D. (2017). Endurance exercise and gut microbiota: A review. *Journal of Sport and Health Science*, *6*, 179–197.

Mahesh Kumar, T., & Misra, A. (2006). Formulation and evaluation of insulin dry powder for inhalation. *Drug Development and Industrial Pharmacy*, *32*, 677–686.

Maiuri, L., Rossi, M., Raia, V., Garipoli, V., Hughes, L. A., Swallow, D., et al. (1994). Mosaic regulation of lactase in human adult-type hypolactasia. *Gastroenterology*, *107*, 54–160.

Maiuri, L., Raia, V., Potter, J., Swallow, D., Ho, M. W., Fiocca, R., et al. (1991). Mosaic pattern of lactase expression by villous enterocytes in human adult-type hypolactasia. *Gastroenterology*, *100*, 359–369.

Martin, C. R., Ling, P. R., & Blackburn, G. ,L. (2016). Review of infant feeding: Key features of breast milk and infant formula. *Nutrients*, *8*(5). Available from https://doi.org/10.3390/nu8050279, pii, E279.

Menini, S., Iacobini, C., Blasetti Fantauzzi, C., Pesce, C. M., & Pugliese, G. (2016). Role of galectin-3 in obesity and impaired glucose homeostasis. *Oxidative Medicine and Cellular Longevity*, *2016*, 9618092. Available from https://doi.org/10.1155/2016/9618092.

Merritt, M. A., Cramer, D. W., Vitonis, A. F., Titus, L. J., & Terry, K. L. (2013). Dairy foods and nutrients in relation to risk of ovarian cancer and major histological subtypes. *International Journal of Cancer*, *132*, 1114–1124.

Merritt, M. A., Poole, E. M., Hankinson, S. E., Willett, W. C., & Tworoger, S. S. (2014). Dairy food and nutrient intake in different life periods in relation to risk of ovarian cancer. *Cancer Causes Control*, *25*, 795–808.

Michaelsen, K. F., Nielsen, A. L., Roos, N., Friis, H., & Mølgaard, C. (2011). Cow's milk in treatment of moderate and severe undernutrition in low-income countries. *Nestle Nutrition Workshop Series: Pediatric Program*, *67*, 99–111.

Mielcarz, D. W., & Kasper, L. H. (2015). The gut microbiome in multiple sclerosis. *Current Treatment Options in Neurology*, *17*, 344. Available from https://doi.org/10.1007/s11940-015-0344-7.

Mills, S., Ross, R. P., Hill, C., Fitzgerald, G. F., & Stanton, C. (2011). Milk intelligence: Mining milk for bioactive substances associated with human health. *International Dairy Journal*, *21*, 377–401.

Minelli, C., Thompson, J. R., & Abrams, K. R. (2004). An integrated approach to the meta-analysis of genetic association studies using Mendelian randomization. *American Journal of Epidemiology*, *160*, 445–452.

Mitra, S. K., Sachan, A., Udupa, V., Seshadri, S. J., & Jayakumar, K. (2003). Histological changes in intestine in semichronic diarrhoea induced by lactose enriched diet in rats: Effect of Diarex-Vet. *Indian Journal of Experimental Biology*, *41*, 211–215.

Mohammadzadeh, M., Faramarzi, E., Mahdavi, R., Nasirimotlagh, B., & Asghari Jafarabadi, M. (2013). Effect of conjugated linoleic acid supplementation on inflammatory factors and matrix metalloproteinase enzymes in rectal cancer patients undergoing chemoradiotherapy. *Integrative Cancer Therapies*, *12*, 496–502.

Morales, E., Azocar, L., Maul, X., Perez, C., Chianale, J., & Miquel, J. F. (2011). The European lactase persistence genotype determines the lactase persistence state and correlates with gastrointestinal symptoms in the Hispanic and Amerindian Chilean population: A case-control and population-based study. *BMJ Open*, *1*, e000125.

Moya, M., Cortes, E., Ballester, M. I., Vento, M., & Juste, M. (1992). Short-term polycose substitution for lactose reduces calcium absorption in healthy term babies. *Journal of Pediatric Gastroenterology and Nutrition*, *14*, 57–61.

Moya, M., Lifschitz, C., Ameen, V., & Euler, A. R. (1999). A metabolic balance study in term infants fed lactose-containing or lactose-free formula. *Acta Paediatrica*, *88*, 1211–1215.

Moynihan, P. J. (2002). Dietary advice in dental practice. *British Dental Journal*, *193*, 563–568.

Mueckler, M., & Thorens, B. (2013). The SLC2 (GLUT) family of membrane transporters. *Molecular Aspects of Medicine*, *34*(2–3), 121–138.

Mundorff-Shrestha, S. A., Featherstone, J. D., Eisenberg, A. D., Cowles, E., Curzon, M. E., Espeland, M. A., et al. (1994). Cariogenic potential of foods. II. Relationship of food composition, plaque microbial counts, and salivary parameters to caries in the rat model. *Caries Research*, *28*, 106–115.

Musso, G., Gambino, R., & Cassader, M. (2011). Interactions between gut microbiota and host metabolism predisposing to obesity and diabetes. *Annual Review of Medicine*, *62*, 361–380.

Naim, H. Y. (2001). Molecular and cellular aspects and regulation of intestinal lactase-phlorizin hydrolase. *Histology & Histopathology*, *16*, 553–561.

Naim, H. Y., Jacob, R., Naim, H., Sambrook, J. F., & Gething, M. J. (1994). The pro region of human intestinal lactase-phlorizin hydrolase. *Journal of Biological Chemistry*, *269*, 26933–26943.

Naim, H. Y., Sterchi, E. E., & Lentze, M. J. (1987). Biosynthesis and maturation of lactase-phlorizin hydrolase in the human small intestinal epithelial cells. *Biochemical Journal*, *241*, 427–434.

Nakamura, H., Moriyama, Y., Makiyama, T., Emori, S., Yamashita, H., Yamazaki, R., et al. (2013). Lactosylceramide interacts with and activates cytosolic phospholipase A2α. *Journal of Biological Chemistry*, *288*(32), 23264–23272. Available from https://doi.org/10.1074/jbc.M113.491431.

Nakamura, H., Moriyama, Y., Watanabe, K., Tomizawa, S., Yamazak, R., Takahashi, H., et al. (2017). Lactosylceramide-induced phosphorylation signaling to group IVA phospholipase A_2 via reactive oxygen species in tumor necrosis factor-α-treated cells. *Journal of Cellular Biochemistry*, *118*(12), 4370–4382. Available from https://doi.org/10.1002/jcb.26091.

Nakate, T., Yoshida, H., Ohike, A., Tokunaga, Y., Ibuki, R., & Kawashima, Y. (2005). Formulation development of inhalation powders for FK888 using the E-haler to improve the inhalation performance at a high dose, and its absorption in healthy volunteers. *European Journal of Pharmaceutics and Biopharmaceutics*, *59*, 25–33.

Narayanan, A., Baskaran, S. A., Amalaradjou, M. A., & Venkitanarayanan, K. (2015). Anticarcinogenic properties of medium chain fatty acids on human colorectal, skin and breast cancer cells in vitro. *International Journal of Molecular Sciences*, *16*, 5014–5027.

Nardone, R., Taylor, A. C., Höller, Y., Brigo, F., Lochner, P., & Trinka, E. (2016). Minimal hepatic encephalopathy: A review. *Neuroscience Research*, *111*, 1–12.

Niki, T., Tsutsui, S., Hirose, S., Aradono, S., Sugimoto, Y., Takeshita, K., et al. (2009). Galectin-9 is a high affinity IgE-binding lectin with anti-allergic effect by blocking IgE-antigen complex formation. *Journal of Biological Chemistry*, *284*, 32344–32352.

Nelson, G., Hoon, M. A., Chandrashekar, J., Zhang, Y., Ryba, N. J., & Zuker, C. S. (2001). Mammalian sweet taste receptors. *Cell*, *106*, 381–390.

Nilsson, Å. (2016). Role of sphingolipids in infant gut health and immunity. *The Journal of Pediatrics*, *173*(Suppl.), S53–S59.

Norén, O., & Sjöström, H. (2001). Structure, biosynthesis and regulation of lactase-phlorizin hydrolase. *Näringsforskning*, *45*(1), 156–160. Available from https://doi.org/10.3402/fnr.v45i0.1798.

Obata, Y., & Pachnis, V. (2016). The effect of microbiota and the immune system on the development and organization of the enteric nervous system. *Gastroenterology*, *151*, 836–844.

Octoratou, M., Merikas, E., Malgarinos, G., Stanciu, C., & Triantafillidis, J. K. A. (2012). Prospective study of pre illness diet in newly diagnosed patients with Crohn's disease. *Revista medico-chirurgicala a Societatii de Medici si Naturalisti din Iasi*, *116*, 40–49.

Oh, E., Jeremian, R., Oh, G., Groot, D., Susic, M., Lee, K., et al. (2017). Transcriptional heterogeneity in the lactase gene within cell-type is linked to the epigenome. *Scientific Reports*, *7*, 41843. Available from https://doi.org/10.1038/srep41843.

O'Hara, J. P., Carroll, S., Cooke, C. B., Morrison, D. J., Preston, T., & King, R. F. (2012). Preexercise galactose and glucose ingestion on fuel use during exercise. *Medicine & Science in Sports & Exercise*, *44*(10), 1958–1967. Available from https://doi.org/10.1249/MSS.0b013e318258bf85.

Ogretmen, B., & Hannun, Y. A. (2004). Biologically active sphingolipids in cancer pathogenesis and treatment. *Nature Reviews Cancer*, *4*(8), 604–616.

Opstelten, J. L., Leenders, M., Dik, V. K., Chan, S. S., van Schaik, F. D., Khaw, K. T., et al. (2016). Results from a European prospective cohort investigation. *Inflammatory Bowel Diseases*, *22*, 1403–1411.

Paasela, M., Kolho, K. L., Vaarala, O., & Honkanen, J. (2014). Lactose inhibits regulatory T-cell-mediated suppression of effector T-cell interferon-γ and IL-17 production. *British Journal of Nutrition*, *112*(11), 1819–1825.

Pan, L. L., Deng, Y. Y., Wang, R., Wu, C., Li, J., Niu, W., et al. (2018). Lactose induces phenotypic and functional changes of neutrophils and macrophages to alleviate acute pancreatitis in mice. *Frontiers in Immunology*, *9*, 751. Available from https://doi.org/10.3389/fimmu.2018.00751, [eCollection 2018].

Park, Y., Leitzmann, M. F., Subar, A., Hollenbeck, A., & Schatzkin, A. (2009). Dairy food, calcium, and risk of cancer in the NIH-AARP Diet and Health Study. *Archives of Internal Medicine*, *169*, 391–401.

Petry, K. G., & Reichardt, J. K. (1998). The fundamental importance of human galactose metabolism: Lessons from genetics and biochemistry. *Trends in Genetics*, *14*(3), 98–102.

Peuhkuri, K., Hukkanen, M., Beale, R., Polak, J. M., Vapaatalo, H., & Korpela, R. (1997). Age and continuous lactose challenge modify lactase protein expression and enzyme activity in gut epithelium in the rat. *Journal of Physiology and Pharmacology*, *48*, 719–729.

Philippou, E., & Al Dhaheri, A. S. (2017). Introduction to dietary carbohydrates and the glycemic index. In E. Philippou (Ed.), *Glycemic index applications in practice* (pp. 1–23). Boca Raton, FL: CRC Press, Taylor & Francis Group.

Pimentel, G., Burton, K. J., Rosikiewicz, M., Freiburghaus, C., von Ah, U., Münger, L. H., et al. (2017). Blood lactose after dairy product intake in healthy men. *British Journal of Nutrition*, *118*(12), 1070–1077. Available from https://doi.org/10.1017/S0007114517003245.

Pitt, P., de Bruijn, K. M., Beeching, M. F., Goldberg, E., & Blendis, L. M. (1980). Studies on breath methane: The effect of ethnic origins and lactulose. *Gut*, *21*(11), 951–954.

Pommergaard, H. C., Burcharth, J., Rosenberg, J., & Raskov, H. (2016). Aspirin, calcitriol, and calcium do not prevent adenoma recurrence in a randomized controlled trial. *Gastroenterology*, *150*, 114–122.

Power, M. L., & Schulkin, J. (2013). Maternal regulation of offspring development in mammals is an ancient adaptation tied to lactation. *Applied & Translational Genomics*, *2*, 55–63.

Prado, E. L., & Dewey, K. G. (2014). Nutrition and brain development in early life. *Nutrition Reviews*, *72*(4), 267–284.

Prokazova, N. V., Samovilova, N. N., Gracheva, E. V., & Golovanova, N. K. (2009). Ganglioside GM3 and its biological functions. *Biochemistry (Moscow)*, *74*, 235–249.

Promrat, K., Longato, L., Wands, J. R., & de la Monte, S. M. (2011). Weight loss amelioration of non-alcoholic steatohepatitis linked to shifts in hepatic ceramide expression and serum ceramide levels. *Hepatology Research*, *41*(8), 754–762.

Qin, L. Q., Xu, J. Y., Wang, P. Y., Hashi, A., Hoshi, K., & Sato, A. (2005). Milk/dairy products consumption, galactose metabolism and ovarian cancer: Meta-analysis of epidemiological studies. *European Journal of Cancer Prevention*, *14*, 13–19.

Rangel, I., Sundin, J., Fuentes, S., Repsilber, D., de Vos, W. M., & Brummer, R. J. (2015). The relationship between faecal-associated and mucosal-associated microbiota in irritable bowel syndrome patients and healthy subjects. *Aliment Pharmacology & Therapeutics*, *42*, 1211–1221.

Rasinpera, H., Kuokkanen, M., Kolho, K.-L., Lindahl, H., Enattah, N. S., Savilahti, E., et al. (2005). Transcriptional downregulation of the lactase (LCT) gene during childhood. *Gut*, *54*, 1660–1661.

Ridel, K. R., Leslie, N. D., & Gilbert, D. L. (2005). An updated review of the long-term neurological effects of galactosemia. *Pediatric Neurology*, *33*, 153–161.

Rios-Covian, D., Gueimonde, M., Duncan, S. H., Flint, H. J., & de los Reyes-Gavilan, C. G. (2015). Enhanced butyrate formation by cross-feeding between *Faecalibacterium prausnitzii* and *Bifidobacterium adolescentis*. *FEMS Microbiology Letters*, *362*(21), fnv176.

Rivera-Huerta, M., Lizárraga-Grimes, V. L., Castro-Torres, I. G., Tinoco-Méndez, M., Macías-Rosales, L., Sánchez-Bartéz, F., et al. (2017). Functional effects of prebiotic fructans in colon cancer and calcium metabolism in animal models. *BioMed Research International*, *2017*, 975–982, ID 9758982.

Rivière, A., Gagnon, M., Weckx, S., Roy, D., & De Vuyst, L. (2015). Mutual cross-feeding interactions between *Bifidobacterium longum* subsp. *longum* NCC2705 and *Eubacterium rectale* ATCC 33656 explain the bifidogenic and butyrogenic effects of arabinoxylan oligosaccharides. *Applied and Environmental Microbiology*, *81*, 7767–7781.

Robayo-Torres, C. C., & Nichols, B. L. (2007). Molecular differentiation of congenital lactase deficiency from adult-type hypolactasia. *Nutrition Reviews*, *65*, 95–98.

Roser, M., Josic, D., Kontou, M., Mosetter, K., Maurer, P., & Reutter, W. (2009). Metabolism of galactose in the brain and liver of rats and its conversion into glutamate and other amino acids. *Journal of Neural Transmission (Vienna)*, *116*(2), 131–139.

Rossi, M., Maiuri, L., Fusco, M. I., Salvati, V. M., Fuccio, A., Auricchio, S., et al. (1997). Lactase persistence versus decline in human adults: Multifactorial events are involved in down-regulation after weaning. *Gastroenterology*, *112*, 1506–1514.

Rozenberg, S., Body, J. J., Bruyère, O., Bergmann, P., Brandi, M. L., & Cooper, C. (2016). Effects of dairy products consumption on health: Benefits and beliefs—a commentary from the Belgian Bone Club and the European Society for Clinical and Economic Aspects of Osteoporosis, Osteoarthritis and Musculoskeletal Diseases. *Calcified Tissue International*, *98*, 1–17.

Ryan, J. M., Rice, G. E., & Mitchell, M. D. (2013). The role of gangliosides in brain development and the potential benefits of perinatal supplementation. *Nutrition Research*, *33*, 877–887.

Saarela, T., Similä, S., & Koivisto, M. (1995). Hypercalcemia and nephrocalcinosis in patients with congenital lactase deficiency. *Journal of Pediatrician*, *127*, 920–923.

Sacerdote, C., Guarrera, S., Smith, G. D., Grioni, S., Krogh, V., Masala, G., et al. (2007). Lactase persistence and bitter taste response: Instrumental variables and Mendelian randomization in epidemiologic studies of dietary factors and cancer risk. *American Journal of Epidemiology*, *166*, 576–581.

Sadre, M., & Karbasi, K. (1979). Lactose intolerance in Iran. *American Journal of Clinical Nutrition*, *32*, 1948–1954.

Salomon, R., Levy, E., Levesque, D., Szilagyi, A., & Seidman, E. (1996). Caco-2 cell disaccharidase activities are unaffected by gestational hormones. *Canadian Journal of Physiology and Pharmacology*, *74*, 1126–1131.

Sandborn, W. J., Feagan, B. G., Wolf, D. C., D'Haens, G., Vermeire, S., Hanauer, S. B., et al. (2016). TOUCHSTONE Study Group. Ozanimod Induction and Maintenance Treatment for Ulcerative Colitis. *New England Journal of Medicine*, *374*, 1754–1762.

Sanders, R. D., Sefton, J. M., Mobe, K. H., & Fridovich-Keil, J. L. (2010). UDP-galactose 4' epimerase (GALE) is essential for development of *Drosophila melanogaster*. *Disease Models & Mechanism*, *3*, 628–638.

Sanz, M. L., Gibson, G. R., & Rastall, R. A. (2005). Influence of disaccharide structure on prebiotic selectivity in vitro. *Journal of Agricultural and Food Chemistry*, *53*, 5192–5199.

Sanz, Y., Olivares, M., Moya-Pérez, Á., & Agostoni, C. (2015). Understanding the role of gut microbiome in metabolic disease risk. *Pediatric Research*, *77*, 236–244.

Schaafsma, G. (2008). Lactose and lactose derivatives as bioactive ingredients in human nutrition. *International Dairy Journal*, *18*, 458–465.

Schaafsma, G., & Visser, R. (1980). Nutritional interrelationships between calcium, phosphorus and lactose in rats. *Journal of Nutrition*, *110*, 1101–1111.

Schadewaldt, P., Kamalanathan, L., Hammen, H. W., & Wendel, U. (2004). Age dependence of endogenous galactose formation in Q188R homozygous galactosemic patients. *Molecular Genetics and Metabolism*, *81*(1), 31–44.

Schweigert, B. S., Shaw, J. H., Zepplin, M., & Elvenhjem, C. (1946). Dental caries in the cotton rat VI. The effect of the amount of protein, fat, and carbohydrate in the diet on the incidence and extent of carious lesions. *Journal of Nutrition*, *31*, 439–447.

Seksik, P., & Landman, C. (2015). Understanding microbiome data: A primer for clinicians. *Digestive Diseases*, *33*(Suppl. 1), 11–16.

Seol, J. E., Cho, C. H., Kim, S. H., & Lee, J. E. (2015). Total and dietary calcium intake and colorectal adenoma in Korean adults. *Journal of Cancer Prevention*, *20*, 153–158.

Shanmugavelan, P., Kim, S. Y., Kim, J. B., Kim, H. W., Cho, S. M., Kim, S. N., et al. (2013). Evaluation of sugar content and composition in commonly consumed Korean vegetables, fruits, cereals, seed plants, and leaves by HPLC-ELSD. *Carbohydrate Research*, *380*, 112–117.

Sharma, M., & Shukla, G. (2016). Metabiotics: one step ahead of probiotics; an insight into mechanisms involved in anticancerous effect in colorectal cancer. *Frontiers in Microbiology*, *7*, 1940, [eCollection 2016].

Sheiham, A. (2001). Dietary effects on dental disease. *Public Health Nutrition*, *4*, 569–591.

Shide, D. J., & Blass, E. M. (1989). Opioidlike effects of intraoral infusions of corn oil and polycose on stress reactions in 10-day-old rats. *Behavioural Neuroscience*, *103*(6), 1168–1175.

Shirreffs, S. M., Watson, P., & Maughan, R. J. (2007). Milk as an effective post-exercise rehydration drink. *British Journal of Nutrition*, *98*(1), 173–180.

Shrier, I., Szilagyi, A., & Correa, J. A. (2008). Impact of lactose containing foods and the genetics of lactase on diseases: An analytical review of population data. *Nutrition and Cancer*, *60*(3), 292–300. Available from https://doi.org/10.1080/01635580701745301.

Skovbjerg, H., Sjöström, H., & Norén, O. (1981). Purification and characterisation of amphiphilic lactase/phlorizin hydrolase from human small intestine. *European Journal of Biochemistry*, *114*(3), 653–661.

Smith, C. E., Coltell, O., Sorlí, J. V., Estruch, R., Martínez-González, M. Á., Salas-Salvadó, J., et al. (2016). Associations of the MCM6-rs3754686 proxy for milk intake in Mediterranean and American populations with cardiovascular biomarkers, disease and mortality: Mendelian randomization. *Science Reports*, *6*, 33188. Available from https://doi.org/10.1038/srep33188.

Stahel, P., Kim, J. J., Xiao, C., & Cant, J. P. (2017). Of the milk sugars, galactose, but not prebiotic galacto oligosaccharide, improves insulin sensitivity in male Sprague-Dawley rats. *PLos One*, *12*(2), e0172260.

Stang, A., Ahrens, W., & Baumgrdt-Elms, C. (2006). Adolescent milk fat and galactose consumption and testicular cell cancer. *Cancer Epidemiology, Biomarkers & Prevention*, *15*, 2189–2195.

Stellaard, F., Koetse, H. A., Elzinga, H., Boverhof, R., Tjoonk, R., Klimp, A., et al. (2000). 13C-carbohydrate breath tests: Impact of physical activity on the rate-limiting step in lactose utilization. *Scandinavian Journal of Gastroenterology*, *35*, 819–823.

Sterchi, E. E., Mills, P. R., Fransen, J. A., Hauri, H. P., Lentze, M. J., Naim, H. Y., et al. (1990). Biogenesis of intestinal lactase-phlorizin hydrolase in adults with lactose intolerance. Evidence for reduced biosynthesis and slowed-down maturation in enterocytes. *Journal of Clinical Investigation*, *86*, 1329–1337.

Struzyńska, L., & Sulkowski, G. (2004). Relationships between glutamine, glutamate, and GABA in nerve endings under Pb-toxicity conditions. *Journal of Inorganic Biochemistry*, *98*(6), 951–958.

Suchy, F. J., Brannon, P. M., Carpenter, T. O., Fernandez, J. R., Gilsanz, V., & Gould, J. B. (2010). National Institutes of Health Consensus Development Conference: Lactose intolerance and health. *Annals of Internal Medicine*, *152*, 792–796.

Suzuki, M., West, C., & Beutler, E. (2001). Large-scale molecular screening for galactosemia alleles in a pan-ethnic population. *Human Genetics*, *109*, 210–215.

Suzuki, T., Mochizuki, K., & Goda, T. (2014). Thyroid and glucocorticoid hormones induce expression of lactase-phlorizin hydrolase gene in CDX-2/HNF-1α co-transfected IEC-6 cells. *Journal of Nutritional Science and Vitaminology (Tokyo)*, *60*, 321–327.

Szilagyi, A. (2004). Redefining lactose as a conditional prebiotic. *Canadian Journal of Gastroenterology*, *18*, 163–167.

Szilagyi, A. (2015). Adaptation to lactose in lactase non persistent people: Effects on intolerance and the relationship between dairy food consumption and evaluation of diseases. *Nutrients*, *7*, 6751–6779, 2015.

Szilagyi, A., Malolepszy, P., Yesovitch, S., Nathwani, U., Vinokuroff, C., Cohen, A., et al. (2005). Inverse dose effect of pretest dietary lactose intake on breath hydrogen results and symptoms in lactase non persistent subjects. *Digestive Diseases and Sciences*, *50*, 2178–2182.

Szilagyi, A., Nathwani, U., Vinokuroff, C., Correa, J. A., & Shrier, I. (2006). The effect of lactose maldigestion on the relationship between dairy food intake and colorectal cancer: A systematic review. *Nutrition and Cancer*, *55*, 141–150.

Tan-Dy, C. R., & Ohlsson, A. (2013). Lactase treated feeds to promote growth and feeding tolerance in preterm infants. *Cochrane Database of Systematic Reviews*. Available from https://doi.org/10.1002/14651858.CD004591.pub3, CD004591.

Thorning, T. K., Bertram, H. C., Bonjour, J. P., de Groot, L., Dupont, D., Feeney, E., et al. (2017). Whole dairy matrix or single nutrients in assessment of health effects: Current evidence and knowledge gaps. *American Journal of Clinical Nutrition*, *105*, 1033–1045.

Thorning, T. K., Raben, A., Tholstrup, T., Soedamah-Muthu, S. S., Givens, I., & Astrup, A. (2016). Milk and dairy products: Good or bad for human health? An assessment of the totality of scientific evidence. *Food & Nutrition Research*, *60*, 32527. Available from https://doi.org/10.3402/fnr.v60.32527, [eCollection 2016].

Timpson, N. J., Greenwood, C. M. T., Soranzo, N., Lawson, D. J., & Richards, J. B. (2018). Genetic architecture: The shape of the genetic contribution to human traits and disease. *Nature Reviews Genetics*, *19*(2), 110–124. Available from https://doi.org/10.1038/nrg.2017.101.

Tishkoff, S. A., Reed, F. A., Ranciaro, A., Voight, B. F., Babbitt, C. C., Silverman, J. S., et al. (2007). Convergent adaptation of human lactase persistence in Africa and Europe. *Nature Genetics*, *39*, 31–40.

Toba, T., Nagashima, S., & Adachi, S. (1991). Is lactose really present in plants? *Journal of the Science of Food and Agriculture*, *54*, 305–308.

Torfadottir, J. E., Steingrimsdottir, L., Mucci, L., Aspelund, T., Kasperzyk, J. L., Olafsson, O., et al. (2012). Milk intake in early life and risk of advanced prostate cancer. *American Journal of Epidemiology*, *175*, 144–153.

Touger-Decker, R., & van Loveren, C. (2003). Sugars and dental caries. *American Journal of Clinical Nutrition*, *78*, 881S–892S.

Travis, R. C., Appleby, P. N., Siddiq, A., Allen, N. E., Kaaks, R., Canzian, F., et al. (2012). Genetic variation in lactase gene, dairy product intake and risk for prostate cancer in the European prospective investigation into cancer and nutrition. *International Journal of Cancer*, *132*, 1901–1910.

Trauner, M., Claudel, T., Fickert, P., Moustafa, T., & Wagner, M. (2010). Bile acids as regulators of hepatic lipid and glucose metabolism. *Digestive Diseases*, *28*, 220–224.

Triantis, V., Bode, L., & Van Neerven, J.J. (2018). Immunological effects of human milk oligosaccharides. Frontiers in Pediatrics, 6, 190:doi:10.3389/fped.2018.00190. eCollection 2018.

Uchida, N., Sakamoto, O., Irie, M., Abukawa, D., Takeyama, J., Kure, S., et al. (2012). Two novel mutations in the lactase gene in a Japanese infant with congenital lactase deficiency. *Tohoku Journal of Experimental Medicine*, *227*, 69–72.

Um, C. Y., Fedirko, V., Flanders, W. D., Judd, S. E., & Bostick, R. ,M. (2017). Associations of calcium and milk product intakes with incident, sporadic colorectal adenomas. *Nutrition and Cancer*, *69*, 416–427.

Uribe, M., Márquez, M. A., García-Ramos, G., Escobedo, V., Murillo, H., Guevara, L., et al. (1980). Treatment of chronic portal-systemic encephalopathy with lactose in lactase-deficient patients. *Digestive Diseases and Sciences*, *25*, 924–928.

Uribe, M., Toledo, H., Perez, F., Vargas, F., Gil, S., Garcia-Ramos, G., et al. (1987). Lactitol, a second-generation disaccharide for treatment of chronic portal-systemic encephalopathy. A double-blind, crossover, randomized clinical trial. *Digestive Diseases and Sciences*, *32*, 1345–1353.

Vállez García, D., Doorduin, J., de Paula Faria, D., Dierckx, R. A., & de Vries, E. F. (2017). Effect of preventive and curative fingolimod treatment regimens on microglia activation and disease progression in a rat model of multiple sclerosis. *Journal of Neuroimmune Pharmacology*. Available from https://doi.org/10.1007/s11481-017-9741-x.

Van Beers, E. H., Al, R. H., Rings, E. H., Einerhand, A. W., Dekker, J., & Büller, H. A. (1995). Lactase and sucrase-isomaltase gene expression during Caco-2 cell differentiation. *Biochemical Journal*, *308*(Pt 3), 769–775.

Van Calcar, S. C., Bernstein, L. E., Rohr, F. J., Scaman, C. H., Yannicelli, S., & Berry, G. T. (2014). A re-evaluation of life-long severe galactose restriction for the nutrition management of classic galactosemia. *Molecular Genetics and Metabolism*, *112*(3), 191–197.

Van Dam, D., & De Deyn, P. P. (2011). Animal models in the drug discovery pipeline for Alzheimer's disease. *British Journal of Pharmacology*, *164*(4), 1285–1300. Available from https://doi.org/10.1111/j.1476-5381.2011.01299.x.

van der Hoeven, N. W., Hollander, M. R., Yıldırım, C., Jansen, M. F., Teunissen, P. F., Horrevoets, A. J., et al. (2016). The emerging role of galectins in cardiovascular disease. *Vascular Pharmacology*, *81*, 31–41.

Varki, A., Cummings, R. D., Esko, J. D., Freeze, H. H., Stanley, P., Bertozzi, C. R., et al. (2009). *Essentials of glycobiology* (2nd ed.). Cold Spring Harbor, NY: Cold Spring Harbor Laboratory Press.

Veettil, S. K., Teerawattanapong, N., Ching, S. M., Lim, K. G., Saokaew, S., Phisalprapa, P., et al. (2017). Effects of chemopreventive agents on the incidence of recurrent colorectal adenomas: Systematic review with network meta-analysis of randomized controlled trials. *OncoTargets and Therapy*, *10*, 2689–2700.

Vesa, T. H., Marteau, P., & Korpela, R. (2000). Lactose intolerance. *Journal of American College of Nutrition*, *19*(2 Suppl.), 165S–175S.

Viguier, M., Advedissian, T., Delacour, D., Poirier, F., & Deshayes, F. (2014). Galectins in epithelial functions. *Tissue Barriers*, *2*, e29103.

Vuong, H. E., & Hsiao, E. Y. (2017). Emerging roles for the gut microbiome in autism spectrum disorder. *Biological Psychiatry*, *81*, 411–423.

Wargovich, M. J., Jimenez, A., McKee, K., Steele, V. E., Velasco, M., Woods, J., et al. (2000). Efficacy of potential chemopreventive agents on rat colon aberrant crypt formation and progression. *Carcinogenesis*, *21*, 1149–1155.

Weaver, L. T., Laker, M. F., & Nelson, R. (1986). Neonatal intestinal lactase activity. *Archives of Disease in Childhood*, *61*, 896–899.

Welling, L., Bernstein, L. E., Berry, G. T., Burlina, A. B., Eyskens, F., & Gautschi, M. (2017). International clinical guideline for the management of classical galactosemia: Diagnosis, treatment, and follow-up. *Journal of Inherited Metabolic Disease*, *40*(2), 171–176.

Wijesinha-Bettoni, R., & Burlingame, B. (2013). In E. Muelhoff, A. Bennet, & D. McMahon (Eds.), *Milk and dairy product composition* (pp. 41–102). Rome, Italy: Food Agriculture Organization (FAO).

Windey, K., Houben, E., Deroover, L., & Verbeke, K. (2015). Contribution of colonic fermentation and fecal water toxicity to the pathophysiology of lactose-intolerance. *Nutrients*, *7*, 7505–7522.

Wolever, T. M. (1990). The glycemic index. *World Review of Nutrition and Dietetics*, *62*, 120–185.

Wu, G. D. (2016). The gut microbiome, its metabolome, and their relationship to health and disease. *Nestle Nutrition Institute Workshop Series*, *84*, 103–110.

Xiao, R., Badger, T. M., & Simme, F. A. (2005). Dietary exposure to soy or whey proteins alters colonic global gene expression profiles during rat colon tumorigenesis. *Molecular Cancer*, *4*, 1.

Yamaguchi, M., Takai, S., Hosono, A., & Seki, T. (2014). Bovine milk-derived α-lactalbumin inhibits colon inflammation and carcinogenesis in azoxymethane and dextran sodium sulfate-treated mice. *Bioscience, Biotechnology and Biochemistry*, *78*, 672–679.

Yeh, K. Y., Yeh, M., & Holt, P. R. (1991). Thyroxine and cortisone cooperate to modulate postnatal intestinal enzyme differentiation in the rat. *American Journal of Physiology*, *260*(3 Pt 1), G371–G378.

Zanabria, R., Tellez, A. M., Griffiths, M., & Corredig, M. (2013). Milk fat globule membrane isolate induces apoptosis in HT-29 human colon cancer cells. *Food & Function*, *4*, 222–230.

Zecca, L., Mesonero, J. E., Stutz, A., Poirée, J. C., Giudicelli, J., Cursio, R., et al. (1998). Intestinal lactase-phlorizin hydrolase (LPH): The two catalytic sites; the role of the pancreas in pro-LPH maturation. *FEBS Letters*, *435*, 225–228.

Ziegler, E. E., & Fomon, S. J. (1983). Lactose enhances mineral absorption in infancy. *Journal of Pediatric Gastroenterology and Nutrition*, *2*, 288–294.

Zittermann, A., Bock, P., Drummer, C., Scheld, K., Heer, M., & Stehle, P. (2000). Lactose does not enhance calcium bioavailability in lactose-tolerant, healthy adults. *American Journal of Clinical Nutrition*, *71*, 931–936.

Zhang, X., Keum, N., Wu, K., Smith-Warner, S. A., Ogino, S., Chan, A. T., et al. (2016). Calcium intake and colorectal cancer risk: Results from the nurses' health study and health professionals follow-up study. *International Journal of Cancer*, *139*, 2232–2242. Available from https://doi.org/10.1002/ijc.30293.

Zhao, G. Q., Zhang, Y., Hoon, M. A., Chandrashekar, J., Erlenbach, I., Ryba, N. J., et al. (2003). The receptors for mammalian sweet and umami taste. *Cell*, *115*(3), 255–266.

Zhu, J., Mu, X., Zeng, J., Xu, C., Liu, J., Zhang, M., et al. (2014). Ginsenoside Rg1 prevents cognitive impairment and hippocampus senescence in a rat model of D-galactose-induced aging. *PLoS One*, *9*, e10129.

Further reading

Abubakar, I., Myhill, D. J., Hart, A. R., Lake, I. R., Harvey, I., Rhodes, J. M., et al. (2007). A case-control study of drinking water and dairy products in Crohn's. Case-control study of drinking water and dairy products in Crohn's disease—further investigation of the possible role of *Mycobacterium avium paratuberculosis*. *American Journal of Epidemiology*, *165*, 776–783.

Ambalam, P., Raman, M., Purama, R. K., & Doble, M. (2016). Probiotics, prebiotics and colorectal cancer prevention. *Best Practice & Research: Clinical Gastroenterology*, *30* (1), 119–131. Available from https://doi.org/10.1016/j.bpg.2016.02.009.

Avena, N. M., Rada, P., & Hoebel, B. G. (2008). Evidence for sugar addiction: Behavioral and neurochemical effects of intermittent, excessive sugar intake. *Neuroscience & Biobehavioral Reviews*, *32*, 20–39.

Dzialanski, Z., Barany, M., Engfeldt, P., Magnuson, A., Olsson, L. A., & Nilsson, T. K. (2016). Lactase persistence versus lactose intolerance: Is there an intermediate phenotype? *Clinical Biochemistry*, *49*, 248–252. Available from https://doi.org/10.1016/j.clinbiochem.2015.11.001.

Fajardo, O., Naim, H. Y., & Lacey, S. W. (1994). The polymorphic expression of lactase in adults is regulated at the messenger RNA level. *Gastroenterology*, *106*, 1233–1241.

Konstantinov, S. R., van der Woude, C. J., & Peppelenbosch, M. P. (2013). Do pregnancy-related changes in the microbiome stimulate innate immunity? *Trends in Molecular Medicine*, *19*, 454–459.

Laxminarayan, S., Reifman, J., Edwards, S. S., Wolpert, H., & Steil, G. M. (2015). Bolus Estimation—rethinking the effect of meal fat content. *Diabetes Technology & Therapeutics*, *17*, 860 866.

Lazzara, F., Szilagyi, A., & Lerman, A. (1985). Lactose intolerance in hyperthyroidism: A possible unmasking effect. *American Journal of Gastroenterology*, *80*, 846, [Abstract].

Le Merrer, J., Becker, J. A., Befort, K., & Kieffer, B. L. (2009). Reward processing by the opioid system in the brain. *Physiological Reviews*, *89*, 1379–1412.

Marton, A., Xue, X., & Szilagyi, A. (2012). Meta-analysis: The diagnostic accuracy of lactose breath hydrogen or lactose tolerance tests for predicting the North European lactase polymorphism C/T -13910. *Alimentary Pharmacology & Therapeutics*, *35*, 429–440.

Menini, S., Iacobini, C., Blasetti Fantauzzi, C., Pesce, C. M., Pugliese, G., Minelli, C., et al. (2004). An integrated approach to the meta-analysis of genetic association studies using Mendelian randomization. *American Journal of Epidemiology*, *160*, 445–452.

Mukherjee, A., Biswas, A., & Das, S. K. (2016). Gut dysfunction in Parkinson's disease. *World Journal of Gastroenterology*, *22*, 5742–5752.

Mulcare, A. C., Weale, M. E., Jones, A. L., Connel, B., Zeitlyn, D., Tarekegn, A., et al. (2004). The T allele of a single-nucleotide polymorphism 13.9 kb upstream of the lactase gene (LCT) (C-13.9kbT) does not predict or cause the lactase-persistence phenotype in Africans. *American Journal of Human Genetics*, *74*, 1102–1110.

Naim, H. ,Y., & Naim, H. (1996). Dimerization of lactase-phlorizin hydrolase occurs in the endoplasmic reticulum, involves the putative membrane spanning domain and is required for an efficient transport of the enzyme to the cell surface. *European Journal of Cellular Biology*, *70*, 198–208.

Szilagyi, A., Lerman, S., Barr, R. G., Stern, J., & MacMullan, S. (1992). Influence of hypothyroidism on gastric emptying, oral cecal transit and lactose absorption. *Clinical and Investigative Medicine*, *15*, A44 [Abstract].

Wattez, J. S., Delmont, A., Bouvet, M., Beseme, O., Goers, S., & Delaha, F. (2015). Maternal perinatal undernutrition modifies lactose and serotranferrin in milk: Relevance to the programming of metabolic diseases? *American Journal of Physiology–Endocrinology and Metabolism*, *308*(5), E393–E401.

Westwater, M. L., Fletcher, P. C., & Ziauddeen, H. (2016). Sugar addiction: The state of the science. *European Journal of Nutrition*, *55*(Suppl. 2), 55–69.

Wray, G. A., & Deloukas, P. (2007). Convergent adaptation of human lactase persistence in Africa and Europe. *Nature Genetics*, *39*, 31–40.

CHAPTER

Lactose intolerance and other related food sensitivities

3

Andrew Szilagyi[1], Catherine Walker[2] and Mark G. Thomas[2]

[1]*Jewish General Hospital, McGill University School of Medicine, Division of Gastroenterology, Montreal, QC, Canada*

[2]*Research Department of Genetics, Evolution and Environment (GEE), University College London, London, United Kingdom*

3.1 Introduction

Milk and dairy food consumption are often implicated in a range of adverse gastrointestinal and other symptoms, something first noted by the Romans (Jones, 1969). These symptoms include cramps, bloating, flatulence, and altered bowel movements, as well as skin and joint problems, mood alterations, and headaches. There is widespread belief that the lactose in milk causes these symptoms. However, there are various other nutrient components in milk and dairy foods, as well as nondairy factors, which can induce similar symptoms. There is also a widely held belief that adverse symptoms caused specifically by the ingestion of lactose—termed lactose intolerance (LI)—are congenital. However, a range of nongenetic factors can also lead to LI. In this chapter we will describe different causes of LI, including the physical and physiological parameters governing the digestion of lactose. We describe the tests of lactose digestion available. We also discuss the confusion relating to the adverse symptoms following dairy consumption not caused by lactose.

In humans lactose is digested into its constituent monosaccharides—glucose and galactose—by the membrane-bound enzyme lactase-phlorizin hydrolase (hereafter referred to as lactase), which is expressed in the brush border of the small intestine (see Chapter 2: Digestion, absorption, metabolism, and physiological effects of lactose for details on lactose digestion). In most humans and all other mammals studied to date, lactase production is downregulated sometime between the end of the weaning period and adulthood (Gerbault et al., 2011; Itan, Powell, Beaumont, Burger, & Thomas, 2009). This trait is known as lactase nonpersistence (LNP) and is considered the ancestral state. However, approximately one-third of the world's adult population continues to produce sufficient quantities of lactase to permit digestion of lactose. This genetically determined and evolutionarily derived trait is called lactase persistence (LP). At least five associated

Lactose. DOI: https://doi.org/10.1016/B978-0-12-811720-0.00003-9

alleles, all within a 100 bp region approximately 14,000 bp upstream of the translation initiation codon of the lactase gene (LCT), have been identified and are likely causative of LP (Enattah et al., 2002; Ingram, Mulcare, Itan, Thomas, & Swallow, 2009; Lewinsky et al., 2005; Tishkoff et al., 2007) (see Chapter 1, The evolution of lactose digestion, for a detailed discussion of the genetic basis for LP). LP is generally considered to be a dominant trait, although reduced lactase expression has been observed in heterozygotes (Swallow, 2003; Swallow & Troelsen, 2016; Troelsen, 2005; Wang et al., 1995).

Lactose maldigestion (LM) is a failure to hydrolyze lactose into its constituent monosaccharides. The major cause of LM is an absence or low levels of lactase expression. In LM individuals, undigested lactose will transit into the ileum and colon (lower small bowel and the large bowel), where it is fermented by bacteria, producing by-products such as fatty acids and various gases, including hydrogen and methane. The production of large quantities of gases can lead to bloating and cramps, two characteristic features of LI. In addition, lactose in the colon can have an osmotic effect, leading to liquefaction of the colonic contents, and consequently diarrhea, a third characteristic feature of LI. LM was reported in 1903 as causing diarrhea in mature dogs following milk consumption (Röhmann & Nagano, 1903). Later the role of intestinal lactase in hydrolyzing lactose was elucidated in humans (Dahlqvist, 1962; Sahi, 1978).

A simple causation chain of LCT ancestral gene → LNP → LM → LI does not represent the full complexity of the relationships between these variables. Genetically LP individuals can fail to express lactase; adult onset secondary hypolactasia—a form of LM—can be caused by diseases in the upper gastrointestinal tract. LM people can often consume modest quantities of lactose without experiencing LI symptoms, while people who express lactase can still experience LM. In addition, symptoms similar to those of LI can follow milk and dairy food consumption which are not caused by lactose (Fig. 3.1).

3.2 Lactose intolerance

Terms related to LI are defined in Table 3.1. Because most of the world's adult population ($\sim$65%) (Itan et al., 2010) are LNP, and so LM, it is expected that the most common cause of LI symptoms is simply the failure to digest the lactose in milk and other dairy foods. However, this is complicated by a number of observations. For example, in East Asian populations with a high prevalence of LNP and LM, the frequency of reported LI is similar to non–East Asian populations, with a poor correlation between LI and LM (Lukito, Malik, Surono, & Wahlqvist, 2015). This may be partly explained by lower exposure to milk and other lactose-containing dairy foods in the diets of East Asian populations, although other factors such as variations in colonic microbiomes may also be important (Fig. 3.2).

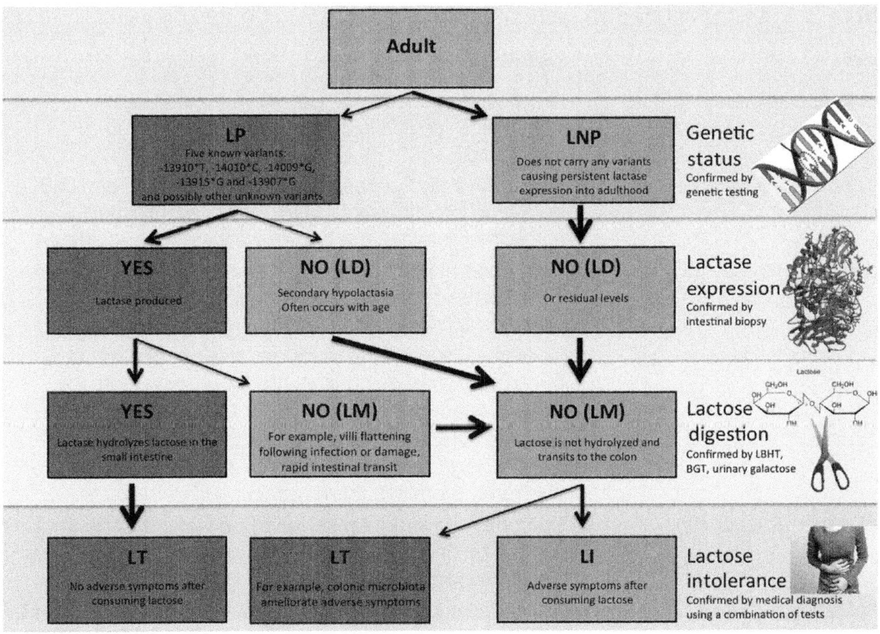

FIGURE 3.1

Diagram relating lactose digestion to lactase gene expression, lactase enzyme activity, and lactose intolerance (LI). The two main outcomes are lactose tolerance (LT) and LI. LP refers to lactase persistence, which is usually associated with one of five known alleles. LNP refers to the lactase nonpersistent phenotype. LD is lactase deficiency, not producing sufficient lactase to hydrolyze lactose or having a condition that prevents lactose from coming into contact with lactase. LM is lactose maldigestion. Arrows indicate the strength and direction of the association. See Table 3.1 for detailed definitions.

LI can be classified into four groups: congenital, neonatal, secondary lactase deficiency, and adult onset lactase deficiency. These classifications are based on the causes for the low levels of intestinal lactase and the symptoms produced by the presence of undigested lactose in the colon.

3.2.1 Congenital lactose intolerance

Virtually all mammals are born with intestinal lactase and are generally able to digest lactose. In humans, congenital lactase deficiency (CLI), or alactasia, is an extremely rare autosomal recessive condition caused by mutations in the LCT (Robayo-Torres & Nichols, 2007; Uchida et al., 2012). Insufficient lactase levels in the small intestine shortly after birth lead to severe symptoms of diarrhea, acidosis, hypercalcemia, and nephrocalcinosis, with its complications (Saarela, Similä, & Koivisto, 1995) upon exposure to dietary lactose. Without treatment involving lactose withdrawal and nutritional support, CLI infants usually fail to thrive.

Table 3.1 Terms related to lactose and its digestion.

Term	Definition
Lactase persistence (LP)	A dominant genetic trait usually associated with continued high levels of lactase production into adulthood.
Lactase nonpersistence (LNP)	A recessive and ancestral genetic trait associated with a decline in intestinal lactase to <10 U/g of tissue sometime between the end of weaning and adulthood.
Lactase deficiency (LD)	Reduction of intestinal lactase enzyme from any cause, either genetic (LNP) or any secondary causes like diseases or injury of the proximal small bowel mucosa.
Lactose maldigestion (LM)	Inability to digest lactose for any reason, primary LNP, but also secondary causes. Most common tests for lactase deficiency are actually for LM.
Lactose intolerance (LI)	Adverse symptoms resulting from the ingestion of lactose, including flatus, gas, bloating, cramps, diarrhea, and rarely, vomiting. LI may occur without LM.
Self-reported LI (SRLI)	Persons believing themselves to be lactose intolerant without testing for LM. Nocebo and psychological characteristics may play a role in milk avoidance.
Lactose sensitivity (LS)	Adverse symptoms with or without symptoms of LI and may also include depression, headache and fatigue, with or without LM. LS symptoms may overlap with irritable bowel syndrome.

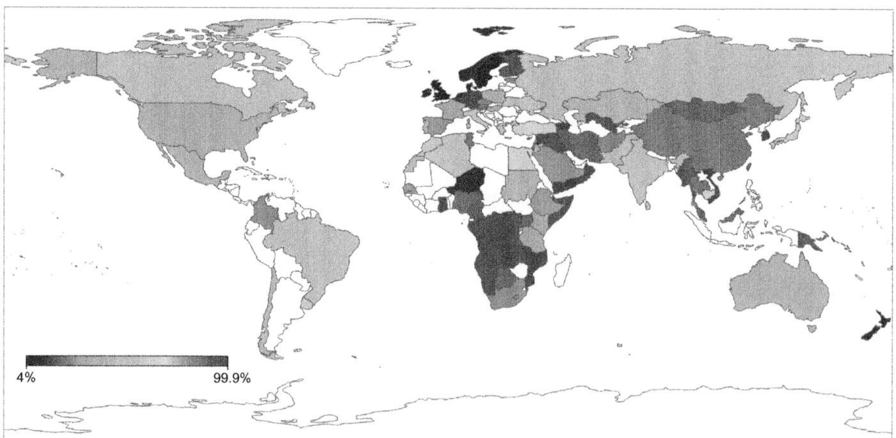

FIGURE 3.2

An approximate distribution of national lactase nonpersistence of populations.

Reprinted with permission from Storhaug, C. L., Fosse, S. K., & Fadnes, L. T. (2017). Country, regional, and global estimates for lactose malabsorption in adults: A systematic review and meta-analysis. Lancet Gastroenterology & Hepatology, *2(10), 738–746 (Storhaug, Fosse, & Fadnes, 2017).*

3.2.2 Neonatal lactose intolerance

Neonatal LI occurs when insufficient levels of lactase are present in the small intestine, and is related to the gestational age of the neonate. There are few studies investigating lactose hydrolysis in neonates. Lactase is first detected around the 12th gestational week and levels remain below those of term neonates through weeks 26–34 of gestation (Antonowicz & Lebenthal, 1977; Nilsson, 2016). Because of this, full lactase activity is often not reached in premature infants, and when it is reached remains open to question.

There is some controversy in the literature over when full-term neonates achieve normal lactase levels. Francavilla et al. (2012) carried out a crossover study in full-term infants to examine the effects of adding 3.8% lactose to formula-fed to infants with cow's milk allergy (CMA). They compared the fecal microbiome of healthy controls without dietary restrictions to each phase of the study, reporting that with added lactose, lactic acid bacteria increased significantly in the CMA cohort, while bacteroides and clostridia decreased. There were no significant differences observed in the control group. While intestinal enzymes were not evaluated, the authors suggested that in the healthy neonate, lactose is incompletely hydrolyzed, allowing some lactose to enter the lower intestine (Francavilla et al., 2012). Similar results were also observed in an earlier study in breastfed neonates where lactose was present in the colon. Due to the normal pH and absence of glucose in the stools, lactose was presumed to be fermented by gut bacteria (Lifschitz, Smith, & Garza, 1983). However, a study by Weaver and colleagues evaluated 40 healthy neonates born between 27 and 42 weeks' gestation for lactose assimilation using urinary lactose: lactulose excretion ratios to assess lactase activity. They found that at 5 days after delivery, 98% of lactose was hydrolyzed by intestinal lactase in full-term and preterm infants. In the subsequent 5–8 days (6–13 days post birth) when the lactose load increases in human breastmilk, they observed that the preterm intestinal epithelium only partially hydrolyzed lactose whereas full-term neonates almost completely hydrolyzed lactose. These findings suggest that lactase levels in pre- and full-term neonates during the first 5 days post-birth are sufficient to meet metabolic demands, but in subsequent days only the full-term infants fully hydrolyzed lactose in step with rising lactose levels in breastmilk (Weaver, Laker, & Nelson, 1986). A more recent Cochrane review evaluated the possible benefit of adding oral lactase to formulae for preterm infants, but the single randomized controlled trial did not find any benefit, suggesting that neonates had achieved their full lactase capacity (Tan-Dy & Ohlsson, 2013). These three studies suggest that further work is needed to clarify if excess lactose ingestion in neonates contributes to bacterial adaptation and at which stages of development.

3.2.3 Primary adult onset hypolactasia

Lactase production decreases with age of the child, and some develop LNP status in early adulthood (Seppo et al., 2008) (See Chapter 1, Table 1.2 for more detailed information on lactase production and age). While primary adult onset hypolactasia generally occurs before the age of 20 years (Flatz, 1987; Sahi, 1994), it was observed in a Finnish study that, on rare occasions, onset may be delayed well beyond the age of 20 years (Seppo et al., 2008). Adult onset lactase deficiency can occur in older LP individuals without injury, illness, or damage to the small intestine. Declining levels of intestinal lactase are genetically determined by the downregulation of LCT. However, in contrast to disease-induced hypolactasia, levels of some other disaccharidases (e.g., sucrase-isomaltase) do not diminish with time (Enattah et al., 2007). During testing for LM, some adults experience symptoms of LI. In older adults (>65) indications of LM may be related to bacterial overgrowth rather than an actual decline in intestinal lactase levels (Almeida et al., 2008).

In 2010 a National Institute of Health single topic conference concluded that LI is "real" but its true prevalence is not known (Suchy et al., 2010). One reason why its prevalence is difficult to quantify is that LM and LI have been used as interchangeable terms despite being distinct conditions. Many individuals describe themselves as LI without a formal diagnosis of LM. This is known as self-reported LI (SRLI) and does not distinguish LI from LM. Systemic symptoms including headache, fatigue, and concentration difficulties attributed to lactose is termed lactose sensitivity (LS). LI and SRLI also merge with the functional gastrointestinal disorders (FGIDs) described in the discussion of irritable bowel syndrome (IBS).

3.2.4 Secondary hypolactasia

Secondary lactase deficiency can occur in both LP and LNP individuals at any age. Diseases, infection, medication, or surgery can damage the integrity of villi in the proximal small intestine, leading to reduced lactase levels. In diseases of the proximal small intestine (e.g., celiac disease), activity of multiple disaccharidases can be reduced, usually as a result of decreased surface area (Mones, Yankah, Duelfer, Bustami, & Mercer, 2011). With treatment the activity of these disaccharidases can return toward normal, although this can take years (Plotkin & Isselbacher, 1964).

Secondary hypolactasia (or lactase deficiency) is the most common cause of LI in children under 5 years of age. It is typically caused by underlying gut conditions, such as viral gastroenteritis, giardiasis, cow's milk enteropathy, celiac disease, or Crohn's disease (Heine et al., 2017). Heine and colleagues concluded that LI in childhood is mostly transient and symptoms of LI will improve when the underlying pathologies are resolved (Heine et al., 2017).

3.3 Diagnostic tests for lactose intolerance

A number of tests are available for LP/LNP and LM. Some of these are presented as tests for LI, although strictly speaking they are not LI tests, as an individual could be LNP and LM without experiencing LI. As a result, most early reports in the literature equated LM with LI. This has caused confusion in the scientific literature and also in public perceptions, with the unintended and negative consequence of contributing to nutritional deficiencies through the dietary avoidance of dairy foods.

3.3.1 Direct test—biopsy

The definitive test for lactase production is assaying for its activity in an intestinal biopsy sample. However, this invasive procedure is not well suited to individual or population-level clinical diagnoses. Furthermore, biopsies can produce false-negative results due to variation in the age-related decline of lactase levels in LNP adults, and the fluctuating nature of this decline providing inaccurate snapshots (Maiuri et al., 1991). More recently, an intestinal biopsy test for lactase deficiency has been reintroduced in the form of a "Quick R" test, which allows rapid diagnosis of enzyme deficiency during gastroscopy (Furnari et al., 2013). This test is convenient when gastroscopy is being conducted for other reasons, allowing the exclusion of secondary causes of LM and lactase deficiency, but is not appropriate for general use due to the invasive nature of gastroscopy.

3.3.2 Indirect tests

Several indirect tests of LI have been developed but for the most part are not standard in clinical practice. Some indirect tests are still occasionally used in pediatric medicine or research. These include stool chromatography (Coppa et al., 2001; Roggero et al., 1986) and sugar acid tests (Robayo-Torres, Quezada-Calvillo, & Nichols, 2006). These tests can be used to test for any sugar malabsorption.

Early indirect tests included capillary galactose or glucose tests following ingestion of large amounts of lactose ($\sim$50 g), a $^{14}CO_2$ test following ingestion of ^{14}C-labeled lactose, or the breath hydrogen tests. Of these, the most common in standard clinical practice and anthropological studies is the lactose breath hydrogen test (LBHT), but the older blood glucose test (BGT, often misleadingly called the lactose tolerance (LT) test) is also used. These indirect tests have been correlated with intestinal biopsies and provide similar results in detecting LM (Newcomer, McGill, Thomas, & Hofmann, 1975).

3.3.3 Blood glucose test

The BGT measures the rise in blood glucose over baseline following an oral lactose load (which can vary between 12.5 and 50 g but is usually 25 g, an equivalent to around half a liter of milk). Blood glucose is measured every 30 minutes, usually for 2 hours. A rise above 1.1–1.4 mmol/L is considered a positive indicator of lactose digestion (Arola, 1994). A rise of less than this is considered a negative outcome, indicating a probable LM status. A variation on the BGT is the measurement of serum galactose, which denotes digester status with a rise above 0.3 mmol/L. Alcohol can be administered before the test to prevent the rapid metabolism of galactose, rendering the test somewhat more reliable (Adam, Rubio-Texeira, & Polaina, 2004). This is because ethanol inhibits epimerase, an enzyme involved in the final step of the Leloir pathway in the liver converting galactose to glucose.

3.3.4 Lactose breath hydrogen test

The LBHT involves measurement of breath H_2 following an oral lactose load, and has become the most widespread diagnostic test of LM because of its simplicity and accuracy (Newcomer et al., 1975). Unlike the BGT and serum galactose test, the LBHT detects the consequences of LM status; lactose not digested by the host reaches the lower intestine where it is fermented by bacteria, releasing gases—especially hydrogen—and short-chain fatty acids (SCFAs) (see Chapter 2: Digestion, absorption, metabolism, and physiological effects of lactose). Hydrogen is highly soluble and diffusible, and crosses the colonic epithelium into the circulatory system, where it is released as a gas in the lung. The amount of hydrogen exhaled is measured, either using an electronic hydrogen sensor or by samples of exhaled air injected at time intervals into a compact gas chromatography apparatus. A rise in hydrogen by >20 parts per million (ppm) above the baseline generally denotes a person with LM. While LM is often equated with LNP, other possible causes of increased H_2 production include excessively rapid intestinal transit, bacterial overgrowth (a consequence of loss of intestinal motility and/or hypochlorhydria; low acid) with aging, which allows usually low titers of bacteria to populate the upper segment of the intestine and ferment lactose, producing H_2, and secondary lactase deficiency (disease-induced loss of intestinal villi and reduction of lactase levels) (Krajicek & Hansel, 2016). By setting the cutoff for the rise in breath H_2 to 10 ppm, the test sensitivity is increased but specificity may be lowered (Strocchi, Corazza, Ellis, Gasbarrini, & Levitt, 1993).

A failure to raise hydrogen levels above 20 ppm is generally consistent with digester status, which in adults is a good proxy for LP status (Metz, Peters, Jenkins, Newman, & Blendis, 1975; Newcomer et al., 1975). However, other interpretations are possible. These include the absence of bacteria that produce H_2 (possible population prevalence about 10%), use of pretest antibiotics (<1 month), drugs that slow intestinal transit including opiates, primary methane production, which may reduce the rise in H_2 (Gasbarrini et al., 2009), or altered microbiome adaptation from regular lactose consumption (Hertzler & Savaiano, 1996). Bacterial adaptation will not affect the BGT because it does not measure bacterial fermentation. The breath tests are thought to be more precise in identifying LM when methane production is incorporated in test results, as the microbiota of some individuals may not produce sufficient hydrogen to mount a positive H_2 response alone, but methane production may be increased instead. Therefore measuring methane output can detect additional LM individuals (Knudsen & Di Palma, 2012; Waud, Matthews, & Campbell, 2008). Another modification is the use of carbon isotopes, an approach best reserved for research purposes (Hiele et al., 1988). The use of isotopes such as ^{13}C, although valid, complicates the general use for clinical work while adding little accuracy. It has been used to determine which patients should undergo intestinal lactase biopsy, but has also been largely replaced by indirect tests (BGT and LBHT).

It may be possible to distinguish LP heterozygote individuals from homozygotes by examining the results of breath tests (Dzialanski et al., 2016). A statistically significant difference in area under the breath hydrogen curve is observed between heterozygote and homozygote digesters, providing indirect information on zygosity. Genetic tests can be used to identify LCT genotypes, but as levels of intestinal lactase can decline with age, these tests may only be weakly informative for clinicians. A possible nongenetic explanation was hypothesized by Enko, Kriegshäuser, Halwachs-Baumann, Mangge, and Schnedl (2017). Lower levels of diamine oxidase (DAO) are associated with more pronounced symptoms of LI. This enzyme is involved in intestinal mucosal proliferation, a common occurrence after intestinal resection where the intestinal villi adapt to reduced surface area (Wolvekamp & De Bruin, 1994). In LM individuals, low DAO levels <10 U/mL predicted higher levels of breath hydrogen with more gastrointestinal symptoms compared with DAO levels >10 U/mL (Enko et al., 2017). However, it remains to be tested if LCT heterozygotes have intermediate levels of DAO and would produce breath hydrogen also at an intermediate level. Whether heterozygotes become less able to digest lactose as they age, and the subsequent clinical impacts, require further examination. An alternative hypothesis is that increased LM in the elderly is related to increased incidence of bacterial overgrowth (Almeida et al., 2008).

3.3.5 Interpreting test results

In the case of the LBHT, considerable variation has been observed in the amount of H_2 produced. Some of this variation can be understood better when the results of other tests are considered. Bacterial overgrowth of the small intestine can give false-positive results through early upper intestinal lactose fermentation by bacteria. Furthermore, LNP individuals who regularly consume milk can stimulate microbial adaptation which mimics digester status in the LBHT, producing false-negative results (Hertzler & Savaiano, 1996). In the case of the BGT there can be discrepancies between capillary and venous blood measurements, with the latter providing more accurate data (Domínguez & Fernández, 2016). In addition, gastric emptying delay or advancement can adversely affect digestion and absorption of digestion products, and medical conditions such as type II diabetes can interfere with the lactose digestion/LM test results. Both the LBHT and the BGT have discrepancies that reduce the utility in comparing the results of these two tests (Romagnuolo, Schiller, & Bailey, 2002; Szilagyi, 2011; Szilagyi et al., 2007).

3.3.6 Urinary galactose tests

Measurement of urinary excretion of galactose following an oral lactose load is also possible. Some authors argue that in addition to the LBHT measuring lactose fermented by intestinal bacteria, the urinary galactose:creatinine ratio could be used to measure the residual absorption of lactose digestion products (Grant et al., 1989). However, the galactose test is rarely used when the LBHT and BGT are used (Sahi, 1978). The additional information of residual lactose fermentation and absorption is rarely useful clinically.

3.3.7 Genetic testing

Genetic tests do not determine whether an individual is LT/LI, but ascertain whether they carry a known variant associated with LP. There are multiple known genetic causes of LP (as discussed in Chapter 1: The evolution of lactose digestion,), including alleles -13910*T, -13915*G, -13907*G, -14009*G, -14010*C (genome positions relative to the start codon of LCT) (Enattah et al., 2008; Ingram et al., 2009; Ranciaro et al., 2014; Tishkoff et al., 2007). A further single nucleotide polymorphism -22018*G is associated with LP in European populations in combination with -13910*T, but this is found in the Hazara of Pakistan without -13910*T whereas in the Chinese Han -13910*T is present without -22018*G, suggesting that the association of the -22018*G allele with LP may be more common in European populations (Ranciaro et al., 2014).

Indirect LM tests can be combined with genetic tests to confirm and refine LI diagnoses. In a recent meta-analysis, a global literature search revealed 19 studies, of which 14 evaluated the LBHT, 2 investigated the BGT, and 3 compared the LBHT and BGT results (Marton, Xue, & Szilagyi, 2012). The overall sensitivity of the LBHT was 0.88 (CI, 0.85–0.90) and specificity was 0.85 (CI, 0.82–0.87). Further analyses revealed that a 50 g lactose (instead of 25 g or less) load and the exclusion of individuals younger than 18 years old increased test sensitivity and minimally changed specificity (lower with the high lactose load and higher in the omission of children). Although only five studies evaluated the BGT, results were consistent with genetic tests. The sensitivity was 0.94 (CI, 0.90–0.97) and specificity was 0.9 (CI, 0.84–0.95). The study supports the observation that LBHT and BGT results can predict the genetic status of the individual reasonably well, with the caveat that in some African populations, variations in microbiomes may harbor different gut bacteria which ferment lactose without significantly raising breath hydrogen output (Ranciaro et al., 2014).

One further caveat in using indirect tests to infer the genetic status relates to age-determined lactase expression levels. While -13910*T explains the majority of LP in Europe (and individuals of European ancestry) and southern Asia, the other known variants together with -13910*T do not sufficiently explain LP in some non-European populations (Ingram et al., 2009). That is to say, LP has been observed in non-European populations in the absence of the five known genetic causes of LP. Furthermore, recent advances in our understanding of how epigenetic factors affect lactase expression suggest that decreasing levels of intestinal lactase may be part of the aging process (Labrie et al., 2016; Swallow & Troelsen, 2016). The -13910*T variant leads to continued expression of lactase into adulthood by increased LCT promoter activity, which in turn increases transcription rates. These increased rates of transcription are achieved by reduced levels of methylation in the -13910*T variant, whereas the -13910*C variant is more prone to methylation over time. DNA methylation is the process of adding a methyl group (CH_3) to the DNA molecule. This does not change the DNA sequence itself but affects transcription regulation and often reduces transcription rates, particularly if methylation occurs in the promoter region. In light of this, LBHT and BGT results need to be considered with respect to age of the individual as there may be an age-related gradient of lactase levels in LP and possibly LNP individuals.

Lastly, heterozygosity for LP-associated alleles should be considered. While one functional copy of an LP variant will maintain sufficient levels of lactase in the small intestine to *meta*bolize most lactose loads, a single functional copy will produce smaller amounts of the enzyme (Weiss, Lee, & Diamond, 1998). It is possible that heterozygotes may experience symptoms of LI earlier than homozygotes during mild illnesses, metabolic challenges, or aging, as they may produce insufficient quantities of lactase, and levels may decline as they age (Swallow, 2003).

Table 3.2 summarizes the tests for LM. Clinically the lactose breath hydrogen and to a lesser extent the BGT are used to assess LM. Other tests are used less often and mainly for research purposes. Lactase genetic tests are used for epidemiological investigations.

3.4 Complexities within lactose intolerance

Although historically LM became intertwined with LI, the complexities of LI now are more accurately characterized and understood.

3.4.1 Single dose of lactose and interaction with residual lactase

Single large doses or suprathreshold doses of lactose can generate LM-positive test results and induce LI symptoms in LNP and LP individuals. When ingesting a large single dose of lactose, the lactase digestion capacity can be overwhelmed, even though normal intestinal levels of lactase are being produced. Using three indirect methods, including LBHT, pulmonary $^{14}CO_2$ output, and ^{14}C levels in stool samples, Bond and Levitt evaluated four LP and six LNP adults. They found that within the LP cohort, 0%–8% of the administered load failed to be digested whereas 42%–75% of the administered lactose failed to be digested in the LNP cohort (Bond & Levitt, 1976).

Table 3.2 Tests for lactose maldigestion.

Test	Method	Interpretation	Comments
Intestinal biopsy	Crosby capsule swallowed by patient	Lactase U/g of tissue Normal 18.3 U/g	Ratio of lactase to sucrose used as sucrase is not usually downregulated May be variable due to age-related fluctuations of lactase levels
Quick test	Endoscopic biopsy With rapid results	Similar to above but more rapid at bedside	Limited use at this time
^{14}C-labeled lactose	Isotopically labeled glucose measured	Similar to measuring rise of glucose	Early method. More labor intensive and not widely used except in research
Blood glucose[a]	Ingestion of lactose 25–50 g Measure blood glucose every 30 min for 2 h	A rise above 1.1–1.4 mmol/L is positive in a lactose digester	Outcome modified by underlying diabetes, small intestinal disorders, or gastrointestinal problems. May not concur with breath hydrogen test results, but agrees with genetic tests
Blood galactose	Same as blood glucose	A rise in galactose >0.3 mmol/L is positive in a lactose digester. Result measured by change in absorbance by the release of galactonic acid and reduced nicotinamide adenine dinucleotide	Rarely used as other tests like the glucose or breath tests more readily available
Breath hydrogen[a]	Measurement of hydrogen and methane after a load of lactose for 3 h	Lactose digesters do not increase breath hydrogen ≥20 ppm Or methane ≥10 ppm	Most common clinical test. Lactose load varies but 25 g best clinical use, 50 g more sensitive for correlation with genetic tests. May disagree with the blood glucose test. Multiple variables can interfere with interpretation
Urinary galactose	Measurement of galactose in urine 5 h after a 50 g lactose load	Normal lactose digesters $\mu = 45$ mg Lactose maldigesters $\mu < 8$ mg 5 h	Surpassed for clinical use by blood glucose or breath hydrogen tests
Genetic tests	DNA from blood, saliva, or other tissue source	Confirms the presence or absence of alleles associated with lactose hydrolysis into adulthood	Most clinical tests assess the C/T -13910 allele

[a]*Breath tests and blood glucose tests are most commonly used in clinical practice.*

In LNP individuals, single-threshold doses have been evaluated in a number of investigations. In general, responses to varying the lactose loads were assessed using the LBHT. Hertzler, Huynh, and Savaiano (1996) noted that breath hydrogen began to rise after a 6 g dose was administered. Vesa et al. reported that gastrointestinal symptoms of LI occurred after increasing doses of lactose added to lactose-free milk, in both LP and LNP cohorts in exposed individuals. They concluded that 0.5–7 g of additional lactose did not cause LI symptoms in LNP individuals (Vesa, Seppo, Marteau, Sahi, & Korpela, 1998). In a Japanese study, Oku et al. found that a maximum single ingestion of 10 g lactose was tolerated but higher doses induced diarrhea (Oku, Nakamura, & Ichinose, 2005). In a double-blind random order study, a lactose dose of 12.5 g (equivalent to about 250 mL of milk) was tolerated either in a single or two half-dose ingestions (Suarez, Savaiano, Arbisi, & Levitt, 1997; Suarez, Savaiano, & Levitt, 1995). Breath hydrogen was found to inversely correlate with lactose pretest intake during breath tests in LNP subjects (Szilagyi et al., 2005).

3.4.2 Lactose intolerance symptoms during tests for lactose maldigestion

LI is confirmed by observing gastrointestinal symptoms following a lactose load test. However, it should be questioned whether symptoms produced by a lactose load confirm true LI or an inability to tolerate the specific lactose load under those test conditions. Houben et al. investigated 1051 clinical outpatients who were referred for lactose digestion evaluation using 50 g lactose load breath testing with standard (no isotope) and carbon isotope ($^{13}CO_2$) analyses. They found that 57% of participants had normal lactose digestion, 29.8% had LM, and 13.2% had either LM or bacterial overgrowth (based on achieving the rise in H_2 to 20 ppm). An additional 16% of subjects without H_2 production were reclassified as LM based on CH_4 levels above 5 ppm (Houben, De Preter, Billen, Van Ranst, & Verbeke, 2015). In this study, 59% reported some symptoms after the lactose load, with about half reporting symptoms even with normal LM outcome. Symptoms were more severe with higher hydrogen output in LM individuals or those with bacterial overgrowth. This discrepancy within LM status and symptoms has been reported previously (Szilagyi et al., 2005). In addition, positive correlations between peak and total hydrogen and total symptoms during tests have been noted (Ladas, Papanikos, & Arapakis, 1982; Szilagyi et al., 2005). It can be argued that aqueous lactose loads (dispersed in water for the purpose of the test) may not always cause LI outside test conditions. A recent review suggests that LI may not be the principal reason for the low levels of dairy food consumption in some East Asian regions. Lukito et al. (2015) suggest that testing for dairy food intolerances rather than LI would be a better choice in determining the true cause of adverse symptoms.

3.4.3 Effect of other nutrients on lactose digestion

Aqueous lactose loads may heighten responses through enhanced osmotic effects, resulting in increased LI symptoms and raised hydrogen output. The corollary is that the type of food eaten with lactose can reduce the rate at which lactose reaches the lower intestine. Adding cereal to milk reduces hydrogen production (Dehkordi, Rao, Warren, & Chawan, 1995). An increased fat content ingested with lactose also delays gastric emptying, reducing the rate of lactose entering the small bowel (Laxminarayan, Reifman, Edwards, Wolpert, & Steil, 2015). Testing lactose digestion status while consuming a standard meal may reflect real life eating habits better than using aqueous lactose challenges. Currently, these types of test conditions are uncommon due to the added complexity of administering these lactose loads.

3.4.4 Effects of pregnancy

Pregnancy is another condition where LT in LNP individuals may improve as pregnancy progresses. Although the continued consumption of lactose can encourage colonic adaptation, the retarding effect of progesterone on intestinal transit times may also contribute to increased LT (Szilagyi, Salomon, Martin, Fokeeff, & Seidman, 1996).

3.4.5 Effects of medication

Various medications also influence the transit times and/or water absorption. As a result, the correct interpretation of test results needs to consider drug effects. The drug loperamide (a common over-the-counter drug used for nonspecific control of diarrhea) slows intestinal transit by activating opioid receptors in the intestine and can improve LT when ingested before a lactose challenge test (Szilagyi, Salomon, Smith, Martin, & Seidman, 1996).

3.4.6 Disease and illness affecting intestinal motility

Similarly, diseases affecting intestinal transit times and gastric emptying can alter the rate of lactose emptying into the small bowel. Hyperthyroidism (a condition where the thyroid gland is overactive) can aggravate LI, possibly by increasing the rate of gastric emptying and intestinal transit (Szilagyi, Lerman, Barr, Colacone, & McMullan, 1991). The result is a rapid increase in lactose concentration with decreased substrate enzyme contact time and more concentrated, rapid delivery of undigested lactose into the colon. Another hypothesis is that thyroxine or triiodothyronine (thyroid hormones) may directly inhibit LCT transcription (Hodin, Chamberlain, & Upton, 1992). While such evidence is available in animal and in vitro studies, as described in Chapter 2, Digestion, absorption, metabolism, and physiological effects of lactose, there are limited studies on the direct effects of thyroid hormones on lactase in humans.

Diseases affecting the villi of the small intestine often reduce lactase levels or lactase exposure to lactose. These diseases include celiac disease, viral, bacterial (e.g., bacterial overgrowth), or parasitic infections (e.g., giardiasis), consequences of radiation therapy, and some medication (e.g., olmesartan for high blood pressure) (Rubio-Tapia et al., 2012).

3.4.7 Microbial adaptation

Early observations in Iran (Sadre & Karbasi, 1979) and Ethiopia (Habte, Sterky, & Hjalmarsson, 1973) showed after a month of regular milk powder consumption, LI symptoms improved in cohorts comprising mostly LNP individuals. Later it was shown that regular consumption of lactose in foods can improve LI symptoms in lactose naïve individuals (nonregular consumers of lactose and therefore nonadapted intestinal microbes). Hertzler and Savaiano (1996) demonstrated that by increasing lactose doses over a 16-day period, LNP individuals produced LBHT results consistent with LP status. This was accompanied by a reduction in some LI symptoms. These results demonstrate the efficacy of lactose as a prebiotic, ingesting therapeutic doses of lactose to reduce gastrointestinal symptoms. Fig. 3.3 indicates the effect of daily lactose consumption on LBHT output in LNP individuals. Symptoms of gas and bloating were reduced, while diarrhea showed no improvement. The reduced hydrogen production is associated with increased fecal β-galactosidase and increased lactic acid producing bacteria as discussed in Chapter 2, Digestion, absorption, metabolism, and physiological effects of lactose.

3.4.8 Pathogenesis of lactose intolerance

When lactose is not broken down into its constituent monosaccharides, it remains in the intestinal lumen. Here it exerts osmotic forces, drawing water into the lumen and increasing intestinal transit. Lactose and water then reach the colon more rapidly causing cramps and possibly diarrhea. He et al. (2008) postulate that the rapid production of SCFA and gas (hydrogen, methane and carbon dioxide) contributes to LI symptoms. Symptoms of LI can be caused by lactose and other carbohydrates, as well as by diseases including celiac disease, inflammatory bowel disease (IBD) (especially Crohn's disease), gynecological neoplasms, neuroendocrine tumors, and bacterial overgrowth. These conditions can be diagnosed relatively easily. However, the most common disorders that can be confused with LI are FGIDs.

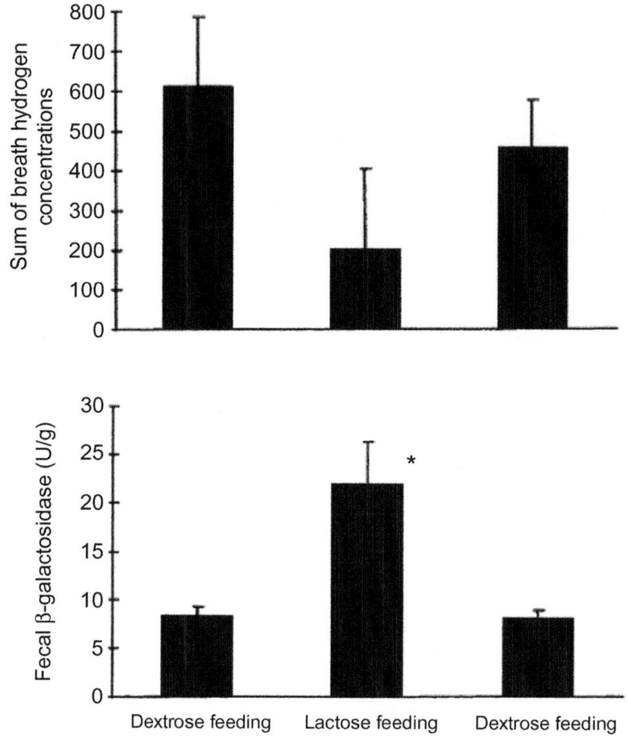

FIGURE 3.3

The reduced hydrogen levels after regular and daily increasing consumption of lactose compared with dextrose. The study was based on 20 participants with previously established lactose maldigestion who were randomly assigned 16 days of increasing weight-adjusted lactose intake (0.6–1.0 g/kg) or similar dose of dextrose as a placebo. After a washout period, participants were given the opposite disaccharide. A lactose challenge test was carried out at the middle of each sugar period and after an additional 2 weeks of any test sugar.

From Hertzler, S. R., & Savaiano, D. A. (1996). Colonic adaptation to daily lactose feeding in lactose maldigesters reduces lactose intolerance. The American Journal of Clinical Nutrition, 64*(2), 232.*

3.4.9 Functional gastrointestinal disorders and irritable bowel syndrome

A group of predominantly gastrointestinal symptoms with no or minimal gross pathology have been classified as FGIDs (Stanghellini et al., 2016). FGIDs are hypothesized to be caused by an interaction of three mechanisms: increased activated intestinal immunity, altered intestinal permeability, and abnormalities in proinflammatory serum cytokines (Collins, 2016).

IBS is one of the most common FGIDs. Symptoms of IBS resemble reactions to maldigested carbohydrates. Symptoms experienced by LNP individuals in response to lactose are not unique and other malabsorbed carbohydrates (fructose, fructo- and galacto-oligosaccharides) can produce similar gastrointestinal effects, such as abdominal cramps, flatus, abdominal distension, and sometimes diarrhea with or without nausea can be manifested after ingestion of other carbohydrates.

IBS is made up of four definable groups, divided by characteristic bowel movement, alterations, and specific time criteria. This includes IBS-D with predominance of diarrhea, IBS-C with predominance of constipation, or IBS-M with mixed symptoms (diarrhea and constipation), and finally undifferentiated IBS (Ford, Lacy, & Talley, 2017). All forms of IBS may be accompanied by abdominal cramps, flatus, and a sensation of bloating. Although there are specific biomarkers for each form of IBS, they are typically not used clinically and there is no specific pathology that explains the varying forms (Camilleri, Halawi, & Oduyebo, 2017). In the cases of IBS-D there is a 15% past history of infectious gastroenteritis preceding development of IBS (Holtmann, Ford, & Talley, 2016). The prevalence of IBS worldwide varies between 10% and 20%, with no distinct geographic patterns (Lovell & Ford, 2012).

3.4.10 The biopsychosocial contribution to lactose intolerance

Daily living stresses and psychological factors such as depression and anxiety are often associated with IBS and other FGIDs. These complex interactions have been labeled as the biopsychosocial model of IBS and FGIDs. Other symptomatic organ involvements such as fibromyalgia, functional dyspepsia, chronic fatigue syndrome, temporomandibular joint pain, and chronic interstitial cystitis with symptoms may also be involved. The pathogenic role or complicity of the intestinal microbiota and their metabolites have been implicated in FGIDs and related disorders as well as other diseases.

3.4.11 Microbiome and functional gastrointestinal disorders

Recent investigations have broadened and deepened our understanding of the microbiome, implicating its involvement in causing both disease and symptoms. The microbiome is sensitive to diet, disease states, antibiotics, stress, and external environmental changes such as air pollution. This has caused researchers to reconsider the pathogenesis of LI, IBS, and other pathological gastrointestinal symptoms. Brain–gut/gut–brain interactions (also called the "gut–brain axis") in extra gastrointestinal manifestations and mood disorders may explain some symptoms and pathology (Holtmann et al., 2016; Liu et al., 2016). There are studies demonstrating cross talk between enteric and central neurons (Cashman, Martin, Dhillon, & Puli, 2016). These findings suggest how stress may precipitate symptoms via changes in the microbiome. In addition to the microbiome, changes in some forms of IBS (associated with diarrhea) have similar features to bacterial changes observed in depressive disorders (Liu et al., 2016). Since many disorders are associated with alterations of the microbiome, lactose digestion in LNP individuals may modify the severity of symptoms.

The association of mood changes like depression (Ledochowski, Sperner-Unterweger, & Fuchs, 1998) and anxiety (Tomba, Baldassarri, Coletta, Cesana, & Basilisco, 2012) with LI and the observation of extraintestinal general symptoms (labeled as LS) (Matthews, Waud, Roberts, & Campbell, 2005; Waud et al., 2008) are similar to symptoms associated with IBS (Borghini, Donato, Alvaro, & Picarelli, 2017).

3.4.12 Is lactose intolerance a cause of irritable bowel syndrome?

When IBS was first described in the 1950s there was a tendency to ascribe it to LI. In IBS, the rate of LM is not increased (Vesa et al., 1998), but LNP patients with IBS may be more sensitive (Dainese et al., 2014; Yang et al., 2013). The reason for this relates to a sense of hypervigilance or hypersensitivity and reduced pain threshold described in IBS patients (Nozu & Kudaira, 2009; Stabell, Stubhaug, Flægstad, & Nielsen, 2013). However, other evidence suggests that LI symptoms are of similar frequency in LP (Farup, Monsbakken, & Vandvik, 2004) and LNP patients with IBS (Yang et al., 2014). In northern Europe the frequency of LNP is 4%–7%, with IBS frequencies higher (10%–19.9% frequencies). However, on examining geographic comparisons, one would not expect a strong positive correlation between the frequency of LNP and IBS because IBS appears to have a less spatially structured distribution than LP/LNP (compare Fig. 3.2 with Fig. 3.4). If IBS were clearly linked with LNP, one would expect high rates

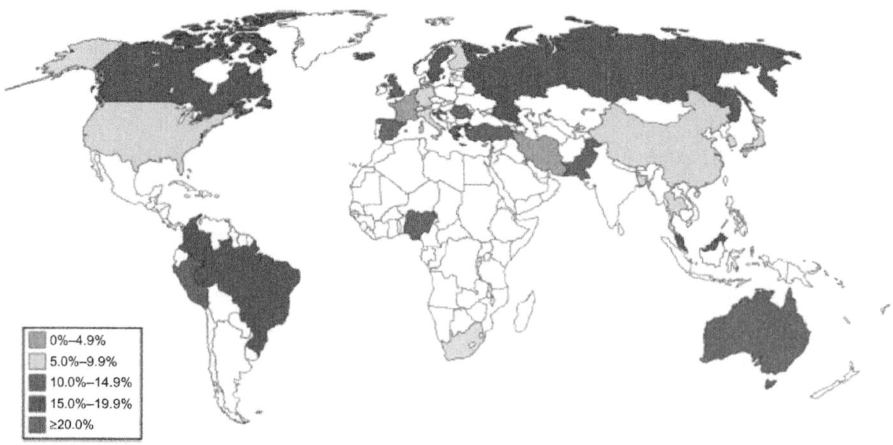

FIGURE 3.4

Global distributions of irritable bowel syndrome (Lovell & Ford, 2012) noting the less spatially structured distribution compared with LP.

of IBS in China ($>90\%$ LNP prevalence) but this is not the case. The correlation between distributions of LNP and IBS is only -0.17 ($P = .4274$) (Szilagyi & Xue, 2017).

3.4.13 The placebo and nocebo effect modifiers

In conditions where no specific therapy exists, such as FGIDs, the placebo effect can alleviate or improve symptoms. A placebo effect is a beneficial or therapeutic effect resulting from a treatment or drug therapy with no known active ingredient. Placebo effects can modify LI. In controlled trials, SRLI individuals were unable to distinguish lactose from the placebo with regard to development of gastrointestinal symptoms (Suarez et al., 1995, 1997). In colonic bacterial adaptation, some studies have hypothesized that symptomatic improvement after regular lactose consumption was perhaps attributable to a placebo effect because both placebo and lactose ingestion improved symptoms (Briet et al., 1997). The correlation of reduced symptoms with lower hydrogen production in challenge tests supports an improvement of symptoms with repeated lactose consumption (Ladas et al., 1982). Increased colonic lactobacilli and bifidobacteria in response to regular lactose consumption in LNP individuals is associated with lower hydrogen production (Hertzler & Savaiano, 1996).

In contrast, a nocebo effect is experiencing adverse or more severe symptoms resulting from consuming a substance with no active or therapeutic ingredient. In response to perceived lactose ingestion, the nocebo effect can elicit adverse symptoms (Vernia, Di Camillo, Foglietta, Avallone, & De Carolis, 2010). Due to the nocebo effect, individuals may limit their dairy food intake to reduce unpleasant symptoms. However, this can inadvertently cause nutritional deficiencies (as described in more detail in Section 3.7).

3.4.14 Lactose in pharmaceutical preparations

It has been debated whether small doses of excipient lactose in medications could cause symptoms. One study described 20 LM individuals of whom 3 experienced significant symptoms with a 5 g lactose load whereas 17 were able to tolerate these doses (Gudmand-Høyer & Simony, 1977). While this study emphasized individual lactose "sensitivity" (not to be confused with the later definition of systemic effects in LS), most drugs contain <100 mg of lactose, with budesonide being the significant outlier, containing 600 mg. Budesonide is a nonsystemic, locally active formulation of cortisone used to treat, for example, asthma and IBD (Eadala, Waud, Matthews, Green, & Campbell, 2009). Perhaps because of the low-lactose content of most drugs (Eadala et al., 2009), there is no evidence that the lactose is causing symptoms even when multiple capsules are taken in one dose. In LNP individuals, as much as 400 mg of lactose does not produce a hydrogen response on breath tests with negligible amounts of lactose reaching the intestinal microbiota (Montalto et al., 2008). Most studies define lactose levels of

>5–6 g as the threshold for eliciting hydrogen production and >10–15 g for the start of symptoms. However, no published controlled trials have evaluated a low dose such as <1 g lactose in producing symptoms (see also Chapter 5: Application of lactose in the pharmaceutical industry).

3.4.15 Food sensitivities

Diet has been implicated in gastrointestinal manifestations of IBS. With the recognition of microbial/host neurone interactions, IBS has been redefined as a gut–brain axis dysfunction (Holtmann et al., 2016; Liu et al., 2016). Research in this area of microbe interactions leading to psychoneurological disorders implicates diet–host interactions in the development of IBS. Lactose was considered as a cause when IBS was first described. Undigested lactose alters the intestinal microbiota, and it is possible that systemic features attributed to LI could be expressed via microbiome complicity (Windey, Houben, Deroover, & Verbeke, 2015). Many foods or food combinations can induce food sensitivities which mimic symptoms attributed to LI (Kitts et al., 1997; Mullin, Shepherd, Roland, Ireton-Jones, & Matarese, 2014). As a result dietary management of IBS has gained wider attention with a number of specific and nonspecific diets.

3.4.16 "Lactose intolerance" as media-driven popular science

LI has captured the imagination of the general public and many dairy and nondairy-related gastrointestinal symptoms are erroneously attributed to lactose without clinical investigation. Food fads often question dairy food consumption, raising objections such as "humans are the only animals who consume other species' milk" and "it is not natural." Furthermore with FGIDs affecting up to 20% of the population, explanations are sought for the causes of these symptoms. Media interest has grown following the increasing frequency of SRLI without defining digestion status. An online search of the term "lactose intolerance" yields over 2.1 million hits, providing an indication of how pervasive this concept is within popular science (Google, 2017). Sites range in topics on dairy foods from helping to harming life. Possibly as a result of this media attention, and self-diagnoses, lactose has been implicated in many unpleasant gastrointestinal symptoms. However, it is likely that other food components are contributing to and/or causing gastrointestinal symptoms. Isolating specific food components causing gastrointestinal symptoms can be difficult, and such symptoms are often multifactorial, requiring clinical investigations to confirm individual cases.

3.5 Allergies, sensitivities, and intolerances to milk

Other nutrients in milk such as fats and proteins can also cause gastrointestinal symptoms similar to LI. This section explores several hypothesized mechanism-inducing symptoms.

3.5.1 Long-chain triacylglycerol

Dairy fats can induce gastrointestinal symptoms that have been confused with LI, including long-chain triacylglycerol in some dairy foods (Mishkin, 1997). Nolan-Clarke found that fat content was more positively correlated with adverse gastrointestinal symptoms than lactose in a cohort of patients with Crohn's disease self-reporting symptoms following dairy food consumption. This suggests that lactose often has been mistakenly identified as the causative nutrient (Nolan-Clark, Tapsell, Hu, Han, & Ferguson, 2011). As the majority of patients in this study were LP (93%), this study reinforces the overlap between LI and other dairy food sensitivities.

3.5.2 Reactions to A1 and A2 dairy proteins

It has been reported that β-casein milk proteins can have physiological effects on the gastrointestinal tract and on modifying inflammation and intestinal transit. There are several types of β-caseins present in cow's milk. The A1 protein, associated with northern European cow breeds such as Friesian, Ayrshire, and British Shorthorn, seems to interact with μ-opioid receptors in the gut and is postulated to decrease the intestinal motility. The A2 β-casein, produced mostly in cows originating in southern France and the Channel Islands including Guernsey, Jersey, Charolais, and Limousin breeds, does not induce these physiological effects. The A1 effects are hypothesized to be mediated via β-casomorphines (mostly BMC-7). In addition, mucin secretion in the gut is increased via interaction of this molecule with the μ-opioid receptor (Pal, Woodford, Kukuljan, & Ho, 2015). Jianqin et al. carried out a 2-week crossover study comparing A1/A2 milk with A2 milk only and found that the A1/A2 combination significantly increased serum biomarkers of inflammation and reduced SCFAs in stool. They also noted changes in stool consistency and measurements of cognitive ability (Jianqin et al., 2015). However, other studies have not found differences related to the consumption of A1/A2 or A2 milk indicating that the current scientific evidence supporting the health benefits of A2 (or more correctly, non-A1) milk is inconclusive (Pasin, 2017). The possible benefits of consuming milk with A2 proteins rather than a combination of A1/A2 proteins merit further evaluation.

3.5.3 Cow's milk allergy

Lactose does not play a role in milk allergy. Allergic reactions are mediated by immunological reactions involving, for example, IgE. Allergies can have very serious outcomes, including death, while carbohydrate intolerances produce uncomfortable but non-life-threatening symptoms. CMA and LI are distinct conditions.

Cow's milk protein allergy (CMA) was first recognized in children with a prevalence of 2%–3% (Høst, Jacobsen, Halken, & Holmenlund, 1995). The main milk proteins–inducing allergies are αs1-, αs2-, β-, and κ-casein and α- and β-lactoglobulin (Wal, 2004). The main manifestations include skin irritation (e.g., dermatitis, urticarial), cardiopulmonary complaints (e.g., asthma), and gastrointestinal symptoms. The gastrointestinal manifestations can be similar to IBD, either ulcerative colitis (bloody diarrhea) or diarrhea without blood (Virta, Kautiainen, & Kolho, 2016). Higher antibody counts to milk protein in IBD have been reported but subsequently it was concluded that these antibodies were unlikely to be pathogenic (Jewell & Truelove, 1972). Without blood in the stool, the symptoms overlap with the IBS.

Adults rarely acquire CMA, but when they do cardiopulmonary and skin symptoms often are more severe and with fewer gastrointestinal symptoms (Lam et al., 2008). Occasionally individuals following a lactose-free (but not dairy-free) diet are diagnosed with CMA through allergy testing (Olivier et al., 2012). Diagnosis relies on a battery of tests including questionnaires, measurement of IgE and IgG, skin tests, and at times food challenge (Liu et al., 2012). CMA plays a minor but important role within milk intolerance/LI in adults (Paajanen et al., 2005). Milk intolerance and CMA are clearly differentiated. In milk intolerance digestive effects are less severe, whereas CMA in children is an allergic reaction with features sometimes resembling colitis (Bahna, 2002). The diagnosis has specific guidelines (Wal, 2004). However, on occasions where CMA in adults suggests LI, an allergic investigation requiring the consultation of an allergist is recommended (Crittenden & Bennett, 2005).

In the last 20 years the number of reported food allergy related diseases has increased, often involving the gastrointestinal tract. One recent example is eosinophilic esophagitis, which has been increasing in incidence. This condition mostly affects male children, but more recently older adolescents and adults are also affected. Restriction of six major foods [milk, wheat, fish/shellfish, nuts (peanuts/pine nuts), egg, and soy] reduces heartburn and difficulty in swallowing (Furuta & Katzka, 2015). Recent studies estimate that 85% of allergies causing eosinophilic esophagitis are attributed to CMA (Kagalwalla et al., 2017). Table 3.3 outlines various nutrients in milk and dairy foods which may produce symptoms.

3.6 Management of lactose intolerance

The growing research interests and public awareness of LI were addressed by a National Institute of Health conference in 2010. Following this conference,

Table 3.3 Table of terms, symptoms, and possible causes.

Terms	Symptoms	Causes
Cow's milk protein allergy (CMA)—children	Skin rashes Anaphylaxis Cardiopulmonary symptoms including tachycardia, altered blood pressure, and asthma Gastrointestinal pain Diarrhea Vomiting	Allergic (IgE, IgG, IgA) reactions to casein, α-lactalbumin and β-lactoglobulin, and bovine serum albumin
Cow's milk protein allergy (CMA)—adults	As above but more severe and fewer gastrointestinal symptoms	As above
Milk sensitivity	Mostly gastrointestinal symptoms Bloating Flatulence Diarrhea Vomiting	Not an immune response (i.e., not an allergy) May be related to lactose fermentation
β-Casein-related sensitivity	Mostly gastrointestinal symptoms Cramps Bloating Altered bowel movements Vomiting	Genetic alteration in β-casein in cows. Mainly European/non-Asian cattle. Casomorphins in milk may alter function of intestinal motility by interacting with μ-opioid receptors to slow transit
Milk fat–related symptoms	Abdominal discomfort, pain, and distension Nausea Vomiting	Possible effect on gastrointestinal motility with a delay in gastric emptying
Lactose as part of FODMAP[a]	Gastrointestinal and systemic symptoms Flatulence Bloating Abdominal pain Constipation Diarrhea	Bacterial fermentation of carbohydrates in the colon Osmotic effects of short-chain fatty acids in the bowel

[a]*Fermentable oligo-, di-, monosaccharide, and polyol sensitivity. The role of lactose in FODMAP may be independent of the process of lactose maldigestion by the host.*

Shaukat et al. published a review of therapeutic options. They found that most LI individuals were able to tolerate 12–15 g of lactose in single-dose ingestion but there was insufficient evidence for other modalities of therapy (Shaukat et al., 2010). The recommendation was made to consume acceptable quantities of lactose in single meal periods (i.e., 1 cup of milk, 12.5 g, or 2 cup equivalents in divided doses).

3.6.1 Fermented dairy products with reduced-lactose content

However, other options may also be effective. These include the consumption of low-lactose products such as cheese, yogurt, other fermented dairy products (kefir), or specially produced reduced-lactose milk. In fermented dairy products the presence of active lactic acid producing bacteria reduces the lactose compared to equivalent amounts of milk (see also Chapter 6: Lactose in the dairy production chain). A list of the lactose content of different dairy foods can be obtained from the Food-Intolerance-Network (2017).

3.6.2 Exogenous lactase

There are a number of lactase-based oral pills which can be taken before consuming dairy foods to reduce or prevent the gastrointestinal symptoms. These lactase medications reduce breath hydrogen production and symptoms in LNP individuals by providing exogenous lactase that cleaves lactose (Francesconi et al., 2016; He, Yang, Bian, & Cui, 1999; Ianiro, Pecere, Giorgio, Gasbarrini, & Cammarota, 2016; Ibba, Gilli, Boi, & Usai, 2014; Montalto et al., 2005). These enzymes are usually derived from fungi or yeast. The quantity of lactase varies but many contain from 1000 to 3000 IU per caplet, and the number of caplets taken can vary depending on consumed lactose load. The dose is taken about 15 minutes before dairy consumption and the efficacy varies depending on quantity of lactose consumed. Most enzymes can predigest 20 g but efficacy is reduced with higher lactose intakes such as 50 g of lactose in single meals (Ianiro et al., 2016).

3.6.3 Bacterial adaptation as therapy

Symptoms of LI may be more severe with intermittent ingestion of dairy. Regular consumption of slightly higher doses of milk (e.g., 12.5–25 g lactose in divided doses) may increase tolerance through encouraging microbial adaptation. Maldigested carbohydrates can promote colonic bacterial adaptation, as hypothesized by Gibson and Roberfroid, who noted that certain carbohydrates promote beneficial host lactic acid bacteria growth. Those carbohydrates were regarded as prebiotics (Gibson & Roberfroid, 1995).

Many bacteria, including lactic acid bacteria used in yogurt production, thrive on regular carbohydrate utilization. Generally, bifidobacteria—the dominant bacterial group in the gut of breastfed infants and those fed with formula enriched with prebiotics—produce less hydrogen gas and therefore cause fewer gas-related symptoms (for details on lactose-fermenting bacteria, see also Chapter 4: Lactose—a conditional prebiotic?). Early studies by Bouhnik et al. showed that the disaccharide lactulose (a disaccharide formed of galactose and fructose) as well as short-chain fructo- and galacto-oligosaccharides promoted the growth of bifidobacteria in adults (Bouhnik, Attar, et al., 2004; Bouhnik, Raskine, et al., 2004). Doses as low as 2.5 g of fructo-oligosaccharides are considered bifidogenic

(Bouhnik et al., 1999). Regular consumption of small doses of lactulose can result in improved symptoms of LI and a reduction of breath hydrogen in response to a lactulose challenge resulting from bacterial adaptation (Hertzler & Savaiano, 1996). A 2-week study administering oral lactulose to participants reported improved symptoms (Szilagyi, Rivard, & Fokeeff, 2001).

In a recent study a galacto-oligosaccharide prebiotic was investigated by the US Food and Drug Administration (FDA), offering another treatment strategy based on microbial adaptation. It was shown that LNP individuals, after using a galacto-oligosaccharide product for 1 month, were symptom-free after ingesting lactose (Azcarate-Peril et al., 2017; Savaiano et al., 2013). This was accompanied by changes in the microbiota composition. The relative abundance of lactose-fermenting *Bifidobacterium*, *Faecalibacterium*, and *Lactobacillus* was shown to be significantly increased in response to galacto-oligosaccharide after the introduction of dairy food. The lactose-fermenting *Roseburia* species were also increased (Azcarate-Peril et al., 2017). This suggests that microbiota adaptation may play a role in alleviating symptoms related to LM. The treatment is likely to be more effective for LNP individuals due to the effect on the microbiome. This product, *RP-G28* (Ritter Pharmaceuticals), has the potential to become the first FDA-approved drug for treatment of LI.

3.6.4 Dietary management—FODMAPs

Gibson and Shepherd hypothesized that certain carbohydrates are not digested well by the host. These carbohydrates then reach the lower intestine and are catabolized by bacteria, producing metabolites and gases responsible for many of the symptoms. The hypothesis is based on the presence of fermentable oligo-, di-, monosaccharides and polyols (FODMAPs) in many food items (Gibson & Shepherd, 2005; see also Chapter 4: Lactose—a conditional prebiotic? for further explanation of this concept). The proposed mechanism is similar to that postulated by He et al.'s for lactose-induced symptoms in LNP populations (He et al., 2008) (see Chapter 2: Digestion, absorption, metabolism, and physiological effects of lactose for discussion). However, in Gibson's hypothesis, LP/LNP status is less relevant, although symptoms may be more pronounced in LNP individuals (Gibson & Shepherd, 2005). The treatment is a restriction of foods containing FODMAPs for a limited time. This is then followed by selective reintroduction of carbohydrates using a diary to record symptoms related to the specific FODMAP. There are a number of clinical trials providing evidence of success with FODMAP-restricted diets in patients with IBS (O'Keeffe et al., 2017) and IBS symptoms in patients with IBD in remission (Pedersen et al., 2017). However, a systematic review of nine trials of FODMAP for IBS found a number of biases, suggesting the results may be attributable to a placebo effect (Krogsgaard, Lyngesen, & Bytzer, 2017).

Prolonged FODMAP diets are generally not recommended because of the potential harm in reduction of SCFAs derived from carbohydrates and an increased metabolism of proteins (Valeur et al., 2016). This combination of reduced SCFA production via altered microbiome fermentation may promote colorectal carcinogenesis (Han et al., 2017), and dieticians recommend the reintroduction of different foods with FODMAP. However, recent studies have used FODMAP without complications for more prolonged periods (O'Keeffe et al., 2017). FODMAP diet may also alleviate fibromyalgia, a nonpathological form of joint symptoms often associated with IBS (Marum, Moreira, Carus, Saraiva, & Guerreiro, 2017).

3.6.5 Dietary management—gluten-restricted diet

Gluten restriction for individuals with nonceliac wheat sensitivity can be an effective dietary management strategy. These individuals find that wheat product consumption is associated with gastrointestinal symptoms and follow a gluten-restricted diet with apparent benefits. The benefits appear to be more pronounced in patients who have the genetic predisposition to celiac disease (HLA-DQ2/8) (Aziz et al., 2016; Vazquez-Roque et al., 2013). Some authors, however, have argued that the main benefit of gluten-free diet is due to its FODMAP restriction and not the gluten itself (El-Salhy, Gunnar Hatlebakk, Helge Gilja, & Hausken, 2015). In the case of secondary hypolactasia caused by celiac disease, a gluten-free diet or other management strategies may encourage the return of lactase to the small intestine although this process may be slow (Plotkin & Isselbacher, 1964).

3.7 Approaches to adults with "lactose intolerance"

The general goal when working with patients is to minimize adverse symptoms while maintaining nutrition. In view of the variety of possible conditions described in this chapter, the traditional approach of taking a full patient history is important in order to accurately link the ingestion of dairy products with symptoms. It is important to note that dairy usually needs to be consumed in order to establish lactose as the cause. Small quantities are usually insufficient to cause symptoms. On the other hand, general food intolerances and some secondary causes of LM may begin any time in life. A large portion of general digestive complaints will fall into the category of functional digestive disorders, each with its own management plan (Fig. 3.5).

Limiting the discussion to lactose or dairy-related symptoms, we start with SRLI. It should be recalled that SRLI is similar in frequency in both LP and LNP populations (Barr, 2013; Zhao et al., 2017). Investigations begin with establishing whether lactose digestion or maldigestion (LM) is occurring, usually with the breath hydrogen test or the BGT. If the patient is LM, lactose withdrawal with lactose-free dairy or exogenous lactase may be tried. Alternatively, the recently

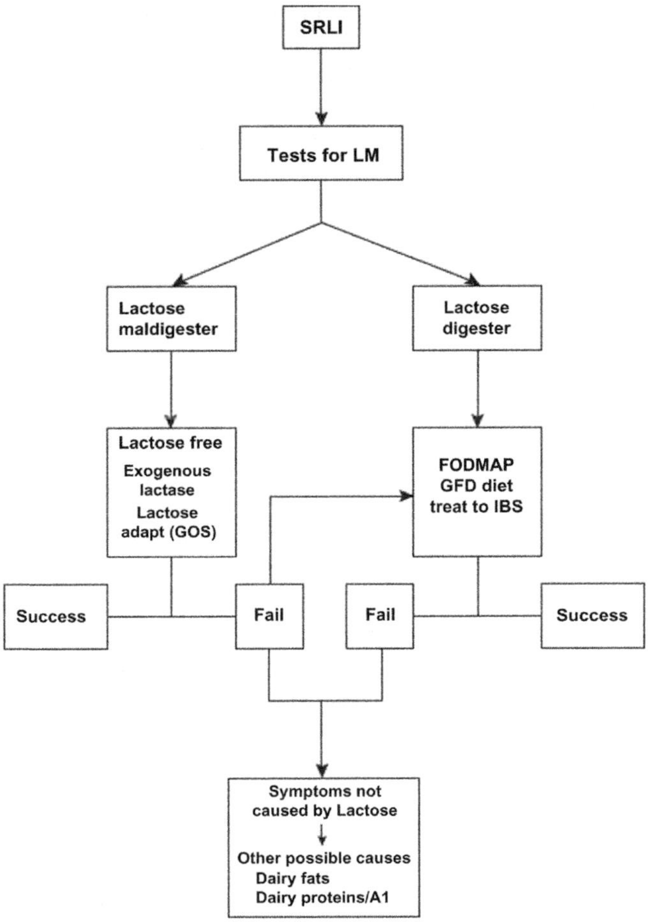

FIGURE 3.5

A schematic decision tree to manage self-reported lactose intolerance, where GOS is galacto-oligosaccharide, FGS is fructo- oligosaccharide, and GFD is gluten-free diet. This is not intended as a clinical guideline.

FDA-approved galacto-oligosaccharide prebiotic may be helpful (Azcarate-Peril et al., 2017). In this latter case, patients are thought to be able to withstand increased lactose consumption by prebiotic modification of intestinal microflora.

In the case of lactose digesters, the techniques outlined earlier may be tried but there is no good evidence of efficacy. However, some patients do report an amelioration of adverse symptoms perhaps attributable to the placebo effect. Alternatively, the carbohydrate elimination, via the outlined FODMAP diet, may also help. A FODMAP diet may also be tried in LM patients where lactose manipulation alone did not achieve the desired effect. Similarly, after celiac disease has been excluded, a gluten-free diet may be used to assess celiac sensitivity in both lactose digesters and LM patients.

If none of these interventions improve symptoms experienced after dairy intake, other dairy product–related causes should be considered. Adult milk allergy is rare and less commonly associated with causing gastrointestinal symptoms. Milk fats, although less well studied, are also possible culprits. Other specific dairy food components may need to be investigated with the help of a dietician. A1 variant milk has recently been investigated for its potential role in causing gastrointestinal disturbances (Pal et al., 2015). However, further studies of A1/A2 milk variants are needed to clarify the clinical implications of this dairy but nonlactose potential cause of symptoms.

3.8 Lactose intolerance in children

Both LP and LNP children can generally tolerate lactose. In early childhood, the immaturity of intestinal lactase can give rise to symptoms of LI, most often due to secondary causes such as gastroenteritis. In a recent survey of pediatricians in Southeast Asia (Thailand, Singapore, and Indonesia) regarding LI in children aged 1–5 years, about half of those questioned estimated that primary LI was <5%. The remainder of estimates ranged from 12% to 20%. Secondary LI estimates ranged from 10.8% to 23.9% (Tan et al., 2018). In cases of secondary LI, restriction of lactose for a limited time until the intestinal mucosa recovers is usually effective. However, it is important to distinguish and rule out CMA, as in these cases lactose need not to be restricted but cow's milk protein should be eliminated from the diet (Heine et al., 2017). In older children, recommendations for adults' LI management apply.

3.9 Nutritional impacts of dairy avoidance

In the 2010 NIH LI conference, the panel concluded that the deficiency of nutrients in dairy food–restricted diets was the most important detrimental consequence of LI. Avoiding dairy foods negatively affects bone health and possibly increases the risk of osteoporosis as dairy foods are a significant source of dietary calcium and vitamin D (Savaiano, 2011). However, the inclusion of dairy foods in the diet may also reduce the risk of hypertension, colorectal cancer, and possibly obesity and metabolic syndrome (Suchy et al., 2010). Hypertension and colon cancer may respond favorably to calcium consumption as it affects cellular communications and promotes antiproliferation behaviors. Other nutrients in milk and dairy foods may also encourage antineoplastic effects. In the case of obesity, diabetes, and the metabolic syndrome, data are conflicting (Bergholdt, Nordestgaard, & Ellervik, 2015; Louie, Flood, Hector, Rangan, & Gill, 2011). Therapeutic consumption of dairy food to promote microbiome health would have a positive health effect for lactose consuming LNP individuals.

The SRLI individuals avoiding or limiting dairy food intake are at risk of nutritional deficiencies through dairy food avoidance strategies. A number of studies have established that in controlled trials these individuals were unable to distinguish lactose from placebo. Nonetheless, these individuals limit their dairy food intake due to a perception that consuming dairy foods will cause unpleasant symptoms (Suarez et al., 1995, 1997). However, the correlation between LM and LI is poor (Lukito et al., 2015) in both high LP and high LNP populations. Other studies have found poor or no correlation between adverse symptoms and LM in patients with IBS, although symptoms may be aggravated by lactose and other poorly absorbed fermentable carbohydrates (Farup et al., 2004; Yang et al., 2014). This phenomenon may persist without IBS. The frequency of SRLI in Canada is 16% (Barr, 2013) while in China it is 15.2% (Zhao et al., 2017). The importance of addressing nutrition in LI remains a serious challenge for clinicians, researchers, and the dairy industries. Even if symptoms are ultimately related to food sensitivities, maintaining dairy foods in the diet for nutrient breadth and microbiome support is important.

In diarrheal illnesses, the use of dairy products has been controversial. In pediatric gastroenteritis, for example, dairy products are assumed to aggravate clinical symptoms through lactose consumption. Intestinal lactase levels decrease temporarily in gastroenteritis due to a flattening of the villi, so avoiding foods containing lactose for about 18 hours reduces gastrointestinal symptoms (MacGillivray, Fahey, & McGuire, 2013). Since this condition is usually time limited, it is unlikely to negatively affect nutrition. However, malnutrition may cause prolonged diarrhea through a blunting of the intestinal villi harboring intestinal lactase. Grenov et al. published recommendations that in pediatric patients with mild to moderate diarrhea, the addition of lactose may improve nutritional intake. The mechanism here may also include intestinal bacterial adaptation (Grenov et al., 2016).

IBD are intestinal disorders often associated with diarrhea. While the role of dairy foods during active phases is still debated, in preillness phases, milk and dairy foods may reduce the risk of Crohn's disease and possibly ulcerative colitis (Opstelten et al., 2016; Szilagyi, Galiatsatos, & Xue, 2016).

LI and other food intolerances may lead to increased rates of comorbidities. Among these are osteoporosis, mental changes, and abdominal pain (Schiffner, Kostev, & Gothe, 2016). Calcium may be important in several specific disorders as outlined at the NIH 2010 conference (Suchy et al., 2010), and may be relevant to growth and maintenance of bones and teeth. It is also involved with regulating muscle contraction (including the heart) and in blood clotting. In addition, dietary calcium may modulate the microbiome (Chaplin, Parra, Laraichi, Serra, & Palou, 2016). In obesity and diabetic control, calcium may modify the microbiome via binding of fats and bile salts (Gomes, Costa, & Alfenas, 2015).

It has been hypothesized that East Asian populations may have a lower metabolic requirement for calcium than northern European populations. A meta-analysis of 12 studies examining the metabolism of calcium in Chinese populations (predominantly LNP) concluded that as little as 300 mg of calcium per day may be adequate to maintain normal functions, whereas 1–2 g per day is recommended by the Institute of Medicine. In China, much of the calcium ingested was derived from nondairy sources, such as soybeans, green vegetables, and fish. More recently, the use of dairy foods in China is being encouraged as an additional source of calcium (Fang, Li, Shi, He, & Li, 2016). Further work is necessary to better define the role of dairy foods in health maintenance and disease risk reduction.

3.10 Conclusion

LI is now better understood by the scientific community to be composed of multiple etiologies related to dairy consumption, with further progress to be made in the public understanding of lactose digestion. Similar symptoms can have nondairy causes, often confused with LI. LI is no longer simply equated with LNP. The adaptive flexibility of the microbiome in responding to regular consumption of dairy foods may encourage individuals, both LP and LNP, to consider adding milk and dairy foods to their diet for health reasons. Recent work on sensitivities and allergies has further demonstrated that dairy components other than lactose can induce symptoms previously attributed to lactose. This highlights the importance of clinical testing to ascertain the cause of LI and how to best manage the LI and non-LI symptoms. Given the nutrient density and nutrient diversity of milk and dairy foods, and their specific value for certain key nutrients such as calcium, nutritional management strategies for their dietary inclusion rather than avoidance offer important avenues for further research. These and other factors raise perceptions and misunderstandings of LI to the level of an important public health issue.

Generally gaps in public understanding of LI and dairy consumption relate to: (1) Increased reporting of different food intolerances and adverse digestive symptoms. (2) "Blanket blame" on LI for all dairy-related symptoms. This notion is promoted by popular media and will require time to sort out. (3) Lack of clear understanding of health benefits of dairy products. This uncertainty may be due both to the lack of awareness by the public of potential benefits and the lack of consistency of findings in nutritional studies. Further studies may clear these discrepancies. (4) Changing demographics in western societies, with increasing proportions of LNP individuals living in traditional milk consuming cultures, but retaining diets that include less milk and dairy foods. Taking into consideration these knowledge gaps may help guide future research and more specific ways to manage true LI-related symptoms.

References

Adam, A. C., Rubio-Texeira, M., & Polaina, J. (2004). Lactose: The milk sugar from a biotechnological perspective. *Critical Reviews in Food Science and Nutrition*, *44*(7–8), 553–557.

Almeida, J. A., Kim, R., Stoita, A., McIver, C. J., Kurtovic, J., & Riordan, S. M. (2008). Lactose malabsorption in the elderly: Role of small intestinal bacterial overgrowth. *Scandinavian Journal of Gastroenterology*, *43*(2), 146–154. Available from https://doi.org/10.1080/00365520701676617.

Antonowicz, I., & Lebenthal, E. (1977). Developmental pattern of small intestinal enterokinase and disaccharidase activities in the human fetus. *Gastroenterology*, *72*(6), 1299–1303.

Arola, H. (1994). Diagnosis of hypolactasia and lactose malabsorption. *Scandinavian Journal of Gastroenterology*, *29*(Suppl. 202), 26–35.

Azcarate-Peril, M. A., Ritter, A. J., Savaiano, D., Monteagudo-Mera, A., Anderson, C., Magness, S. T., & Klaenhammer, T. R. (2017). Impact of short-chain galactooligosaccharides on the gut microbiome of lactose-intolerant individuals. *Proceedings of the National Academy of Sciences*, *114*(3), E367–E375.

Aziz, I., Trott, N., Briggs, R., North, J. R., Hadjivassiliou, M., & Sanders, D. S. (2016). Efficacy of a gluten-free diet in subjects with irritable bowel syndrome-diarrhea unaware of their HLA-DQ2/8 genotype. *Clinical Gastroenterology and Hepatology*, *14*(5), 696–703.

Bahna, S. L. (2002). Cow's milk allergy vs cow milk intolerance. *Annals of Allergy, Asthma & Immunology*, *89*(6), 56–60.

Barr, S. I. (2013). Perceived lactose intolerance in adult Canadians: A national survey. *Applied Physiology, Nutrition, and Metabolism*, *38*(8), 830–835.

Bergholdt, H. K., Nordestgaard, B. G., & Ellervik, C. (2015). Milk intake is not associated with low risk of diabetes or overweight-obesity: A Mendelian randomization study in 97,811 Danish individuals. *American Journal of Clinical Nutrition*, *102*(2), 487–496.

Bond, J. H., & Levitt, M. D. (1976). Quantitative measurement of lactose absorption. *Gastroenterology*, *70*(6), 1058–1062.

Borghini, R., Donato, G., Alvaro, D., & Picarelli, A. (2017). New insights in IBS-like disorders: Pandora's box has been opened; review. *Gastroenterology and Hepatology From Bed to Bench*, *10*(2), 79–89.

Bouhnik, Y., Attar, A., Joly, F., Riottot, M., Dyard, F., & Flourie, B. (2004). Lactulose ingestion increases faecal bifidobacterial counts: A randomised double-blind study in healthy humans. *European Journal of Clinical Nutrition*, *58*(3), 462–466.

Bouhnik, Y., Raskine, L., Simoneau, G., Vicaut, E., Neut, C., Flourié, B., ... Bornet, F. R. (2004). The capacity of nondigestible carbohydrates to stimulate fecal bifidobacteria in healthy humans: A double-blind, randomized, placebo-controlled, parallel-group, dose-response relation study. *American Journal of Clinical Nutrition*, *80*(6), 1658–1664.

Bouhnik, Y., Vahedi, K., Achour, L., Attar, A., Salfati, J., Pochart, P., & Rambaud, J.-C. (1999). Short-chain fructo-oligosaccharide administration dose-dependently increases fecal bifidobacteria in healthy humans. *The Journal of Nutrition*, *129*(1), 113–116.

Briet, F., Pochart, P., Marteau, P., Flourie, B., Arrigoni, E., & Rambaud, J. (1997). Improved clinical tolerance to chronic lactose ingestion in subjects with lactose intolerance: A placebo effect? *Gut*, *41*(5), 632–635.

Camilleri, M., Halawi, H., & Oduyebo, I. (2017). Biomarkers as a diagnostic tool for irritable bowel syndrome: Where are we? *Expert Review of Gastroenterology & Hepatology*, *11*(4), 303–316.

Cashman, M. D., Martin, D. K., Dhillon, S., & Puli, S. R. (2016). Irritable bowel syndrome: A clinical review. *Current Rheumatology Reviews*, *12*(1), 13–26.

Chaplin, A., Parra, P., Laraichi, S., Serra, F., & Palou, A. (2016). Calcium supplementation modulates gut microbiota in a prebiotic manner in dietary obese mice. *Molecular Nutrition & Food Research*, *60*(2), 468–480.

Collins, S. (2016). The intestinal microbiota in the irritable bowel syndrome. *International Review of Neurobiology*, *131*, 247–261.

Coppa, G., Pierani, P., Zampini, L., Bruni, S., Carloni, I., & Gabrielli, O. (2001). Characterization of oligosaccharides in milk and feces of breast-fed infants by high-performance anion-exchange chromatography. *Bioactive Components of Human Milk*, *501*, 307–314.

Crittenden, R. G., & Bennett, L. E. (2005). Cow's milk allergy: A complex disorder. *Journal of the American College of Nutrition*, *24*(Suppl. 6), 582S–591S.

Dahlqvist, A. (1962). Specificity of the human intestinal disaccharidases and implications for hereditary disaccharide intolerance. *Journal of Clinical Investigation*, *41*(3), 463.

Dainese, R., Casellas, F., Mariné-Barjoan, E., Vivinus-Nébot, M., Schneider, S. M., Hébuterne, X., & Piche, T. (2014). Perception of lactose intolerance in irritable bowel syndrome patients. *European Journal of Gastroenterology & Hepatology*, *26*(10), 1167–1175.

Dehkordi, N., Rao, D., Warren, A., & Chawan, C. (1995). Lactose malabsorption as influenced by chocolate milk, skim milk, sucrose, whole milk, and lactic cultures. *Journal of the American Dietetic Association*, *95*(4), 484–486.

Domínguez, J. J., & Fernández, S. A. (2016). Correlation between capillary and venous blood glucose in the lactose tolerance test. *Digestive Diseases and Sciences*, *61*(1), 208.

Dzialanski, Z., Barany, M., Engfeldt, P., Magnuson, A., Olsson, L. A., & Nilsson, T. K. (2016). Lactase persistence vs lactose intolerance: Is there an intermediate phenotype? *Clinical Biochemistry*, *49*(3), 248–252.

Eadala, P., Waud, J., Matthews, S., Green, J., & Campbell, A. (2009). Quantifying the 'hidden' lactose in drugs used for the treatment of gastrointestinal conditions. *Alimentary Pharmacology & Therapeutics*, *29*(6), 677–687.

El-Salhy, M., Gunnar Hatlebakk, J., Helge Gilja, O., & Hausken, T. (2015). The relation between celiac disease, nonceliac gluten sensitivity and irritable bowel syndrome. *Nutrition Journal*, *14*(1), 92.

Enattah, N. S., Jensen, T. G., Nielsen, M., Lewinski, R., Kuokkanen, M., Rasinpera, H., & Khalil, I. F. (2008). Independent introduction of two lactase-persistence alleles into human populations reflects different history of adaptation to milk culture. *The American Journal of Human Genetics*, *82*(1), 57–72.

Enattah, N. S., Sahi, T., Savilahti, E., Terwilliger, J. D., Peltonen, L., & Jarvela, I. (2002). Identification of a variant associated with adult-type hypolactasia. *Nature Genetics*, *30* (2), 233–238.

Enattah, N. S., Trudeau, A., Pimenoff, V., Maiuri, L., Auricchio, S., Greco, L., & Rahgozar, S. (2007). Evidence of still-ongoing convergence evolution of the lactase persistence T-13910 alleles in humans. *The American Journal of Human Genetics*, *81* (3), 615–625.

Enko, D., Kriegshäuser, G., Halwachs-Baumann, G., Mangge, H., & Schnedl, W. J. (2017). Serum diamine oxidase activity is associated with lactose malabsorption phenotypic variation. *Clinical Biochemistry*, *1*(50), 50–53.

Fang, A.-P., Li, K.-J., Shi, H.-Y., He, J.-J., & Li, H. (2016). Habitual dietary calcium intakes and calcium metabolism in healthy adults Chinese: A systematic review and meta-analysis. *Asia Pacific Journal of Clinical Nutrition*, *25*(4), 776–784.

Farup, P., Monsbakken, K., & Vandvik, P. (2004). Lactose malabsorption in a population with irritable bowel syndrome. *Scandinavian Journal of Gastroenterology*, *39*(7), 645–649.

Food-Intolerance-Network. (2017). <https://www.food-intolerance-network.com/food-intolerances/lactose-intolerance/tables-of-lactose-content.html> Accessed 15.08.2017.

Flatz, G. (1987). Genetics of lactose digestion in humans. In *Advances in Human Genetics* (pp. 1–77). Boston, MA: Springer.

Ford, A. C., Lacy, B. E., & Talley, N. J. (2017). Irritable bowel syndrome. *New England Journal of Medicine*, *376*(26), 2566–2578.

Francavilla, R., Calasso, M., Calace, L., Siragusa, S., Ndagijimana, M., Vernocchi, P., & Guerzoni, E. (2012). Effect of lactose on gut microbiota and metabolome of infants with cow's milk allergy. *Pediatric Allergy and Immunology*, *23*(5), 420–427.

Francesconi, C. F., Machado, M. B., Steinwurz, F., Nones, R. B., Quilici, F. A., Catapani, W. R., ... Bafutto, M. (2016). Oral administration of exogenous lactase in tablets for patients diagnosed with lactose intolerance due to primary hypolactasia. *Arquivos de Gastroenterologia*, *53*(4), 228–234.

Furnari, M., Bonfanti, D., Parodi, A., Franzè, J., Savarino, E., Bruzzone, L., & Savarino, V. (2013). A comparison between lactose breath test and quick test on duodenal biopsies for diagnosing lactase deficiency in patients with self-reported lactose intolerance. *Journal of Gastroenterology*, *47*(2), 148–152.

Furuta, G. T., & Katzka, D. A. (2015). Eosinophilic esophagitis. *New England Journal of Medicine*, *373*(17), 1640–1648.

Gasbarrini, A., Corazza, G., Gasbarrini, G., Montalto, M., Di Stefano, M., Basilisco, G., & Anania, C. (2009). 1st Rome H2-Breath Testing Consensus Conference Working Group. Methodology and indications of H2-breath testing in gastrointestinal diseases: The Rome Consensus Conference. *Alimentary Pharmacology & Therapeutics*, *29* (Suppl. 1), 1–49.

Gerbault, P., Liebert, A., Itan, Y., Powell, A., Currat, M., Burger, J., & Thomas, M. G. (2011). Evolution of lactase persistence: An example of human niche construction. *Philosophical Transactions of the Royal Society of London B: Biological Sciences*, *366* (1566), 863–877.

Gibson, G., & Roberfroid, M. (1995). Dietary modulation of the human colonic microbiota: Introducing the concept of prebiotics. *The Journal of Nutrition*, *125*(6), 1401–1412.

Gibson, P., & Shepherd, S. (2005). Personal view: Food for thought–western lifestyle and susceptibility to Crohn's disease. The FODMAP hypothesis. *Alimentary Pharmacology & Therapeutics*, *21*(12), 1399–1409.

Gomes, J., Costa, J., & Alfenas, R. (2015). Could the beneficial effects of dietary calcium on obesity and diabetes control be mediated by changes in intestinal microbiota and integrity? *British Journal of Nutrition*, *114*(11), 1756–1765.

Google. (2017). Google<https://www.google.co.uk/search?q = lactose + intolerance&ie = utf-8&oe = utf-8&client = firefox-b&gws_rd = cr&dcr = 0&ei = xSkAWtThEpCja7-yj4AH> Accessed 06.11.2017.

Grant, J. D., Bezerra, J. A., Thompson, S. H., Lemen, R. J., Koldovsky, O., & Udall, J. N. (1989). Assessment of lactose absorption by measurement of urinary galactose. *Gastroenterology*, *97*(4), 895–899.

Grenov, B., Briend, A., Sangild, P. T., Thymann, T., Rytter, M. J. H., Nielsen, A.-L. H., & Michaelsen, K. F. (2016). Undernourished children and milk lactose. *Food and Nutrition Bulletin*, *37*(1), 85–99.

Gudmand-Høyer, E., & Simony, K. (1977). Individual sensitivity to lactose in lactose malabsorption. *The American Journal of Digestive Diseases*, *22*(3), 177–181.

Habte, D., Sterky, G., & Hjalmarsson, B. (1973). Lactose malabsorption in Ethiopian children. *Acta Paediatrica Scandinavica*, *62*(6), 649.

Han, M., Wang, C., Liu, P., Li, D., Li, Y., & Ma, X. (2017). Dietary fiber gap and host gut microbiota. *Protein and Peptide Letters*, *24*(5), 388–396.

He, M., Yang, Y., Bian, L., & Cui, H. (1999). Effect of exogenous lactase on the absorption of lactose and its intolerance symptoms. *Journal of Hygiene Research*, *28*(5), 309.

He, T., Venema, K., Priebe, M., Welling, G., Brummer, R., & Vonk, R. (2008). The role of colonic metabolism in lactose intolerance. *European Journal of Clinical Investigation*, *38*(8), 541–547.

Heine, R. G., AlRefaee, F., Bachina, P., De Leon, J. C., Geng, L., Gong, S., . . . Rogacion, J. M. (2017). Lactose intolerance and gastrointestinal cow's milk allergy in infants and children—common misconceptions revisited. *World Allergy Organization Journal*, *10* (1), 41.

Hertzler, S. R., & Savaiano, D. A. (1996). Colonic adaptation to daily lactose feeding in lactose maldigesters reduces lactose intolerance. *The American Journal of Clinical Nutrition*, *64*(2), 232.

Hertzler, S. R., Huynh, B.-C. L., & Savaiano, D. A. (1996). How much lactose is low lactose? *Journal of the American Dietetic Association*, *96*(3), 243–246.

Hiele, M., Ghoos, Y., Rutgeerts, P., Vantrappen, G., Carchon, H., & Eggermont, E. (1988). $^{13}CO_2$ breath test using naturally ^{13}C-enriched lactose for detection of lactase deficiency in patients with gastrointestinal symptoms. *The Journal of Laboratory and Clinical Medicine*, *112*(2), 193–200.

Hodin, R. A., Chamberlain, S. M., & Upton, M. P. (1992). Thyroid hormone differentially regulates rat intestinal brush border enzyme gene expression. *Gastroenterology*, *103*(5), 1529–1536.

Holtmann, G. J., Ford, A. C., & Talley, N. J. (2016). Pathophysiology of irritable bowel syndrome. *The Lancet Gastroenterology & Hepatology*, *1*(2), 133–146.

Høst, A., Jacobsen, H., Halken, S., & Holmenlund, D. (1995). The natural history of cow's milk protein allergy/intolerance. *European Journal of Clinical Nutrition*, *49*(Suppl. 1), S13–8.

Houben, E., De Preter, V., Billen, J., Van Ranst, M., & Verbeke, K. (2015). Additional value of CH_4 measurement in a combined $^{13}C/H_2$ lactose malabsorption breath test: A retrospective analysis. *Nutrients*, *7*(9), 7469–7485.

Ianiro, G., Pecere, S., Giorgio, V., Gasbarrini, A., & Cammarota, G. (2016). Digestive enzyme supplementation in gastrointestinal diseases. *Current Drug Metabolism*, *17*(2), 187.

Ibba, I., Gilli, A., Boi, M. F., & Usai, P. (2014). Effects of exogenous lactase administration on hydrogen breath excretion and intestinal symptoms in patients presenting lactose malabsorption and intolerance. *BioMed Research International*, *2014*, 680196.

Ingram, C. J., Mulcare, C. A., Itan, Y., Thomas, M. G., & Swallow, D. M. (2009). Lactose digestion and the evolutionary genetics of lactase persistence. *Human Genetics*, *124*(6), 579–591.

Itan, Y., Powell, A., Beaumont, M. A., Burger, J., & Thomas, M. G. (2009). The origins of lactase persistence in Europe. *PLoS Computational Biology*, *5*(8), e1000491.

Itan, Y., Jones, B. L., Ingram, C. J., Swallow, D. M., & Thomas, M. G. (2010). A worldwide correlation of lactase persistence phenotype and genotypes. *BMC Evolutionary Biology*, *10*(1), 36.

Jewell, D., & Truelove, S. (1972). Circulating antibodies to cow's milk proteins in ulcerative colitis. *Gut*, *13*(10), 796–801.

Jianqin, S., Leiming, X., Lu, X., Yelland, G. W., Ni, J., & Clarke, A. J. (2015). Effects of milk containing only A2 beta casein vs milk containing both A1 and A2 beta casein proteins on gastrointestinal physiology, symptoms of discomfort, and cognitive behavior of people with self-reported intolerance to traditional cows' milk. *Nutrition Journal*, *15*, 35.

Jones, H. L. (1969). The geography of Strabo with an English translation in eight Volumes: *V: Books X-XII*. London–Cambridge: William Heinemann LTD–Harvard University Press.

Kagalwalla, A. F., Wechsler, J. B., Schwartz, S., Makhija, M. M., Olive, A. P., Davis, C. M., & Ross, J. (2017). Diagnostic utility of allergy testing in children undergoing four-food elimination diet for eosinophilic esophagitis. *Gastroenterology*, *152*(5), S856.

Kitts, D., Yuan, Y., Joneja, J., Scott, F., Szilagyi, A., Amiot, J., & Zarkadas, M. (1997). Adverse reactions to food constituents: Allergy, intolerance, and autoimmunity. *Canadian Journal of Physiology and Pharmacology*, *75*(4), 241–254.

Knudsen, C. D., & Di Palma, J. A. (2012). Carbohydrate challenge tests: Do you need to measure methane? *Southern Medical Journal*, *105*(5), 251–253.

Krajicek, E. J., & Hansel, S. L. (2016). Small intestinal bacterial overgrowth: A primary care review. *Mayo Clinic Proceedings*, *91*(12), 1828–1833, [Elsevier].

Krogsgaard, L., Lyngesen, M., & Bytzer, P. (2017). Systematic review: Quality of trials on the symptomatic effects of the low FODMAP diet for irritable bowel syndrome. *Alimentary Pharmacology & Therapeutics*, *45*(12), 1506–1513.

Labrie, V., Buske, O. J., Oh, E., Jeremian, R., Ptak, C., Gasiūnas, G., & Adamonis, K. (2016). Lactase nonpersistence is directed by DNA-variation-dependent epigenetic aging. *Nature Structural & Molecular Biology*, *23*(6), 566–573.

Ladas, S., Papanikos, J., & Arapakis, G. (1982). Lactose malabsorption in Greek adults: Correlation of small bowel transit time with the severity of lactose intolerance. *Gut*, *23* (11), 968–973.

Lam, H. Y., Van Hoffen, E., Michelsen, A., Guikers, K., Van Der Tas, C., Bruijnzeel-Koomen, C., & Knulst, A. (2008). Cow's milk allergy in adults is rare but severe: Both casein and whey proteins are involved. *Clinical & Experimental Allergy*, *38*(6), 995–1002.

Laxminarayan, S., Reifman, J., Edwards, S. S., Wolpert, H., & Steil, G. M. (2015). Bolus estimation—rethinking the effect of meal fat content. *Diabetes Technology & Therapeutics*, *17*(12), 860.

Ledochowski, M., Sperner-Unterweger, B., & Fuchs, D. (1998). Lactose malabsorption is associated with early signs of mental depression in females (a preliminary report). *Digestive Diseases and Sciences*, *43*(11), 2513–2517.

Lewinsky, R. H., Jensen, T. G., Møller, J., Stensballe, A., Olsen, J., & Troelsen, J. T. (2005). T-13910 DNA variant associated with lactase persistence interacts with Oct-1 and stimulates lactase promoter activity in vitro. *Human Molecular Genetics*, *14*(24), 3945–3953.

Lifschitz, C. H., Smith, E. O., & Garza, C. (1983). Delayed complete lactase sufficiency in breast-fed infants. *Journal of Pediatric Gastroenterology and Nutrition*, *2*, 478–482.

Liu, L.-L., Yao, H., Zhang, X.-L., Zhang, H.-L., Chao, P.-L., Tong, M.-L., & Yang, T.-C. (2012). Characteristics of patients suffering from cow milk allergy. *International Immunopharmacology*, *1*(14), 94–98.

Liu, Y., Zhang, L., Wang, X., Wang, Z., Zhang, J., Jiang, R., & Xia, Z. (2016). Similar fecal microbiota signatures in patients with diarrhea-predominant irritable bowel syndrome and patients with depression. *Clinical Gastroenterology and Hepatology*, *14* (11), 1602–1611. e1605.

Louie, J. C. Y., Flood, V., Hector, D., Rangan, A., & Gill, T. (2011). Dairy consumption and overweight and obesity: A systematic review of prospective cohort studies. *Obesity Reviews*, *12*(7), e582–92.

Lovell, R. M., & Ford, A. C. (2012). Global prevalence of and risk factors for irritable bowel syndrome: A meta-analysis. *Clinical Gastroenterology and Hepatology*, *10*(7), 712–721. e714.

Lukito, W., Malik, S. G., Surono, I. S., & Wahlqvist, M. L. (2015). From lactose intolerance to lactose nutrition. *Asia Pacific Journal of Clinical Nutrition*, *24*(Suppl.), S1.

MacGillivray, S., Fahey, T., & McGuire, W. (2013). Lactose avoidance for young children with acute diarrhoea. *Cochrane Database of Systematic Reviews, Oct. 31*(10), CD005433.

Maiuri, L., Raia, V., Potter, J., Swallow, D., Ho, M. W., Fiocca, R., & Quaroni, A. (1991). Mosaic pattern of lactase expression by villous enterocytes in human adult-type hypolactasia. *Gastroenterology*, *100*(2), 359–369.

Marton, A., Xue, X., & Szilagyi, A. (2012). Meta-analysis: The diagnostic accuracy of lactose breath hydrogen or lactose tolerance tests for predicting the North European lactase polymorphism C/T-13910. *Alimentary Pharmacology & Therapeutics*, *4*(35), 429–440.

Marum, A. P., Moreira, C., Carus, P. T., Saraiva, F., & Guerreiro, C. S. (2017). A low fermentable oligo-di-mono saccharides and polyols (FODMAP) diet is a balanced therapy for fibromyalgia with nutritional and symptomatic benefits. *Nutrición Hospitalaria*, *34*(3), 667–674.

Matthews, S. B., Waud, J., Roberts, A. G., & Campbell, A. K. (2005). Systemic lactose intolerance: A new perspective on an old problem. *Postgraduate Medical Journal*, *81* (953), 167–173.

Metz, G., Peters, T., Jenkins, D. A., Newman, A., & Blendis, L. (1975). Breath hydrogen as a diagnostic method for hypolactasia. *The Lancet*, *305*(7917), 1155–1157.

Mishkin, S. (1997). Dairy sensitivity, lactose malabsorption, and elimination diets in inflammatory bowel disease. *The American Journal of Clinical Nutrition*, *65*(2), 564–567.

Mones, R. L., Yankah, A., Duelfer, D., Bustami, R., & Mercer, G. (2011). Disaccharidase deficiency in pediatric patients with celiac disease and intact villi. *Scandinavian Journal of Gastroenterology*, *46*(12), 1429–1434.

Montalto, M., Gallo, A., Santoro, L., D'onofrio, F., Curigliano, V., Covino, M., & Gasbarrini, G. (2008). Low-dose lactose in drugs neither increases breath hydrogen excretion nor causes gastrointestinal symptoms. *Alimentary Pharmacology & Therapeutics*, *28*(8), 1003–1012.

Montalto, M., Nucera, G., Santoro, L., Curigliano, V., Vastola, M., Covino, M., & Gasbarrini, G. (2005). Effect of exogenous b-galactosidase in patients with lactose malabsorption and intolerance: A crossover double-blind placebo-controlled study. *European Journal of Clinical Nutrition*, *59*, 489–493.

Mullin, G. E., Shepherd, S. J., Roland, B. C., Ireton-Jones, C., & Matarese, L. E. (2014). Irritable bowel syndrome. *Journal of Parenteral and Enteral Nutrition*, *38*(7), 781–799.

Newcomer, A. D., McGill, D. B., Thomas, P. J., & Hofmann, A. F. (1975). Prospective comparison of indirect methods for detecting lactase deficiency. *New England Journal of Medicine*, *293*(24), 1232–1236.

Nilsson, Å. (2016). Role of sphingolipids in infant gut health and immunity. *The Journal of Pediatrics*, *173*, S53–S59.

Nolan-Clark, D., Tapsell, L. C., Hu, R., Han, D. Y., & Ferguson, L. R. (2011). Effects of dairy products on Crohn's disease symptoms are influenced by fat content and disease location but not lactose content or disease activity status in a New Zealand population. *Journal of the American Dietetic Association*, *111*(8), 1165–1172.

Nozu, T., & Kudaira, M. (2009). Altered rectal sensory response induced by balloon distention in patients with functional abdominal pain syndrome. *Biopsychosocial Medicine*, *3*, 13.

O'Keeffe, M., Jansen, C., Martin, L., Williams, M., Seamark, L., Staudacher, H., & Lomer, M. (2017). Long-term impact of the low-FODMAP diet on gastrointestinal symptoms, dietary intake, patient acceptability, and healthcare utilization in irritable bowel syndrome. *Neurogastroenterology & Motility*, *30*(1), e13154.

Oku, T., Nakamura, S., & Ichinose, M. (2005). Maximum permissive dosage of lactose and lactitol for transitory diarrhea, and utilizable capacity for lactose in Japanese female adults. *Journal of Nutritional Science and Vitaminology*, *51*(2), 51–57.

Olivier, C. E., Silva Lorena, S. L., Pavan, C. R., Pereira Goncalves dos Santos, R. A., dos Santos Lima, R. P., Pinto, D. G., Zollner, R. D. L. (2012). Is it just lactose intolerance? In: *Paper presented at the allergy and asthma proceedings*.

Opstelten, J. L., Leenders, M., Dik, V. K., Chan, S., van Schaik, F. D., Khaw, K.-T., & Lindgren, S. (2016). Dairy products, dietary calcium and the risk of inflammatory bowel disease: Results from a European prospective cohort investigation. *Inflammatory Bowel Diseases*, *22*(6), 1403–1411.

Paajanen, L., Korpela, R., Tuuren, T., Honkanen, J., Järvelä, I., Ilonen, J., & Kokkonen, J. (2005). Cow milk is not responsible for most gastrointestinal immune-like syndromes—evidence from a population-based study. *American Journal of Clinical Nutrition*, *82*(6), 1327–1335.

Pal, S., Woodford, K., Kukuljan, S., & Ho, S. (2015). Milk intolerance, beta-casein and lactose. *Nutrients*, *7*(9), 7285–7297.

Pasin, G. (2017). What is A2 milk? Retrieved from <http://cdrf.org/2017/02/09/a2-milk-facts/> Accessed 26.06.2018.

Pedersen, N., Ankersen, D. V., Felding, M., Wachmann, H., Végh, Z., Molzen, L., & Munkholm, P. (2017). Low-FODMAP diet reduces irritable bowel symptoms in patients with inflammatory bowel disease. *World Journal of Gastroenterology*, *23*(18), 3356–3366.

Plotkin, G. R., & Isselbacher, K. J. (1964). Secondary disaccharidase deficiency in adult celiac disease (nontropical sprue) and other malabsorption states. *New England Journal of Medicine*, *271*(20), 1033–1037.

Ranciaro, A., Campbell, M. C., Hirbo, J. B., Ko, W.-Y., Froment, A., Anagnostou, P., & Omar, S. A. (2014). Genetic origins of lactase persistence and the spread of pastoralism in Africa. *The American Journal of Human Genetics*, *94*(4), 496–510.

Robayo-Torres, C. C., & Nichols, B. L. (2007). Molecular differentiation of congenital lactase deficiency from adult-type hypolactasia. *Nutrition Reviews*, *65*(2), 95.

Robayo-Torres, C. C., Quezada-Calvillo, R., & Nichols, B. L. (2006). Disaccharide digestion: Clinical and molecular aspects. *Clinical Gastroenterology and Hepatology*, *4*(3), 276–287.

Roggero, P., Mosca, F., Motta, G., Mangiaterra, V., Perazzani, M., Offredi, M., & Careddu, P. (1986). Sugar absorption in healthy preterm and full-term infants. *Journal of Pediatric Gastroenterology and Nutrition*, *5*(2), 214–219.

Röhmann, F., & Nagano, J. (1903). Ueber die Resorption und die fermentative Spaltung der Disaccharide im Dünndarm des ausgewachsenen Hundes. *Pflügers Archiv European Journal of Physiology*, *95*(11), 533–605.

Romagnuolo, J., Schiller, D., & Bailey, R. J. (2002). Using breath tests wisely in a gastroenterology practice: An evidence-based review of indications and pitfalls in interpretation. *The American Journal of Gastroenterology*, *97*(5), 1113–1126.

Rubio-Tapia, A., Herman, M. L., Ludvigsson, J. F., Kelly, D. G., Mangan, T. F., Wu, T..T., & Murray, J. A. (2012). Severe spruelike enteropathy associated with olmesartan. *Mayo Clinic Proceedings*, *87*(8), 732–738.

Saarela, T., Similä, S., & Koivisto, M. (1995). Hypercalcemia and nephrocalcinosis in patients with congenital lactase deficiency. *The Journal of Pediatrics*, *6*(127), 920–923.

Sadre, M., & Karbasi, K. (1979). Lactose intolerance in Iran. *American Journal of Clinical Nutrition*, *32*(9), 1948–1954.

Sahi, T. (1978). Dietary lactose and the aetiology of human small-intestinal hypolactasia. *Gut*, *19*(11), 1074.

Sahi, T. (1994). Genetics and epidemiology of adult-type hypolactasia. *Scandinavian Journal of Gastroenterology*, *29*(Suppl. 202), 7–20.

Savaiano, D. (2011). Lactose intolerance: An unnecessary risk for low bone density. *Nestle Nutrition Workshop Series. Paediatric Programme*, *67*, 161–171.

Savaiano, D. A., Ritter, A. J., Klaenhammer, T. R., James, G. M., Longcore, A. T., Chandler, J. R., & Foyt, H. L. (2013). Improving lactose digestion and symptoms of lactose intolerance with a novel galacto-oligosaccharide (RP-G28): A randomized, double-blind clinical trial. *Nutrition Journal*, *12*, 160.

Schiffner, R., Kostev, K., & Gothe, H. (2016). Do patients with lactose intolerance exhibit more frequent comorbidities than patients without lactose intolerance? An analysis of routine data from German medical practices. *Annals of Gastroenterology*, *29*(2), 174.

Seppo, L., Tuure, T., Korpela, R., Järvelä, I., Rasinperä, H., & Sahi, T. (2008). Can primary hypolactasia manifest itself after the age of 20 years? A two-decade follow-up study. *Scandinavian Journal of Gastroenterology*, *43*(9), 1082–1087.

Shaukat, A., Levitt, M. D., Taylor, B. C., Macdonald, R., Shamliyan, T. A., Kane, R. L., & Wilt, T. J. (2010). Systematic review: Effective management strategies for lactose intolerance. *Annals of Internal Medicine*, *152*(12), 797–803.

Stabell, N., Stubhaug, A., Flægstad, T., & Nielsen, C. S. (2013). Increased pain sensitivity among adults reporting irritable bowel syndrome symptoms in a large population-based study. *Pain*, *3*(154), 385–392.

Stanghellini, V., Chan, F. K., Hasler, W. L., Malagelada, J. R., Suzuki, H., Tack, J., & Talley, N. J. (2016). Gastroduodenal disorders. *Gastroenterology*, *150*(6), 1380–1392.

Storhaug, C. L., Fosse, S. K., & Fadnes, L. T. (2017). Country, regional, and global estimates for lactose malabsorption in adults: A systematic review and meta-analysis. *Lancet Gastroenterology & Hepatology*, *2*(10), 738–746.

Strocchi, A., Corazza, G., Ellis, C. J., Gasbarrini, G., & Levitt, M. D. (1993). Detection of malabsorption of low doses of carbohydrate: Accuracy of various breath H_2 criteria. *Gastroenterology*, *105*(5), 1404–1410.

Suarez, F. L., Savaiano, D., Arbisi, P., & Levitt, M. D. (1997). Tolerance to the daily ingestion of two cups of milk by individuals claiming lactose intolerance. *The American Journal of Clinical Nutrition*, *65*(5), 1502–1506.

Suarez, F. L., Savaiano, D. A., & Levitt, M. D. (1995). A comparison of symptoms after the consumption of milk or lactose-hydrolyzed milk by people with self-reported severe lactose intolerance. *New England Journal of Medicine*, *333*(1), 1–4.

Suchy, F., Brannon, P., Carpenter, T., Fernandez, J., Gilsanz, V., Gould, J., & Mennella, J. (2010). National Institutes of Health Consensus Development Conference: Lactose intolerance and health. *Annals of Internal Medicine*, *152*(12), 792–796.

Swallow, D. M. (2003). Genetics of lactase persistence and lactose intolerance. *Annual Review of Genetics*, *37*(1), 197–219.

Swallow, D. M., & Troelsen, J. T. (2016). Escape from epigenetic silencing of lactase expression is triggered by a single-nucleotide change. *Nature Structural & Molecular Biology*, *23*(6), 505–507.

Szilagyi, A. (2011). Lactase phenotype and relationship to diseases. In K. Urbano (Ed.), *Advances in genetics research* (Vol. 6, pp. 287–308). New York, NY: Nova Science Publishers Inc.

Szilagyi, A., Galiatsatos, P., & Xue, X. (2016). Systematic review and meta-analysis of lactose digestion, its impact on intolerance and nutritional effects of dairy food restriction in inflammatory bowel diseases. *Nutrition Journal*, *15*(1), 67.

Szilagyi, A., Lerman, S., Barr, R., Colacone, A., & McMullan, S. (1991). Reversible lactose malabsorption and intolerance in Graves' disease. *Clinical and Investigative Medicine*, *14*(3), 188–197.

Szilagyi, A., Malolepszy, P., Yesovitch, S., Nathwani, U., Vinokuroff, C., Cohen, A., & Xue, X. (2005). Inverse dose effect of pretest dietary lactose intake on breath hydrogen results and symptoms in lactase nonpersistent subjects. *Digestive Diseases and Sciences*, *50*(11), 2178–2182.

Szilagyi, A., Malolepszy, P., Hamard, E., Xue, X., Hilzenrat, N., Ponniah, M., & Chong, G. (2007). Comparison of a real-time polymerase chain reaction assay for lactase genetic polymorphism with standard indirect tests for lactose maldigestion. *Clinical Gastroenterology and Hepatology*, *5*(2), 192–196.

Szilagyi, A., Rivard, J., & Fokeeff, K. (2001). Improved parameters of lactose maldigestion using lactulose. *Digestive Diseases and Sciences*, *46*(7), 1509–1519.

Szilagyi, A., Salomon, R., Martin, M., Fokeeff, K., & Seidman, E. (1996). Lactose handling by women with lactose malabsorption is improved during pregnancy. *Clinical and Investigative Medicine*, *19*(6), 416.

Szilagyi, A., Salomon, R., Smith, B. E., Martin, M., & Seidman, E. (1996). Determinants of prolonged oral cecal transit time during late phase pregnancy. *Clinical and Investigative Medicine*, *19*(1), 20.

Szilagyi, A., & Xue, X. (2017). Comparison of geographic distributions of irritable bowel syndrome with inflammatory bowel disease fail to support common evolutionary roots. *Medical Hypotheses*, *110*, 31–37.

Tan, M. L. N., Muhardi, L., Osatakul, S., Hegar, B., Vandenplas, Y., Ludwig, T., . . . Quak, S. H. (2018). An electronic questionnaire survey evaluating the perceived prevalence and practices of lactose intolerance in 1 to 5 year old children in South East Asia. *Pediatric Gastroenterology, Hepatology & Nutrition*, *21*(3), 170–175.

Tan-Dy, C., & Ohlsson, A. (2013). Lactase treated feeds to promote growth and feeding tolerance in preterm infants. *The Cochrane Database of Systematic Reviews*, *3*, CD004591.

Tishkoff, S. A., Reed, F. A., Ranciaro, A., Voight, B. F., Babbitt, C. C., Silverman, J. S., & Osman, M. (2007). Convergent adaptation of human lactase persistence in Africa and Europe. *Nature Genetics*, *39*(1), 31–40.

Tomba, C., Baldassarri, A., Coletta, M., Cesana, B., & Basilisco, G. (2012). Is the subjective perception of lactose intolerance influenced by the psychological profile? *Alimentary Pharmacology & Therapeutics*, *7*(36), 660–669.

Troelsen, J. T. (2005). Adult-type hypolactasia and regulation of lactase expression. *Biochimica et Biophysica Acta (BBA)—General Subjects*, *1723*(1), 19–32.

Uchida, N., Sakamoto, O., Irie, M., Abukawa, D., Takeyama, J., Kure, S., & Tsuchiya, S. (2012). Two novel mutations in the lactase gene in a Japanese infant with congenital lactase deficiency. *Tohoku Journal of Experimental Medicine*, *227*(1), 69–72.

Valeur, J., Røseth, A. G., Knudsen, T., Malmstrøm, G. H., Fiennes, J. T., Midtvedt, T., & Berstad, A. (2016). Fecal fermentation in irritable bowel syndrome: Influence of dietary restriction of fermentable oligosaccharides, disaccharides, monosaccharides and polyols. *Digestion*, *94*(1), 50–56.

Vazquez-Roque, M. I., Camilleri, M., Smyrk, T., Murray, J. A., Marietta, E., O'Neill, J., & Eckert, D. (2013). A controlled trial of gluten-free diet in patients with irritable bowel syndrome-diarrhea: Effects on bowel frequency and intestinal function. *Gastroenterology*, *144*(5), 903–911. e903.

Vernia, P., Di Camillo, M., Foglietta, T., Avallone, V. E., & De Carolis, A. (2010). Diagnosis of lactose intolerance and the "nocebo" effect: The role of negative expectations. *Digestive and Liver Disease*, *42*(9), 616–619.

Vesa, T. H., Seppo, L. M., Marteau, P. R., Sahi, T., & Korpela, R. (1998). Role of irritable bowel syndrome in subjective lactose intolerance. *The American Journal of Clinical Nutrition*, *67*(4), 710–715.

Virta, L. J., Kautiainen, H., & Kolho, K. L. (2016). Symptoms suggestive of cow's milk allergy in infancy and pediatric inflammatory bowel disease. *Pediatric Allergy and Immunology*, *27*(4), 361–367.

Wal, J.-M. (2004). Bovine milk allergenicity. *Annals of Allergy, Asthma & Immunology*, *93*(5), S2–S11.

Wang, Y., Harvay, C. B., Pratt, W. S., Sams, V., Sarner, M., Rossi, M., & Swallow, D. M. (1995). The lactase persistence/non-persistence polymorphism is controlled by a cis-acting element. *Human Molecular Genetics*, *4*(4), 657–662.

Waud, J. P., Matthews, S. B., & Campbell, A. K. (2008). Measurement of breath hydrogen and methane, together with lactase genotype, defines the current best practice for investigation of lactose sensitivity. *Annals of Clinical Biochemistry*, *45*(1), 50–58.

Weaver, L., Laker, M., & Nelson, R. (1986). Neonatal intestinal lactase activity. *Archives of Disease in Childhood*, *61*(9), 896.

Weiss, S. L., Lee, E. A., & Diamond, J. (1998). Evolutionary matches of enzyme and transporter capacities to dietary substrate loads in the intestinal brush border. *Proceedings of the National Academy of Sciences*, *95*(5), 2117–2121.

Windey, K., Houben, E., Deroover, L., & Verbeke, K. (2015). Contribution of colonic fermentation and fecal water toxicity to the pathophysiology of lactose-intolerance. *Nutrients*, *7*(9), 7505–7522.

Wolvekamp, M., & De Bruin, R. (1994). Diamine oxidase: An overview of historical, biochemical and functional aspects. *Digestive Diseases*, *12*(1), 2–14.

Yang, J., Deng, Y., Chu, H., Cong, Y., Zhao, J., Pohl, D., & Fox, M. (2013). Prevalence and presentation of lactose intolerance and effects on dairy product intake in healthy subjects and patients with irritable bowel syndrome. *Clinical Gastroenterology and Hepatology*, *3*(11), 262–268. e261.

Yang, J., Fox, M., Cong, Y., Chu, H., Zheng, X., Long, Y., & Dai, N. (2014). Lactose intolerance in irritable bowel syndrome patients with diarrhoea: The roles of anxiety, activation of the innate mucosal immune system and visceral sensitivity. *Alimentary Pharmacology & Therapeutics*, *3*(39), 302–311.

Zhao, A., Szeto, I. M.-Y., Wang, Y., Li, C., Pan, M., Li, T., & Zhang, Y. (2017). Knowledge, attitude, and practice (KAP) of dairy products in Chinese urban population and the effects on dairy intake quality. *Nutrients*, *9*(7), 668.

Further reading

Szilagyi, A., Leighton, H., Burstein, B., & Xue, X. (2014). Latitude, sunshine, and human lactase phenotype distributions may contribute to geographic patterns of modern disease: The inflammatory bowel disease model. *Clinical Epidemiology*, *6*, 183.

CHAPTER

Lactose—a conditional prebiotic?

4

Michael G. Gänzle[1,2]

[1]*Department of Agricultural, Food and Nutritional Science, University of Alberta, Edmonton, AB, Canada*

[2]*College of Bioengineering and Food Science, Hubei University of Technology, Wuhan, P.R. China*

4.1 Introduction

Lactose is the major carbohydrate in bovine milk. Because lactose is a by-product of the fractionation of proteins and fat from bovine milk (Fox, Guinee, Cogan, & McSweeney, 2017; Gänzle, Haase, & Jelen, 2008; Jelen, 1979), it is a relatively cheap carbohydrate resource. Lactose valorization in the dairy industry includes food applications and the use as excipient for pharmacological applications (see Chapter 5: Application of lactose in the pharmaceutical industry). In addition, enzymatic conversion of lactose to functional food and feed ingredients with prebiotic properties is increasingly employed (Chen & Gänzle, 2017; Gänzle et al., 2008; Torres, do Pilar, Gonçalves, Teixeira, & Rodrigues, 2010). Prebiotic oligosaccharides derived from lactose include galacto-oligosaccharides (GOS), lactulose, and lactosucrose (Gänzle et al., 2008). Human milk oligosaccharides (HMOs) found in human breast milk are based on lactose and have well-established prebiotic properties' however, their metabolism by bifidobacteria not only requires β-galactosidases but is also dependent on fucosidases, sialidases, and β-hexoaminidases, and thus differs fundamentally from lactose and GOS (Sela et al., 2008; Smilowitz, Lebrilla, Mills, German, & Freeman, 2014).

Prebiotic carbohydrates provide health benefits that are mediated by the metabolism of intestinal microbiota. Intestinal microbiota impact multiple aspects of human physiology and a restricted functionality of intestinal microbiota as imposed by the Western lifestyle predisposes to multiple chronic diseases (Sonnenburg & Sonnenburg, 2014). Intestinal microbiota contribute to carbohydrate digestion; increase the energy harvest by hydrolysis and fermentation of dietary components that are not hydrolyzed by host enzymes (Hamaker & Tuncil, 2014); improve barrier properties of the intestinal epithelium; increase resistance to infection by enteric pathogens (Vogt & Finlay, 2017); and modulate the risk of colon cancer (Cho & Blaser, 2012; Zhang et al., 2017). Extraintestinal effects mediated by intestinal microbiota relate not only to immunity and inflammation

Lactose. DOI: https://doi.org/10.1016/B978-0-12-811720-0.00004-0

(Cho & Blaser, 2012; Postler & Gosh, 2017) but also to brain development and behavior (Collins, Surette, & Bercik, 2012). Composition and activity of intestinal microbiota are strongly determined by the host diet (Carmody et al., 2015; Sonnenburg & Sonnenburg, 2014) and the bacterial metabolites that mediate intestinal and extraintestinal effects are increasingly understood (Postler & Gosh, 2017). Physiological benefits from dietary glycans including prebiotic carbohydrates are mainly attributed to bacterial fermentation of carbohydrates to short chain fatty acids (Bindels, Delzenne, Cani, & Walter, 2015; Hamaker & Tuncil, 2014; Postler & Gosh, 2017).

Prebiotic lactose derivatives are structurally highly related to lactose (Fig. 4.1) and the indigestibility of lactose in lactase nonpersisters (humans whose lactase gene has been switched off in adulthood) suggests that lactose may also have a comparable physiological functionality as prebiotic GOS (see also Chapter 2: Digestion, absorption, metabolism, and physiological effects of lactose and Chapter 3: Lactose intolerance and other related food sensitivities). This chapter will address the question whether lactose itself is a prebiotic carbohydrate, which is conditional on the lactase nonpersistence of the consumer (see Chapter 3: Lactose intolerance and other related food sensitivities). To answer this question, definitions related to the nutritional functionality of dietary carbohydrates are revisited and the contribution of host enzymes and intestinal microbiota to

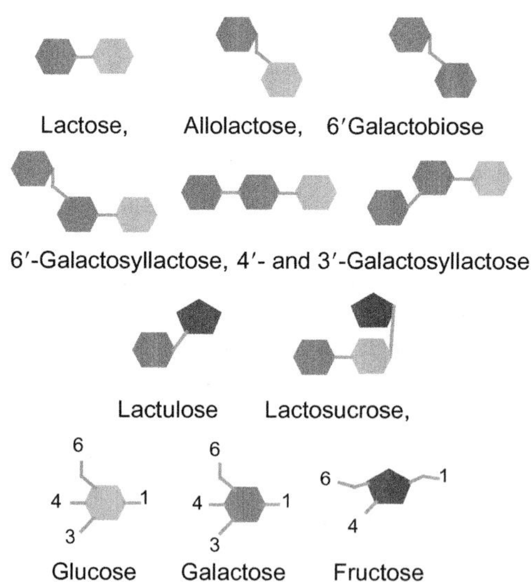

FIGURE 4.1

Schematic representation of the structure of lactose in comparison to prebiotic lactose derivatives.

Drawn according to Chen, X. Y., & Gänzle, M. G. (2017). Lactose and lactose-derived oligosaccharides: More than prebiotics? International Dairy Journal, 67, *61–72.*

digestion of lactose and prebiotic lactose derivatives are discussed. Subsequent sections summarize the arguments in favor and against the classification of lactose as a conditional prebiotic.

4.2 Prebiotics, dietary fiber, and FODMAPs: definitions

The term prebiotic was introduced in 1995 on the basis of functional properties but has evolved in the past decades and lacks clearly defined criteria for assessing prebiotic activity. Moreover, the term prebiotic is not recognized by regulatory agencies or the Codex Alimentarius, and the scientific definition of the term remains controversial (for reviews, see Bindels et al., 2015; Hutkins et al., 2016; Yan, Hu, & Gänzle, 2018). Definitions published in the scientific literature were based on the criteria (1) (lack of) small intestinal hydrolysis and absorption, (2) selective stimulation of human microbiota, (3) metabolism by intestinal microbiota, and (4) beneficial effects on human health. Prebiotic applications for extraintestinal body sites were considered; however, the limited evidence for health beneficial effects of prebiotic carbohydrates in extraintestinal body sites (Coste, Judlin, Lepargneur, & Bou-Antoun, 2012) and the wide recognition of the term prebiotic as nutritional concept argue strongly in favor of restricting the term to the intestinal tract (Bindels et al., 2015). The use of high-throughput sequencing of 16S rRNA genes for characterization of the impact of prebiotics on intestinal microbiota since 2010 (Martínez, Kim, Duffy, Schlegel, & Walter, 2010) demonstrated that changes in the composition of intestinal microbiota that are induced by dietary oligosaccharides are not restricted to specific bacterial taxa. Moreover, current knowledge on the mechanisms by which prebiotic carbohydrates benefit human health demonstrates that these beneficial physiological effects relate to modulation of the metabolism of intestinal microbiota, in particular the formation of short chain fatty acids, rather than its taxonomic composition (Bindels et al., 2015; Chassard & Lacroix, 2013; Louis, Flint, & Michel, 2016; Reichardt et al., 2018; Yan et al., 2018). The definition of prebiotics as compounds that escape small intestinal hydrolysis and absorption and exert health benefits that are mediated by microbial metabolites, thus best fits the current scientific knowledge on the effect of dietary carbohydrates on host health (Bindels et al., 2015; Yan et al., 2018). This definition also encompasses disaccharides including lactose that are excluded from the dietary fiber definition and will be used as a basis for further discussion in this chapter.

Health beneficial effects of dietary glycans are predominantly recognized for plant-derived dietary fiber (Hamaker & Tuncil, 2014). Dietary fiber is defined by regulatory agencies worldwide and the Codex Alimentarius on the basis of structural and functional properties. The Codex Alimentarius definition includes carbohydrate polymers with 10 or more monomers that are resistant to digestion and absorption in the small intestine of humans, and have a physiological benefit to human health (Codex Alimentarius Commission, 2008; Cummings & Stephen, 2007). In addition, several regulatory agencies such as the European Food Safety Agency (EFSA), the US Food and Drug Administration, and Health Canada include oligosaccharides with a degree of polymerization (DP) of 3–9 (EU, 2011; FDA, 2016; Hamaker & Tuncil, 2014; Health Canada, 2012). Thus lactosucrose and GOS with a DP of 3 or more, but not lactulose and lactose, are included in this definition.

The term "FODMAPs" was coined to designate adverse health benefits of oligosaccharides and has gained substantial traction since its introduction in 2012. Fermentable, oligosaccharides, disaccharides, monosaccharides, and polyols (FODMAPs) are carbohydrates and polyols that are not hydrolyzed and absorbed in the small intestine (Staudacher & Whelan, 2016). The term was defined in relationship with factors that contribute to symptoms in people suffering from the irritable bowel syndrome (IBS). The specific contribution of the diverse compounds that are included in the term FODMAP to the development of adverse symptoms remains to be elucidated; however, the FODMAP concept is increasingly used by the food industry for the development of low-FODMAP products and guides food choices of consumers (De Giorgio, Volta, & Gibson, 2016; Laatikainen et al., 2016). Lactose, GOS, and other lactose-derived prebiotics are included in the FODMAP definition; however, experimental evidence for negative consequences of FODMAP in IBS is particularly linked to dietary fructans, including sucrose and fructose malabsorption in IBS patients (Latulippe & Skoog, 2011; Murray et al., 2014). The mechanisms related to the adverse effects of FODMAPs in IBS, however, also link to the mechanisms leading to adverse digestive symptoms in lactose intolerance or the adverse effects of overconsumption of prebiotic compounds including fructo-oligosaccharides (FOS) or lactulose (see below).

4.3 Hydrolysis, absorption, and fermentation of lactose and galacto-oligosaccharides in lactase-persistent and lactase-nonpersistent individuals

GOS are β-linked oligosaccharides with a DP of 2–9 that consist of glucose or galactose at the reducing end and are extended by β-linked galactose (Fig. 4.1; Gänzle, 2012; van Leeuwen, Kuipers, Dijkhuizen, & Kamerling, 2016). International Union of Pure and Applied Chemistry (IUPAC) nomenclature includes disaccharides in the term oligosaccharides. Lactose is thus a GOS; however, due to its different and conditional digestibility, it is often considered separately. The conditional expression of brush border lactase is discussed in detail in Chapters 1–3. In short, human infants and approximately 35% of human adults (lactase persisters) express brush border β-glycosidase with activity on lactose. However, preterm infants and a majority of human adults (lactase nonpersisters) do not express brush border β-galactosidase (Gerbault et al., 2011; Kien, McClead, & Cordero, 1998). In vitro studies with rat small intestinal extracts that contain glycosyl hydrolases with activities that are comparable to the activity of human brush border enzymes demonstrate that GOS other than lactose are also hydrolyzed by brush border lactase (Ferreira-Lazarte, Olano, Villamiel, & Moreno, 2017). β-(1→4)-Linked GOS are hydrolyzed faster than β-(1→6)-linked GOS and disaccharides are hydrolyzed faster than trisaccharides or higher oligosaccharides (Ferreira-Lazarte et al., 2017; Hernández-Hernández et al., 2012). Because intestinal lactase is localized in the brush border, however, transport of digesta is faster than hydrolysis and only a small proportion of GOS are hydrolyzed in the small intestine (Ferreira-Lazarte et al., 2017).

Dietary di- and trisaccharides are metabolized by intestinal microbiota in the terminal ileum (Booijink et al., 2010; Zoetendal et al., 2012) (Fig. 4.2). While proximal parts of the small intestine harbor only low numbers of bacteria, cell counts in the terminal ileum reach 10^9 cfu/g (Booijink et al., 2010). Ileal microbiota are highly variable between individuals and within one individual over time; nevertheless, coliforms, *Veillonella*, and lactic acid bacteria including *Enterococcus*, *Weissella*, and *Streptococcus* spp. were identified as core members

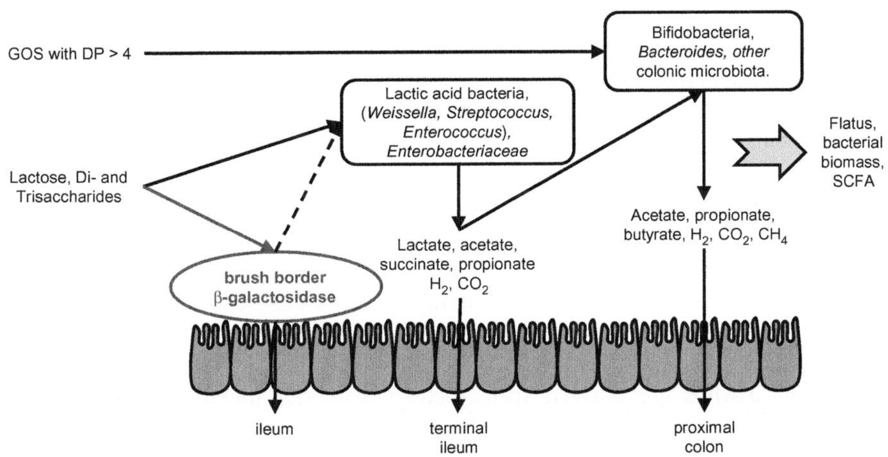

FIGURE 4.2

Overview of metabolism of lactose and GOS by intestinal microbiota. Lactose and dietary GOS with DP of 2 and 3 are hydrolyzed by brush border lactase in the ileum, or metabolized by intracellular β-galactosidases of ileal microbiota. Brush border lactase hydrolyzes β-(1→4)-linked GOS faster than β-(1→6)-linked GOS and disaccharides are hydrolyzed faster than trisaccharides (Ferreira-Lazarte et al., 2017; Hernández-Hernández et al., 2012). Brush border β-galactosidases are not expressed in preterm infants, or in lactase-nonpersistent adults, restricting lactose and GOS digestion to metabolism by intestinal microbiota. Metabolites of ileal microbiota, particularly lactate, succinate, and acetate, are further converted to butyrate or acetate by colonic microbiota.

of the ileal microbiome (Booijink et al., 2010; Zoetendal et al., 2012). Lactic acid bacteria and Enterobacteriaceae metabolize GOS with intracellular β-galactosidases or β-phospho-galactosidases (Gänzle & Follador, 2012). β-Phospho-galactosidases hydrolyze disaccharides after transport and phosphorylation by phosphotransferase systems (PTS) that catalyze simultaneous transport and phosphorylation. The LacEF PTS system in lactobacilli is specific for lactose and does not transport other GOS (Gänzle & Follador, 2012). The lactose permease LacS, which is present in lactic acid bacteria, bifidobacteria, and *Escherichia coli*, transports di- and trisaccharides; tetrasaccharides or higher oligosaccharides are not transported, or transported much more slowly (Andersen et al., 2011; Gänzle & Follador, 2012; Gopal, Sullivan, & Smart, 2001). Therefore ileal microbiota preferentially metabolize mono- and disaccharides (Zoetendal et al., 2012). Metabolism of oligosaccharides with a DP of 4 or more depends on extracellular glycosyl hydrolase 42 family β-galactosidases which are expressed by bifidobacteria and other members of colonic microbiota (Louis et al., 2016; O'Connell Motherway, Kinsella, Fitzgerald, & van Sinderen, 2013). The DP thus strongly influences the site and rate of bacterial metabolism of GOS. The influence of the DP in in vivo fermentation of GOS has not been investigated; however, it is reasonable to assume that, in analogy to observations with β-fructans differing in their molecular weight (Murray et al., 2014), a higher DP of GOS shifts bacterial metabolism from the terminal ileum to the proximal colon (Fig. 4.2). Bacterial β-galactosidases generally hydrolyze β-(1→4)- and β-(1→6)-linked GOS at an equal rate (Torres et al., 2010); major contributors to lactose and GOS metabolism in humans include, but are not limited to, bifidobacteria and lactic acid bacteria (Davis, Martínez, Walter, & Hutkins, 2010; Krumbeck et al., 2018; Maathuis, van den Heuvel, Schoterman, & Venema, 2012). Major metabolites of small intestinal microbiota include acetate, butyrate, lactate, formate, and succinate; their formation is associated with microbial production of H_2 and CO_2 (Fig. 4.2; Zoetendal et al., 2012). Ileal metabolites are absorbed in the small intestine, or converted to acetate, propionate, and butyrate by colonic microbiota (Fig. 4.2).

Although HMOs are derived from lactose, bacterial metabolic enzymes for HMO metabolism differ from lactose metabolic pathways. HMOs carry lactose at the reducing end and are elongated with galactose, *N*-acetylglucosamine, fucose, or sialic acid to generate a large structural diversity of oligosaccharides (Smilowitz et al., 2014). HMOs are not hydrolyzed by lactose-metabolizing enzymes of bifidobacteria, Enterobacteriaceae, or lactic acid bacteria (Hoeflinger, Davis, Chow, & Miller, 2015; James, Motherway, Bottacini, & van Sinderen, 2016; Schwab & Gänzle, 2011). Bifidobacteria, but not lactic acid bacteria, or Enterobacteriaceae harbor multiple dedicated hydrolases and transport enzymes for metabolism of the diverse HMOs (James et al., 2016; Sela & Mills, 2010) conferring a clear growth advantage for bifidobacteria in the presence of HMOs. HMOs reduce the risk of Enterobacteriaceae overgrowth and thus necrotizing enterocolitis. The unique, diverse, and dedicated array of metabolic enzymes for

HMO utilization that is present in some species of bifidobacteria is key for the dominance of *Bifidobacterium* spp. in breastfed infants (Matsuki et al., 2016; Sela & Mills, 2010).

4.4 Lactose is not a prebiotic

Two main points argue against the classification of lactose as prebiotic. First, lactose is hydrolyzed and absorbed in a substantial part of adult humans and prebiotic activity is thus conditional on the genetic background and nutritional status. Second, beneficial effects on health are unanimously required for recognition of compounds as prebiotics. Reports on health effects of lactose in lactase-nonpersistent individuals, however, overwhelmingly describe negative health effects and, in case of premature infants, may favor intestinal overgrowth of Enterobacteriaceae.

Sugar alcohols, monosaccharides, and oligosaccharides that are not hydrolyzed and absorbed in small intestine cause osmotic diarrhea in humans when a tolerated dose is exceeded (Florent et al., 1985; Hammer & Hammer, 2012; Oku & Nakamura, 2002; Oku & Nakamura, 2007). Studies with diverse compounds that escape absorption in the small intestine consistently report a tolerated dose of about 0.3 g/kg body weight in adults (Oku & Nakamura, 2002, 2007). The value of about 0.3 g/kg body weight corresponds well to the tolerance of about 15 g per day of prebiotic GOS and FOS (Davis et al., 2010; Torres et al., 2010). Diarrhea is dependent on the osmotic load of the carbohydrates, which prevents absorption of water from the intestinal lumen. In addition, rapid microbial conversion of the glycans to organic acids, H_2, and CO_2 contributes to adverse effects (Hammer & Hammer, 2012; He et al., 2008; Venema, 2012). Adverse effects of osmotic diarrhea are transient and cease as soon as the glycan is cleared by microbial fermentation. Microbial lactose fermentation in preterm infants, however, has been related to overgrowth of pathogenic Enterobacteriaceae and necrotic enterocolitis (Lin, 2004). These consequences may relate to the immature microbiome of preterm infants and the capacity of some Enterobacteriaceae for rapid lactose metabolism.

While the glycan load causing adverse symptoms is largely independent of the glycan structure (Oku & Nakamura, 2002), high DP glycans are less likely to elicit adverse effects (Murray et al., 2014). The reduction of adverse symptoms with increasing DP relates to the reduced osmotic load of oligosaccharides when compared to monosaccharides and also to the reduced rate of fermentation (see earlier).

Differences in the small intestinal digestibility of lactose between individuals make prebiotic properties of lactose dependent on lactase expression in the small intestine. In lactase-nonpersistent adults the capacity to digest lactose without adverse symptoms is dependent on the fermentation capacity of the intestinal microbiota (Azcarate-Peril et al., 2017; He et al., 2008; Venema, 2012). Adaptation of intestinal microbiota in lactase-nonpersistent individuals enhances the capacity to digest lactose without adverse symptoms (Azcarate-Peril et al., 2017; He et al., 2008). Changes in composition and activity of intestinal microbiota as a result of, for example, antibiotic treatment or intestinal inflammation, however, may also reduce the tolerance for lactose (He et al., 2008; Venema, 2012). In addition, dietary β-galactosidase can substitute for the lack of brush border lactase activity (Kolars, Levitt, Aouji, & Savaiano, 1984). Dietary β-galactosidases are provided by lactic acid bacteria in yoghurt and make prebiotic properties of lactose and GOS conditional on the host diet. Of note, the conditional absorption of dietary carbohydrates is also reported for fructose and sucrose. Approximately 15% of human adults have a reduced capacity to absorb dietary fructose; because small intestinal fermentation of fructose contributes to IBS, the prevalence of fructose malabsorption among IBS patients is much higher compared to the healthy population (Latulippe & Skoog, 2011; Murray et al., 2014). Sucrose hydrolysis is also conditional on the genetic background. Congenital sucrase–isomaltase deficiency is rare but predisposes to IBS (Henström et al., 2018).

4.5 Lactose is a prebiotic

Two main points argue in favor of including lactose in the definition of prebiotic compounds as dietary compounds that exert health benefits that are mediated by bacterial fermentation. First, the effects of lactose on composition and activity of intestinal microbiota are highly similar to the effects of GOS and the other lactose-derived oligosaccharides, lactosucrose, and lactulose (see later), which are broadly recognized as prebiotics with physiological benefits. Mechanisms underlying the health beneficial effects of prebiotics are mainly related to a reduction of the colonic pH and the formation of short chain fatty acids by intestinal microbiota. Since intestinal fermentation of lactose generates the same metabolites as fermentation of GOS and other prebiotic compounds, lactose fermentation is expected to exert comparable health benefits. Second, a majority of clinical studies demonstrating health beneficial effects of GOS were conducted with commercial GOS preparations that include considerable amounts of lactose, which—dependent on dosage and lactase activity in the study population—may also escape small intestinal hydrolysis to be fermented in the lower intestine.

The health beneficial effects of GOS have been demonstrated in multiple clinical studies (Paganini et al., 2017; for reviews, see Macfarlane, Steed, & Macfarlane, 2008; Torres et al., 2010). Most commercial GOS preparations, however, contain lactose as a considerable component (Table 4.1). Most clinical studies do not document whether participants are lactase-persistent, lactose-intolerant, or lactase-nonpersistent with high tolerance for lactose (Davis et al., 2010;

Table 4.1 Composition of nine commercial galacto-oligosaccharide preparations.

	A	B[a]	C	D[a]	E	F	G	H	I[b]	K
DP 1	38	38	24	20	20	9	7	3	1	0.3
DP 2	36	26	33	43	32	46	39	47	0.2	20
DP 3	17	24	35	24	34	32	32	36	57	46
DP 4	7	10	7	10	11	11	14	12	28	22
DP > 5	2	2	1	3	3	2	7	2	13	12

[a]*Designates the commercial GOS preparations which have been used in a majority of clinical and animal studies.*
[b]*The product is not marketed as food ingredient but as a pharmaceutical used to alleviate lactose intolerance.*
Compiled with information from Azcarate-Peril, M. A., Ritter, A. J., Savaiano, D., Monteagudo-Mera, A., Andersson, C., Magnoso, S., & Klaenhammer, T. R. (2017). Impact of short chain galactooligosaccharides on the gut microbiome of lactose-intolerant individuals. Proceedings of the National Academy of Science of the United States of America, 114, *E367–E375; Chen, X. Y., & Gänzle, M. G. (2017). Lactose and lactose-derived oligosaccharides: More than prebiotics?* International Dairy Journal, 67, *61–72; Guo, Q., Goff, H. D., & Cui, S. W. (2017). Structural characterisation of galacto-oligosaccharides (VITAGOS™) synthesized by transgalactosylation of lactose.* Bioactive Carbohydrates and Dietary Fibre. *https://doi.org/10.1016/j.bcdf.2017.07.007; van Leeuwen, S. S., Kuipers, B. J., Dijkhuizen, L., & Kamerling, J. P. (2016). Comparative structural characterization of 7 commercial galacto-oligosaccharide (GOS) products.* Carbohydrate Research, *425, 48–58 (Guo, Goff, & Cui, 2017; Ito et al., 1990).*

Macfarlane et al., 2008; Paganini et al., 2017; Torres et al., 2010). A study that screened participants for lactose intolerance as exclusion criterion found that some of the lactose-tolerant participants were lactase-nonpersistent—in these participants, the adaptation of intestinal microbiota allowed lactose digestion without adverse effects (Krumbeck et al., 2018). Lactose also had a bifidogenic effect in these lactose-tolerant and lactase-nonpersistent individuals (Krumbeck et al., 2018). Lactose may thus contribute to the health beneficial effects of commercial GOS preparations. Similarly, the adverse effects of excessive ingestion of GOS are comparable to the symptoms of lactose consumption in lactase nonpersisters (Oku & Nakamura, 2002). A dose of GOS of more than 0.3 g/kg body weight, corresponding to approximately 15 g/person per day, can cause intestinal bloating and osmotic diarrhea. However, an increase in the capacity of the intestinal microbiota for GOS fermentation reduces the adverse effects over time (Davis et al., 2010; Krumbeck et al., 2018; Mego et al., 2017). Notably, GOS consumption increases the tolerance to dietary lactose in lactase nonpersisters (Savaiano et al., 2013), consistent with the findings that bacterial lactose hydrolyzing enzymes also metabolize nonlactose GOS with DP 2 and 3 (see earlier).

Only few studies report the response of intestinal microbiota to dietary lactose. A crossover study comparing the microbiome of 16 infants with cow allergy reported increased cell counts of enterococci, streptococci, and bifidobacteria in response to inclusion of lactose in the infant formula (Francavilla et al., 2012).

Lactose content in the infant formula, 3.8%, was comparable to that of cow's milk. The response of the microbiome to lactose in healthy infants indicates that lactose is not fully hydrolyzed in the small intestine even if brush border lactase is expressed—the slow hydrolysis by brush border enzymes allows in transporting digesta to the terminal ileum and the colon before all lactose is hydrolyzed (Francavilla et al., 2012; Lifschitz, O'Brian Smith, & Garza, 1983). The abundance of enterococci and staphylococci was increased in lactase-nonpersistent adults ($n=24$) in response to lactose dose of 15 g per day for 6 days (Ito & Kimura, 1993). A daily dose of 25 g lactose over 2 weeks induced a moderate increase of bifidobacteria in lactase-nonpersistent individuals ($n=23$) when compared to lactase persisters (Szilagyi et al., 2010). Lactose has comparable beneficial effects as lactulose in lactase-nonpersistent patients of chronic portal-systemic encephalopathy (Uribe et al., 1987). Shifts in colonic microbiota upon infusion of lactose into the colon of two lactase-persistent adults were dependent on the individual composition of colonic microbiota and included an increased abundance of bifidobacteria, *Akkermansia*, and *Faecalibacterium* (Venema, 2012). An in vitro study with fecal microbiota reported an increased abundance of bifidobacteria upon addition of lactose to the fermentation system (Mäkivuokko, Saarinen, Ouwehand, & Rautonen, 2006). The number and statistical power of intervention studies with lactose in lactase nonpersisters is low but these data are supported by epidemiological evidence. The correlation of the genotype of 1514 individuals enrolled in three independent Dutch population cohorts with the composition of the intestinal microbiota revealed an association of lactase nonpersistence and the abundance of bifidobacteria (Bonder et al., 2016). This correlation was observed only in lactase nonpersisters consuming dairy products, indicating that lactose contributes to this effect. Overall, in vivo studies on lactose fermentation conform to the findings reported earlier that disaccharides are rapidly fermented by diverse members of ileal microbiota, including, but not limited to, bifidobacteria and lactic acid bacteria. A detailed analysis of the fate of lactose in lactase-persistent and lactase-nonpersistent individuals, however, remains subject to future studies.

Health effects of GOS are summarized in Table 4.2. Oligosaccharides with low small intestinal digestibility are generally noncariogenic and low glycemic (Table 4.2). A moderate dose of nondigestible and fermentable oligosaccharides will improve intestinal motility and relieve constipation while an excessive dose results in osmotic diarrhea (Table 4.2). Because these effects appear to be independent on the glycan structure, lactose effects in lactase nonpersisters are equivalent to the effects of other oligosaccharides in all humans (Oku & Nakamura, 2007; Yan et al., 2018).

Major health benefits of prebiotic carbohydrates are derived from intestinal conversion to short chain fatty acids (Table 4.2) (for reviews, see Bindels et al., 2015; Hamaker & Tuncil, 2014; Postler & Ghosh, 2017; Yan et al., 2018). Lactose and GOS conversion in the terminal ileum generates lactate, acetate, and succinate as major fermentation products (Booijink et al., 2010; Zoetendal et al.,

Table 4.2 Beneficial and adverse health effects of GOS and lactose in lactase-nonpersistent individuals.

Beneficial health effects	Adverse health effects
General health effects of oligosaccharides with low small intestinal digestibility	
Noncariogenic Low glycemic and reduced caloric content (1) Intestinal motility and relief of constipation (2)	Osmotic diarrhea (2)
Health effects related to microbial metabolism of oligosaccharides	
Production of short chain fatty acids contributing to (3): – energy supply of colonic mucosa and improved barrier properties – satiety, glucose homeostasis, and insulin sensitivity – antiinflammatory effects – reduced colonic pH resulting in reduced proteolytic fermentation and pathogen exclusion	Production of gas and intestinal discomfort (2)

1. *Oku and Nakamura (2007).*
2. *He et al. (2008), Venema (2012), Hammer and Hammer (2012).*
3. *Cherbut et al. (1998), Sakata (1987), Canfora, Jocken, and Blaak (2015), Byrne, Chambers, Morrison, and Frost (2015), Tolhurst et al. (2012), Smith et al. (2013), Kelly et al. (2015), Holtug, Clausen, Hove, Christiansen, and Mortensen (1992), Wutzke and Scholübbers (2013).*

Modified from Yan, Y. L., Hu, Y., & Gänzle, M. G. (2018). Prebiotics, FODMAPs and dietary fibre—conflicting concepts in development of functional food products? Current Opinion in Food Science, 20, *30–37.*

2012); these metabolic intermediates are converted to butyrate and propionate by colonic microbiota (Louis, Scott, Duncan, & Flint, 2007; Reichardt et al., 2018; Russell, Hoyles, Flint, & Dumas, 2014). Short chain fatty acids resulting from saccharolytic fermentation act locally by reducing the luminal pH and by providing butyrate as the preferred energy source of the colonic mucosa. A reduced luminal pH improves the bioavailability of minerals, inhibits Bacteroidetes, and reduces proteolytic fermentation in the proximal colon (Korpela, 2018; Reichardt et al., 2018). Colonic fermentation of amino acid generates a multitude of metabolites which have beneficial but also detrimental effects on human health, including indole, NH_4^+, H_2S, aromatic amines, and polyamines (for review, see Portune et al., 2016). Saccharolytic fermentation is therefore generally regarded as more beneficial than proteolytic fermentation (Korpela, 2018).

Butyrate is the preferred energy source of colonic enterocytes, and butyrate production by colonic microbiota thus improves the proliferation of the colonic mucosa as well as barrier integrity (Postler & Ghosh, 2017; Sakata, 1987). Systemic effects of short chain fatty acids are mediated by G-protein-coupled receptors GPR41, GPR43, and GPR109A (Kim, Kang, Park, Yanagisawa, & Kim, 2013; Thangaraju et al., 2009; Tolhurst et al., 2012). These receptors are involved in regulation of satiety, insulin, and glucose homeostasis, and mediate antiinflammatory effects and improved immunity (for reviews, see Bindels et al., 2015; Hamaker & Tuncil, 2014; Postler & Ghosh, 2017; Yan et al., 2018). Because many of the health benefits of dietary fiber and prebiotic carbohydrates are mediated by microbial conversion to short chain fatty acids, lactose fermentation to short chain fatty acids in lactase nonpersisters (Bonder et al., 2016; Venema, 2012) can be assumed to have comparable health beneficial effects.

4.6 Open questions

Health beneficial effects of prebiotic oligosaccharides and adverse health effects of FODMAPs are clearly the two sides of the same coin (Fig. 4.2 and Table 4.2); microbial oligosaccharide fermentation generates health beneficial short chain fatty acids, but too vigorous fermentation and gas formation also contribute to adverse outcomes (Yan et al., 2018). Knowledge on how the DP of oligosaccharides and their molecular structure influences the rate of fermentation and the balance of beneficial and adverse effects, however, is only beginning to emerge (Murray et al., 2014) and needs further studies in well-characterized target populations. In particular, the differential effects of lactose, low-molecular-weight GOS, and high-molecular-weight GOS on composition and activity of human intestinal microbiota need more detailed investigation. Likewise, the comparison of the effects of lactose and GOS on intestinal microbiota and host health in lactase nonpersisters is required to confirm a role of lactose as a conditional prebiotic carbohydrate. Of note, isomalto-oligosaccharides are substrates for brush border hydrolases and also have prebiotic effects, as the hydrolysis by brush border enzymes is limited by diffusion and allows for transport of digesta to the distal intestinal tract and microbial fermentation (Hu, Winter, Chen, & Gänzle, 2017). Similarly, a portion of the dietary lactose may be fermented by intestinal microbiota even in lactase-persistent individuals; however, the balance of small intestinal absorption and large intestinal fermentation is poorly documented for slowly digestible oligosaccharides.

4.7 Conclusion

Human digestion of lactose encompasses a more or less continuous transition from lactose tolerance with the full ability for lactose hydrolysis and absorption in the small intestine to lactose maldigestion in the absence of sufficient lactase activity (lactase nonpersistence) with manifest lactose intolerance. The metabolic activity of the intestinal microbiota and its adaptation to lactose fermentation is a strong modulator of factors leading to symptoms of lactose intolerance. Lactose digestion and its role as a potential prebiotic with beneficial health effects or as FODMAP with potential adverse health effects thus depends on genetic factors and is also subject to temporal variability in individuals. Diet, medication, or disease-induced changes in composition and function of the microbiome strongly influence the ability to digest lactose without adverse symptoms. A sufficient dose of undigested lactose can cause osmotic diarrhea and bloating—a property shared with other FODMAPs—but lactose may also have positive health effects. Beneficial health effects are mediated by the increased metabolic activity of the lactose-fermenting intestinal microbiota and match in a broad-sense prebiotic effect of other prebiotic oligosaccharides. However, these effects are highly dependent on the extent of small intestinal lactose hydrolysis and the adaptation of intestinal microbiota in individual subjects. Because studies on health benefits of lactose were not performed in subjects that were sufficiently characterized with respect to their microbiota and actual intestinal lactase activity, currently lactose cannot be designated as prebiotic. However, recognition of the multiple dosage and host-dependent roles of lactose as digestible and glycemic carbohydrate, as a FODMAP with possible adverse health effects, and potentially as a health beneficial prebiotic carbohydrate will allow its use as functional component in personalized nutrition.

References

Andersen, J. M., Barrangou, R., Abou Hachem, M., Lahtinen, S., Goh, Y. J., Svensson, B., & Klaenhammer, T. R. (2011). Transcriptional and functional analysis of galactooligosaccharide uptake by lacS in *Lactobacillus acidophilus*. *Proceedings of the National Academy of Science of the United States of America*, *108*, 17785–17790.

Azcarate-Peril, M. A., Ritter, A. J., Savaiano, D., Monteagudo-Mera, A., Anderson, C., Magness, S., & Klaenhammer, T. R. (2017). Impact of short-chain galactooligosaccharides on the gut microbiome of lactose-intolerant individuals. *Proceedings of the National Academy of Science of the United States of America*, *114*, E367–E375.

Bindels, L. B., Delzenne, N. M., Cani, P. D., & Walter, J. (2015). Towards a more comprehensive concept for prebiotics. *Nature Reviews Gastroenterology and Hepatology*, *12*, 303–310.

Bonder, M. J., Kurilshikov, A., Tigchelaar, E. F., Mujagic, Z., Imhann, F., Vila, A. V., et al. (2016). The effect of host genetics on the gut microbiome. *Nature Genetics*, *48*, 1407–1412.

Booijink, C. C., El-Aidy, S., Rajilić-Stojanović, M., Heilig, H. G., Troost, F. J., Smidt, H., ... Zoetendal, E. G. (2010). High temporal and inter-individual variation detected in the human ileal microbiota. *Environmental Microbiology*, *12*, 3213–3227.

Byrne, C. S., Chambers, E. S., Morrison, D. J., & Frost, G. (2015). The role of short chain fatty acids in appetite regulation and energy homeostasis. *International Journal of Obesity (London)*, *39*, 1331–1338.

Canfora, E. E., Jocken, J. W., & Blaak, E. E. (2015). Short-chain fatty acids in control of body weight and insulin sensitivity. *Nature Reviews Endocrinology*, *11*, 577–591.

Carmody, R. N., Gerber, G. K., Luevano, J. M., Jr., Gatti, D. M., Somes, L., Svenson, K. L., et al. (2015). Diet dominates host genotype in shaping the murine gut microbiota. *Cell Host Microbe*, *17*, 72–84.

Chassard, C., & Lacroix, C. (2013). Carbohydrates and the human gut microbiota. *Current Opinion in Clinical Nutrition and Metabolic Care*, *16*, 453–460.

Chen, X. Y., & Gänzle, M. G. (2017). Lactose and lactose-derived oligosaccharides: More than prebiotics? *International Dairy Journal*, *67*, 61–72.

Cherbut, C., Ferrier, L., Rozé, C., Anini, Y., Blottière, H., Lecannu, G., et al. (1998). Short-chain fatty acids modify colonic motility through nerves and polypeptide YY release in the rat. *American Journal of Physiology*, *275*, G1415–G1422.

Cho, I., & Blaser, M. J. (2012). The human microbiome: At the interface of health and disease. *Nature Reviews Genetics*, *13*, 260–270.

Codex Alimentarius Commission. (2008). *Report of the 30th session of the codex committee on nutrition and foods for special dietary uses*. Cape Town, South Africa: ALINORM.

Collins, S. M., Surette, M., & Bercik, P. (2012). The interplay between the intestinal microbiota and the brain. *Nature Reviews Microbiology*, *10*, 735–742.

Coste, I., Judlin, P., Lepargneur, J. P., & Bou-Antoun, S. (2012). Safety and efficacy of an intravaginal prebiotic gel in the prevention of recurrent bacterial vaginosis: A randomized double-blind study. *Obstetrics and Gynecology International*, *2012*, 147867. Available from https://doi.org/10.1155/2012/147867.

Cummings, J. H., & Stephen, A. M. (2007). Carbohydrate terminology and classification. *European Journal of Clinical Nutrition*, *61*(Suppl. 1), S5–S18.

Davis, L. M., Martínez, I., Walter, J., & Hutkins, R. (2010). A dose dependent impact of prebiotic galactooligosaccharides on the intestinal microbiota of healthy adults. *International Journal of Food Microbiology*, *144*, 285–292.

De Giorgio, R., Volta, U., & Gibson, P. R. (2016). Sensitivity to wheat, gluten and FODMAPs in IBS: Facts or fiction? *Gut*, *65*, 169–178.

EU regulation 1169/2011 (2011). *Official Journal of the European Union* (2011), L. 304, 18–63.

FDA. (2016). Food and Drug Administration, Revision of the Nutrition and Supplement Facts Label, United States Federal Register (2016), Vol. 81, No. 103.

Ferreira-Lazarte, A., Olano, A., Villamiel, M., & Moreno, F. J. (2017). Assessment of in vitro digestibility of dietary carbohydrates using rat small intestinal extract. *Journal of Agricultural and Food Chemistry*, *65*, 8046–8053.

Florent, C., Flourie, B., Leblond, A., Rautureau, M., Bernier, J. J., & Rambaud, J. C. (1985). Influence of chronic lactulose ingestion on the colonic metabolism of lactulose in man (an in vivo study). *Journal of Clinical Investigation*, *75*, 608–613.

Fox, P. F., Guinee, T. P., Cogan, T. M., & McSweeney, P. L. (2017). *Whey and whey products. Fundamentals of cheese science* (2nd ed., pp. 755–769)New York, NY: Springer United States.

Francavilla, R., Calasso, M., Calace, L., Siragusa, S., Ndagijimana, M., Vernocchi, P., ... De Angelis, M. (2012). Effect of lactose on gut microbiota and metabolome of infants with cow's milk allergy. *Pediatric Allergy and Immunology*, *23*, 420–427.

Gänzle, M. G. (2012). Enzymatic synthesis of galactooligosaccharides and other lactose derivatives (hetero-oligosaccharides) from lactose. *International Dairy Journal*, *22*, 116–122.

Gänzle, M. G., & Follador, R. (2012). Metabolism of oligosaccharides in lactobacilli: A review. *Frontiers in Microbiology*, *3*, 340.

Gänzle, M. G., Haase, G., & Jelen, P. (2008). Lactose—crystallisation, hydrolysis and value-added derivatives. *International Dairy Journal*, *18*, 685–694.

Gerbault, P., Liebert, A., Itan, Y., Powell, A., Currat, M., Burger, J., . . . Thomas, M. G. (2011). Evolution of lactase persistence: An example of human niche construction. *Philosophical Transactions of the Royal Society London B. Biological Sciences.*, *366*, 863–877.

Gopal, P. K., Sullivan, P. A., & Smart, J. B. (2001). Utilisation of galacto-oligosaccharides as selective substrates for growth by lactic acid bacteria including *Bifidobacterium lactis* DR10 and *Lactobacillus rhamnosus* DR20. *International Dairy Journal*, *11*, 19–25.

Guo, Q., Goff, H. D., & Cui, S. W. (2017). Structural characterisation of galacto-oligosaccharides (VITAGOS™) synthesized by transgalactosylation of lactose. *Bioactive Carbohydrates and Dietary Fibre*. Available from https://doi.org/10.1016/j.bcdf.2017.07.007.

Hamaker, B. R., & Tuncil, Y. E. (2014). A perspective on the complexity of dietary fiber structures and their potential effect on the gut microbiota. *Journal of Molecular Biology*, *426*, 3838–3850.

Hammer, H. F., & Hammer, J. (2012). Diarrhea caused by carbohydrate malabsorption. *Gastroenterology Clinics of North America*, *41*, 611–627.

He, T., Venema, K., Priebe, M. G., Welling, G. W., Brummer, R. J., & Vonk, R. J. (2008). The role of colonic metabolism in lactose intolerance. *European Journal of Clinical Investigation*, *38*, 541–547.

Health Canada. (2012). Policy for labelling and advertising of dietary fibre-containing food products. <https://www.canada.ca/en/health-canada/services/food-nutrition/legislation-guidelines/policies/policy-labelling-advertising-dietary-fibre-containing-food-products-2012.html>.

Henström, M., Diekmann, L., Bonfiglio, F., Hadizadeh, F., Kuech, E. M., von Köckritz-Blickwede, M., et al. (2018). Functional variants in the sucrase-isomaltase gene associate with increased risk of irritable bowel syndrome. *Gut*, *67*, 263–270.

Hernández-Hernández, O., Marín-Manzano, M. C., Rubio, L. A., Moreno, F. J., Sanz, M. L., & Clemente, A. (2012). Monomer and linkage type of galacto-oligosaccharides affect their resistance to ileal digestion and prebiotic properties in rats. *Journal of Nutrition*, *142*, 1232–1239.

Hoeflinger, J. L., Davis, S. R., Chow, J., & Miller, M. J. (2015). In vitro impact of human milk oligosaccharides on Enterobacteriaceae growth. *Journal of Agricultural and Food Chemistry*, *63*, 3295–3302.

Holtug, K., Clausen, M. R., Hove, H., Christiansen, J., & Mortensen, P. B. (1992). The colon in carbohydrate malabsorption: Short-chain fatty acids, pH, and osmotic diarrhoea. *Scandinavian Journal of Gastroenterology*, *27*, 545–552.

Hu, Y., Winter, V., Chen, X. Y., & Gänzle, M. G. (2017). Effect of acceptor carbohydrates on oligosaccharide- and polysaccharide synthesis by dextransucrase DsrM from *Weissella cibaria*. *Food Research International*, *99*, 603–611.

Hutkins, R. W., Krumbeck, J. A., Bindels, L. B., Cani, P. D., Fahey, G., Jr., Goh, Y. J., . . . Sanders, M. E. (2016). Prebiotics: Why definitions matter. *Current Opinion in Biotechnology*, *37*, 1–7.

Ito, M., & Kimura, M. (1993). Influence of lactose on faecal microflora in lactose maldigestors. *Microbial Ecology in Health and Disease*, *6*, 73–76.

Ito, M., Deguchi, Y., Miyamori, A., Matsumoto, K., Kikuchi, H., Matsumoto, K., et al. (1990). Effects of administration of galactooligosaccharides on the human faecal microflora, stool weight and abdominal sensation. *Microbial Ecology in Health and Disease*, *3*, 285–292.

James, K., Motherway, M. O., Bottacini, F., & van Sinderen, D. (2016). *Bifidobacterium breve* UCC2003 metabolises the human milk oligosaccharides lacto-*N*-tetraose and lacto-*N*-neo-tetraose through overlapping, yet distinct pathways. *Scientific Reports*, *6*, 38560.

Jelen, P. (1979). Industrial whey processing technology: An overview. *Journal of Agricultural and Food Chemistry*, *27*, 658–661.

Kelly, C. J., Zheng, L., Campbell, E. L., Saeedi, B., Scholz, C. C., Bayless, A. J., et al. (2015). Crosstalk between microbiota-derived short-chain fatty acids and intestinal epithelial HIF augments tissue barrier function. *Cell Host Microbe*, *17*, 662–671.

Kien, C. L., McClead, R. E., & Cordero, L., Jr. (1998). Effects of lactose intake on lactose digestion and colonic fermentation in preterm infants. *The Journal of Pediatrics*, *133*, 401–415.

Kim, M. H., Kang, S. G., Park, J. H., Yanagisawa, M., & Kim, C. H. (2013). Short-chain fatty acids activate GPR41 and GPR43 on intestinal epithelial cells to promote inflammatory responses in mice. *Gastroenterology*, *145*, 396–406.

Kolars, J. C., Levitt, M. D., Aouji, M., & Savaiano, D. A. (1984). Yogurt—an autodigesting source of lactose. *New England Journal of Medicine*, *310*, 1–3.

Korpela, L. (2018). Diet, microbiota, and metabolic health: Trade-off between saccharolytic and proteolytic fermentation. *Annual Reviews in Food Science and Technology*, *9*, 65–84.

Krumbeck, J. A., Rasmussen, H. E., Hutkins, R. H., Clarke, J., Shawron, K., Keshavarzian, A., et al. (2018). Probiotic *Bifidobacterium* strains and galactooligosaccharides improve intestinal barrier function in obese adults but show no synergism when used together as synbiotics. *Microbiome*, *2*, 121.

Laatikainen, R., Koskenpato, J., Hongisto, S. M., Loponen, J., Poussa, T., Hillilä, M., & Korpela, R. (2016). Randomised clinical trial: Low-FODMAP rye bread vs. regular rye bread to relieve the symptoms of irritable bowel syndrome. *Alimentary Pharmacology & Therapeutics*, *44*, 460–470.

Latulippe, M. E., & Skoog, S. M. (2011). Fructose malabsorption and intolerance: Effects of fructose with and without simultaneous glucose ingestion. *Critical Reviews in Food Science and Nutrition*, *51*, 583–592.

Lifschitz, C. H., O'Brian Smith, E., & Garza, C. (1983). Delayed complete lactase sufficiency in breast-fed infants. *Journal of Pediatric Gastroenterology and Nutrition*, *2*, 478–482.

Lin, J. (2004). Too much short chain fatty acids cause neonatal necrotizing enterocolitis. *Medical Hypotheses*, *62*, 291–293.

Louis, P., Flint, H. J., & Michel, C. (2016). How to manipulate the microbiota: Prebiotics. *Advances in Experimental Medical Biology*, *902*, 119–142.

Louis, P., Scott, K. P., Duncan, S. H., & Flint, H. J. (2007). Understanding the effects of diet on bacterial metabolism in the large intestine. *Journal of Applied Microbiology*, *102*, 1197–1208.

Macfarlane, G. T., Steed, H., & Macfarlane, S. (2008). Bacterial metabolism and health-related effects of galacto-oligosaccharides and other prebiotics. *Journal of Applied Microbiology*, *104*, 305–344.

Mäkivuokko, H. A., Saarinen, M. T., Ouwehand, A. C., & Rautonen, N. E. (2006). Effects of lactose on colon microbial community structure and function in a four-stage semi-continuous culture system. *Bioscience, Biotechnology, and Biochemistry*, *70*, 2056–2063.

Martínez, I., Kim, J., Duffy, P. R., Schlegel, V. L., & Walter, J. (2010). Resistant starches types 2 and 4 have differential effects on the composition of the fecal microbiota in human subjects. *PLoS One.*, *5*, e15046.

Maathuis, A. J., van den Heuvel, E. G., Schoterman, M. H., & Venema, K. (2012). Galacto-oligosaccharides have prebiotic activity in a dynamic in vitro colon model using a ^{13}C-labeling technique. *Journal of Nutrition*, *142*, 1205–1212.

Matsuki, T., Yahagi, K., Mori, H., Matsumoto, H., Hara, T., Tajima, S., ... Kurokawa, K. (2016). A key genetic factor for fucosyllactose utilization affects infant gut microbiota development. *Nature Communications*, *7*, 11939.

Mego, M., Manichanh, C., Accarino, A., Campos, D., Pozuelo, M., Varela, E., ... Azpiroz, F. (2017). Metabolic adaptation of colonic microbiota to galactooligosaccharides: A proof of concept study. *Alimentary Pharmacology and Therapeutics*, *45*, 670–680.

Murray, K., Wilkinson-Smith, V., Hoad, C., Costigan, C., Cox, E., Lam, C., ... Spiller, R. C. (2014). Differential effects of FODMAPs (fermentable oligo-, di-, mono-saccharides and polyols) on small and large intestinal contents in healthy subjects shown by MRI. *American Journal of Gastroenterology*, *109*, 110–119.

O'Connell Motherway, M., Kinsella, M., Fitzgerald, G. F., & van Sinderen, D. (2013). Transcriptional and functional characterization of genetic elements involved in galacto-oligosaccharide utilization by *Bifidobacterium breve* UCC2003. *Microbial Biotechnology*, *6*, 67–79.

Oku, T., & Nakamura, S. (2007). Threshold for transitory diarrhea induced by ingestion of xylitol and lactitol in young male and female adults. *Journal of Nutritional Science and Vitaminology (Tokyo)*, *53*, 13–20.

Oku, T., & Nakamura, S. (2002). Digestion, absorption, fermentation, and metabolism of functional sugar substitutes and their available energy. *Pure and Applied Chemistry*, *74*, 1253–1261.

Paganini, D., Uyoga, M. A., Kortman, G. A. M., Cercamondi, C. I., Moretti, D., Barth-Jaeggi, T., et al. (2017). Prebiotic galacto-oligosaccharides mitigate the adverse effects of iron fortification on the gut microbiome: A randomised controlled study in Kenyan infants. *Gut*, *66*, 1956–1967.

Portune, K. J., Beaumont, M., Davila, A.-M., Tomé, D., Blachier, F., & Sanz, Y. (2016). Gut microbiota role in dietary protein metabolism and health-related outcomes: The two sides of the coin. *Trends in Food Science & Technology*, *57*, 213–232.

Postler, T. S., & Ghosh, S. (2017). Understanding the holobiont: How microbial metabolized affect human health and shape the immune system. *Cell Metabolism*, *26*, 110–130.

Reichardt, N., Vollmer, M., Holtrop, G., Farquharson, F. M., Wefers, D., Bunzel, M., ... Louis, P. (2018). Specific substrate-driven changes in human faecal microbiota composition contrast with functional redundancy in short-chain fatty acid production. *The ISME Journal*, *12*, 610–622.

Russell, W. R., Hoyles, L., Flint, H. J., & Dumas, M.-E. (2014). Colonic bacterial metabolites and human health. *Current Opinion in Microbiology*, *16*, 246–254.

Sakata, T. (1987). Stimulatory effect of short-chain fatty acids on epithelial cell proliferation in the rat intestine: A possible explanation for trophic effects of fermentable fibre, gut microbes and luminal trophic factors. *British Journal of Nutrition*, *58*, 95–103.

Savaiano, D. A., Ritter, A. J., Klaenhammer, T. R., James, G. M., Longcore, A. T., Chandler, J. R., et al. (2013). Improving lactose digestion and symptoms of lactose intolerance with a novel galacto-oligosaccharide (RP-G28): A randomized, double-blind clinical trial. *Nutrition Journal*, *12*, 160. Available from https://doi.org/10.1186/1475-2891-12-160.

Schwab, C., & Gänzle, M. G. (2011). Lactic acid bacteria fermentation of human milk oligosaccharide components, human milk oligosaccharides and galactooligosaccharides. *FEMS Microbiology Letters*, *315*, 141–148.

Sela, D. A., Chapman, J., Adeuya, A., Kim, J. H., Chen, F., Whitehead, T. R., et al. (2008). The genome sequence of *Bifidobacterium longum* subsp. *infantis* reveals adaptations for milk utilization within the infant microbiome. *Proceedings of the National Academy of Science of the United States*, *105*, 18964–18969.

Sela, D. A., & Mills, D. A. (2010). Nursing our microbiota: Molecular linkages between bifidobacteria and milk oligosaccharides. *Trends in Microbiology*, *18*, 298–307.

Smilowitz, J. T., Lebrilla, C. B., Mills, D. A., German, J. B., & Freeman, S. L. (2014). Breast milk oligosaccharides: Structure-function relationships in the neonate. *Annual Review of Nutrition*, *34*, 143–169.

Smith, P. M., Howitt, M. R., Panikov, N., Michaud, M., Gallini, C. A., Bohlooly, Y. M., et al. (2013). The microbial metabolites, short-chain fatty acids, regulate colonic Treg cell homeostasis. *Science*, *341*, 569–573.

Sonnenburg, E. D., & Sonnenburg, J. L. (2014). Starving our microbial self: The deleterious consequences of a diet deficient in microbiota-accessible carbohydrates. *Cell Metabolism*, *20*, 779–786.

Staudacher, H. M., & Whelan, K. (2016). Altered gastrointestinal microbiota in irritable bowel syndrome and its modification by diet: Probiotics, prebiotics and the low FODMAP diet. *Proceedings of the Nutritional Society*, *75*, 306–318.

Szilagyi, A., Shrier, I., Heilpern, D., Je, J., Park, S., Chong, G., ... Lee, B. (2010). Differential impact of lactose/lactase phenotype on colonic microflora. *Canadian Journal of Gastroenterology*, *24*, 373–379.

Thangaraju, M., Cresci, G. A., Liu, K., Ananth, S., Gnanaprakasam, J. P., Browning, D. D., et al. (2009). GPR109A is a G-protein-coupled receptor for the bacterial fermentation product butyrate and functions as a tumor suppressor in colon. *Cancer Research*, *69*, 2826–2832.

Tolhurst, G., Heffron, H., Lam, Y. S., Parker, H. E., Habib, A. M., Diakogiannaki, E., et al. (2012). Short-chain fatty acids stimulate glucagon-like peptide-1 secretion via the G-protein-coupled receptor FFAR2. *Diabetes*, *61*, 364–371.

Torres, D. P. M., do Pilar, M., Gonçalves, F., Teixeira, J. A., & Rodrigues, L. R. (2010). Galacto-oligosaccharides: Production, properties, applications, and significance as prebiotics. *Comprehensive Reviews in Food Science and Food Safety*, *9*, 438–454.

Uribe, M., Toledo, H., Perez, F., Vargas, F., Gil, S., Garcia-Ramos, G., et al. (1987). Lactitol, a second-generation disaccharide for treatment of chronic portal-systemic encephalopathy. A double-blind, crossover, randomized clinical trial. *Digestive Diseases and Sciences*, *32*, 1345–1353.

van Leeuwen, S. S., Kuipers, B. J., Dijkhuizen, L., & Kamerling, J. P. (2016). Comparative structural characterization of 7 commercial galacto-oligosaccharide (GOS) products. *Carbohydrate Research*, *425*, 48–58.

Venema, K. (2012). Intestinal fermentation of lactose and prebiotic lactose derivatives, including human milk oligosaccharides. *International Dairy Journal*, *22*, 123–140.
Vogt, S. L., & Finlay, B. B. (2017). Gut microbiota-mediated protection against diarrheal infections. *Journal of Travel Medicine*, *24*(Suppl. 1), S39–S43.
Wutzke, K. D., & Scholübbers, D. (2013). The metabolic effect of resistant starch and yoghurt on the renal and faecal nitrogen and ammonia excretion in humans as measured by lactose-[$^{15}N_2$]ureide. *Isotopes in Environmental and Health Studies*, *49*, 464–470.
Yan, Y. L., Hu, Y., & Gänzle, M. G. (2018). Prebiotics, FODMAPs and dietary fibre–conflicting concepts in development of functional food products? *Current Opinion in Food Science*, *20*, 30–37.
Zhang, J., Empl, M. T., Schwab, C., Fekry, M. I., Engels, C., Schneider, M., et al. (2017). Gut microbial transformation of the dietary imidazoquinoxaline mutagen MeIQx reduces its cytotoxic and mutagenic potency. *Toxicological Science*, *159*, 266–276.
Zoetendal, E. G., Raes, J., van den Bogert, B., Arumugam, M., Booijink, C. C., Troost, F. J., … Kleerebezem, M. (2012). The human small intestinal microbiota is driven by rapid uptake and conversion of simple carbohydrates. *The ISME Journal*, *6*, 1415–1426.

Further reading

Fukuda, S., Toh, H., Hase, K., Oshima, K., Nakanishi, Y., Yoshimura, K., et al. (2011). Bifidobacteria can protect from enteropathogenic infection through production of acetate. *Nature*, *469*, 543–547.

CHAPTER

Application of lactose in the pharmaceutical industry

5

Gerald A. Hebbink and Bastiaan H.J. Dickhoff
DFE Pharma, Wageningen, The Netherlands

5.1 Introduction

Lactose is used as an excipient in the pharmaceutical industry. It is one of the most common of all excipients, present in 60%–70% of registered oral solid dose formulations. Historically, an excipient was defined as an inert substance or ingredient used to prepare pharmaceutical dosage forms (Pifferi & Restani, 2003). Today, however, it is understood that excipients are more than inert substances, as they also add functionality to the formulation and dosage form. Excipients are now used to facilitate the production of the optimal dosage form, while they are needed to take the active ingredient in a drug to exactly where they are needed in a patient's body.

The excipient should be safe to use, which means the materials used need to be inherently, and they also needed to be treated and controlled carefully. Processes used to produce the excipient, for example, must prioritize safety by controlling impurities and ensuring that the product is stable. An excipient should be widely and easily available to prevent problems in drug manufacture caused by a shortage of excipients. And an excipient should be cost-effective, to keep drug dosage forms as affordable as possible.

The most important applications of lactose in pharmaceuticals can be found in oral solid dosage formulations, like tablets and inhalation. The web-based RxList was consulted to identify and quantify the use of the most common forms of excipient in tablet and capsule formulations (Fig. 5.1).

Lactose and microcrystalline cellulose (MCC) are the most widely used for registered tablet formulations, between them covering around 60% of all drug products available in the RXList.com. About 45% of drug product formulations are made with various combinations of lactose and MCC. The other most commonly found excipients are the superdisintegrants: sodium starch glycolate and croscarmellose sodium. In general, these are only present in small amounts (around 10%) in a formulation and have a clear defined role, which is ensuring disintegration of the dosage form to make the active ingredient more easily available.

Lactose. DOI: https://doi.org/10.1016/B978-0-12-811720-0.00005-2

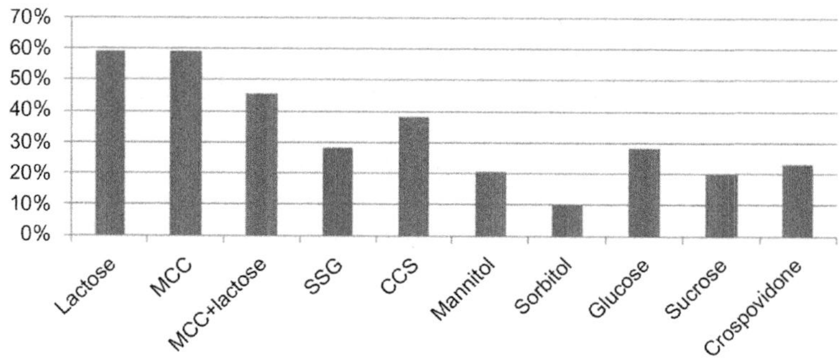

FIGURE 5.1

Excipient usage in tablets and capsules, according RXList 2017.[1]

There are several types of dosage form, each of them posing different challenges in production and usage.

Lactose in pharmaceutics is normally found in oral solid dosage forms, including tablets, capsules, and sachets, and in dry powder inhalers (DPIs). Excipients can be used in oral solid dosage forms as fillers and binders, disintegrants, lubricants, and flow agents. Lactose has an important role as a filler/binder: it fills a dosage form, enabling the formulation to flow more easily, as well, and it provides the binding needed in manufacturing tablets and compacts.

The major pulmonary inhalation dosage forms are pressurized metered dosage inhalators (pMDIs) and DPIs. In both cases, an excipient is needed both to prepare the dosage form and efficiently deliver the drug to exactly where it is needed.

The pharmaceutical industry has several methods for preparing solid dosage formulations. These approaches are all concerned with processing the individual powdery ingredients, converting them into the final dosage form, such as a tablet.

The oldest and most common technique is wet granulation (WG). Powders are mixed with a liquid to form a dough, which is then processed further by drying and milling. This leaves the final powder, which is then ready for use in a dosage form. This processing method involves a whole series of steps, and that makes the process relatively costly. There is another concern with this process, which is the stability of the active pharmaceutical ingredient (API) when it interacts with the liquid. For WG, ingredients must be stable all the way through the process, during every step, while the final wet-granulated powder must be capable of being tableted. Where lactose is concerned, materials fall into two categories: milled or sieved grades, with the finer grades having the highest tableting capacity. The flow of lactose is of less interest here, because during granulation the consistency of the powder changes anyway, so the optimal flow can be achieved by good process design.

[1] www.RXlist.com, a medical online source that lists information on US prescription medications.

An alternative, faster tableting approach is direct compression (DC). This involves only one step in blending the dry ingredients before moving directly to tablet manufacture. This method is simpler and less-costly than WG but it does have a drawback. There is sometimes a concern that the active ingredient, the drug itself, may not be sufficiently or evenly dispersed throughout the powder. This can lead to uneven spread of the drug in the powder, resulting in dosages, that is, tablets, having different content levels.

Lactose used in this process needs to have a good flow to counteract this potential issue and ensure even spread of the drug. In addition, the lactose used should be capable of forming a stable blend with a drug. Finally, during tablet production, the lactose should have good tableting capability to form stable blends when compacted. Special types of lactose have been developed for this purpose. The functional properties of these types of lactose need to balance between processing characteristics, such as powder flow and tableting capabilities, like compactibility.

A third method is dry granulation (DG) or roller compaction (RC) of a formulation. In this technique a blend is first compacted by pressing it in-between two rollers, followed by breaking or milling of the formed, compacted material. Excipients must have the ability to be recompacted, which means the powder obtained after the first roll-compaction process should still be capable of being compacted into the final dosage form.

Because there are several different processing techniques in the pharmaceutical industry, different grades of lactose have been developed to fit them, and a list of lactose types and suppliers can be found in Table 5.1.

Lactose is a versatile excipient that can be used flexibly for the development of different products. The most widely used forms being crystalline α-lactose

Table 5.1 Overview of global and local pharmaceutical lactose suppliers.

	Manufacturer	Milled and sieved	Spray-dried	Granulated	Anhydrous	Inhalation quality
Global suppliers	DFE Pharma	x	x	x	x	x
	Kerry-Sheffield	x	x		x	x
	Meggle	x	x	x	x	x
Local suppliers	Alpavit	x		x		
	Armor	x		x		x
	BASF	x	x			
	Bhole Baba	x				
	Dawning	x	x			
	Danone	x				
	Lactose India	x				
	Milkaut	x				

monohydrate and crystalline anhydrous lactose, consisting mainly of β-lactose. Other forms of lactose are either amorphous forms, which are generated by rapid drying of a lactose solution or milling of lactose, or different polymorphic forms of the α-lactose molecule. In Section 5.2 we will describe the different forms of lactose and their production in more detail.

Many specific excipient properties are directly related to the chemical and surface properties of the different excipients, and in particular to their interaction with moisture. The ability to react with moisture is an important requirement for excipients in the pharmaceutical industry. The crystalline α-monohydrate and β-anhydrous forms are not hygroscopic and have a very limited ability to interact with moisture. Amorphous lactose, on the other hand, is hygroscopic and dissolves easily in liquids.

Both α-monohydrate and β-anhydrous are commercially available in different forms. The properties of those materials are strongly related to the chemical and physical properties of their surface and material structure. Other lactose forms are not commercially available in their pure forms, but small fractions of them at surfaces of commercial forms play a significant role in the properties of the materials.

A wide range of commercial grade lactose exists to suit the many different applications and production methods. Crystallization of α-monohydrate lactose is achieved by double crystallization from a supersaturated aqueous solution below 93.5°C, followed by removal of liquids and drying of the remaining material. The result is crystalline lactose with particles sizes ranging from small to large, with a variety of shapes. The crystalline form of α-lactose monohydrate typically has a tomahawk shape, though processing conditions have a profound effect and a distribution of different forms is obtained as well.

The material obtained is milled, sieved, or goes through a combination of both processes to develop the plethora of products with varying particle size distributions (PSD), which can be extremely useful for WG processes.

For DPI applications, the most common lactose products are also milled and sieved grades. Lactose used for DPI has to meet stringent requirements. Because drugs are delivered directly to the lungs, microbiological requirement is very strict to avoid the infection of the patient. Another important parameter in DPI is PSD. This is an important factor in determining the flow of a material and in delivering the drug. Variations in surface properties also affect DPI. All these parameters should be understood and carefully managed to provide safe and predictable drug delivery.

Spray drying can deliver a product with excellent flow and high compactibility, achieved by using suspensions of a fine grade of milled α-monohydrate lactose. Good flow properties are due to the size and shape of the obtained particles, while high compactibility is caused by a combination of the fine grade of milled lactose with a ductile amorphous fraction, generated by the rapid drying of the liquid suspension in lactose solution. Small particles give stronger materials after compacting and this, in combination with the binding capacity of the amorphous fraction, makes it possible to manufacture strong tablets. This type of product leads to a stable blend with uniform distribution of the active ingredient. That is because the product is porous, which improves the binding of drug particles and good dispersion and stability.

Crystallization of a lactose solution above 93.5°C results in the formation of β-lactose crystals. On a commercial scale this is achieved by drum drying, also referred to as roller drying. This is a rapid process resulting in a coarse material, built up from many small β-lactose crystals in combination with an anhydrous form of α-lactose and mixed crystal forms of α- and β-lactose. Due to the polymorphism and the presence of many small particles, the material performs extremely well in DC processes, while the material is also recompactible, making it the lactose of choice for RC applications.

In DC applications, lactose can provide all the requirements for compaction processes on its own.

Developing a coarse material with fine primary particles can also be achieved through agglomeration or granulation. A fine spray of a lactose solution is sprayed on a fine grade of lactose in, for example, a fluid bed or a high-shear granulator. The resulting product does not include an amorphous fraction, but is still able to form stable compacted materials. Due to the way it is produced, the product is porous, giving it the ability to disperse drug products very effectively, while leading to a stable product.

The market for pharmaceutical grade lactose is currently dominated by just a couple of global suppliers (Table 5.2).

Table 5.2 General recommended usage of lactose in oral solid dosages forms.

	Tablets			Other solid dosage forms		
	Wet granulation	Dry granulation	Direct compression	Capsules	Sachets	Spheres
Milled lactose	+++	+	o	+	o	+++
Sieved lactose	o	o	+	+++	+++	o
Spray-dried lactose	+	o	+++	++	+++	+
Granulated lactose	+	o	+++	+++	+++	+
Anhydrous lactose	+	+++	+++	+++	+	+

+++ Highly recommended.
++ Recommended.
+ Possible but not recommended.
o Not advised.
Dry granulation in this overview includes roller compaction and slugging. Spheres in this overview are made by extrusion- spheronization.

The use of lactose in pharmaceutical dosage forms leads to one or two concerns, which are mainly related to the compatibility of lactose with certain APIs, such as materials containing amine. Reactions between these materials require the presence of a solubilizing medium, like water, and heat to overcome the prevalent energy barriers of the reaction. Lactose is a nonhygroscopic material, which makes the presence of water unlikely. This factor, in combination with prevention of heat intensive steps, means this concern can be mitigated. Several successful formulations of amine-containing APIs with lactose exist including, for example, formulations with the amines trandolapril (Roumeli et al., 2013), pregabalin (Lovdahl, Hurley, Tobias, & Priebe, 2002), and alendronate.

Second, concerns related to lactose intolerance are often raised. Approximately 67% of the world population has reduced lactase levels, sometime after weaning (see Chapters 1–3). After consumption of lactose by a person lacking lactase activity, the lactose is fermented in the gut by the gut flora as a natural process as is described in detail in Chapter 4, Lactose—a conditional prebiotic? When there is a high amount of lactose, this might result in the typical intolerance symptoms, such as flatulence and painful stomach cramps (Chapter 3: Lactose intolerance and other related food sensitivities).

It is generally recognized and accepted that a daily intake of 10–12 g does not result in physical problems. To put this into perspective, if a glass of milk (250 mL) delivers the total acceptable dose of lactose (12 g), two tablets that use lactose as an excipient deliver between 0.2 and 1.4 g of lactose, which is at most 12% of the acceptable limit and normally much less (Silanikove, Leitner, & Merin, 2015).

Drug safety is a critically important requirement, as patients should be able to take a drug without fear of side effects, receiving only the prescribed, calculated benefits of the drug, itself. As excipients form a large and essential part of any drug formulation, it is important to guarantee the safety of these materials, as well. Lactose production and products are therefore continuously tested for quality and purity. Many of these tests and requirements are described in regulatory documents. Product release and certificates are issued to demonstrate and prove that quality standards are being maintained.

Lactose is used in a large number of applications in many different ways, and is considered a safe excipient that has limited negative interactions with the drug substances, even though there are some issues to consider and manage. In the following sections, the different aspects of lactose production methods, products, and applications will be discussed in more detail.

5.2 Types of lactose and production methods

This section begins with a description of the different forms of lactose available. Before describing commercial manufacture of pharmaceutical grade lactose, we will describe crystallization of pharmaceutical grade α-lactose monohydrate, as all commercial products start with this as its essential raw material.

5.2.1 Introduction

Excipients to be used for pharmaceutical applications must have the right level of quality and also the functional properties needed to make safe and effective dosage forms. Where quality is concerned, the excipient should be safe in use, stable both in terms of its own properties and stable in combination with other materials, such as drugs. A number of functional properties can be defined. Flow of a dosage form is a key parameter in manufacturing many different types of formulation. This is typically a function brought by the excipients, which defines flow rates for the formulation as a whole, and this can be achieved by modification of particle size and shape.

For tableting applications, compactibilty of the formulation is a key parameter. For tableting applications, it has been shown that the smaller the particle size of lactose, the stronger the tablets that can be made (Vromans, De Boer, Bolhuis, Lerk, & Kussendrager, 1986), which is illustrated in Fig. 5.2.

Although a small lactose particle size is beneficial for tableting properties, the flow capability of those powders is limited (Shah, Karde, Ghoroi, & Heng, 2017). The challenge is therefore to create a material that flows well but also possesses good flow tableting properties.

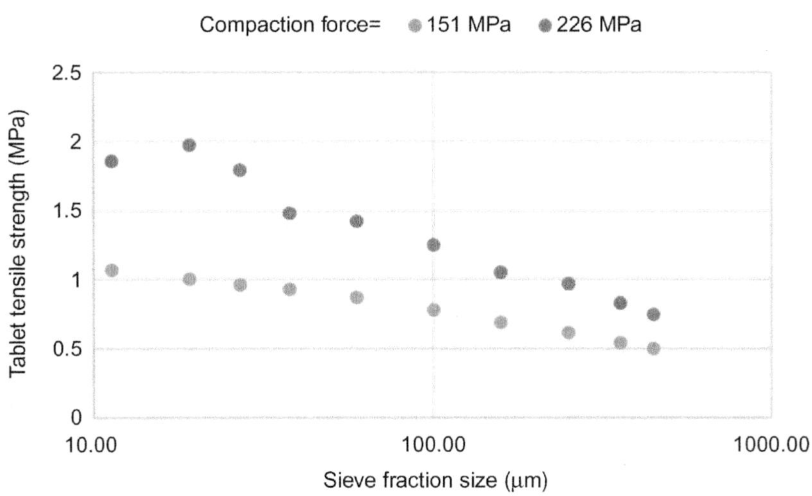

FIGURE 5.2

Correlation of tensile strength with particle size of α-lactose monohydrate (Vromans et al., 1986). The particle sizes in the original report are designated as in-between two sieve size fractions. This has been converted to the geometric mean of upper and lower sieve size.

5.2.2 Lactose polymorphic form and crystallization

Lactose is obtained from whey by concentration and evaporation. This is followed by refining to produce pharmaceutical grade lactose, which complies with quality and regulatory requirements. To comply with specific applications in pharmacy, such as tableting or inhalation applications, postcrystallization processes are required to deliver a specific functionality.

5.2.2.1 Forms of lactose

A number of different lactose forms and polymorphs are known for lactose. The different conditions to achieve the various forms are illustrated in Fig. 5.3

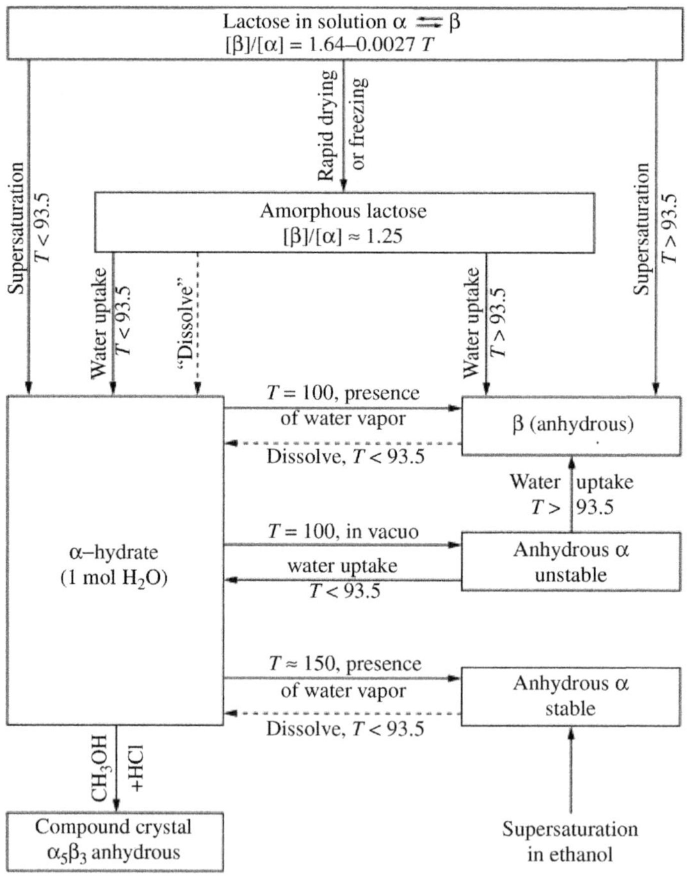

FIGURE 5.3

Conversion scheme for lactose

Taken from Walstra, P., Wouters, J. T. M., Geurts, T. J. (Eds.). (2005). Dairy science & technology *(2nd ed., 782 pp.). Abingdon, Oxford, UK: CRC Press, Taylor & Francis Group.*

(Hourigan, Lifran, Vu, Listiohadi, & Sleigh, 2013; Walstra, Wouters, & Geurts, 2005). This is a conversion scheme that starts with lactose dissolved in water to yield all different known solid forms.

Supersaturation of a lactose solution at temperatures below 93.5°C will result in crystallization of α-lactose monohydrate, as long as sufficient time is given for mutarotation between α- and β-lactose before crystallization. At supersaturation above 93.5°C, anhydrous β-lactose will crystallize. Rapid drying, for example, spray drying, or freeze drying of a lactose solution will result in amorphous lactose. Under the right conditions, this can convert to either α-lactose monohydrate or anhydrous β-lactose. Anhydrous α-lactose forms are obtained by removing water from the α-lactose monohydrate crystal. At relatively low temperatures an unstable type of anhydrous α-lactose is formed and at higher temperature, a stable anhydrous α-lactose is formed. The difference between the two anhydrous α-lactose forms has been proven by several different experimental methods, including crystal form–dependent techniques like X-ray diffraction (Kirk, Dann, & Blatchford, 2007). The industrial production of stable anhydrous α-lactose has been known for a long time already (Sharp, 1940) and involves thermal dehydration of α-lactose monohydrate (Heikonen & Lallukka, 1985; Kussendrager & Andreae, 1984).

5.2.2.2 Industrial crystallization of pharmaceutical grade lactose

Due to the wide variety and natural origin of raw materials (Durham, 2009), there may be wide variations in the properties of lactose products. This creates a challenge for applications that rely strongly on the functionality of the lactose, such as for pharmaceutical applications. Particle size and shape are in general strongly related to functionality, so control of these attributes via crystallization and the downstream processes is extremely important. Pharmaceutical applications also demand strong control on the key quality attributes, including impurities and microbiological contamination.

Pharmaceutical grade lactose is obtained from sour or sweet cheese whey or from cheese whey or milk permeate streams by double crystallization. A general description of this process can be found in the *Ullmann's Encyclopedia of Industrial Chemistry* (Westhoff, Kuster, Heslinga, Pluim, & Verhage, 2014) and a publication by Walstra et al. (2005).

Crystallization starts with concentration of liquid raw materials, for example, by evaporation, to produce a solid material content of 55%–65%. As crystallization cools, it yields a yellowish edible grade of α-lactose monohydrate crystals that are harvested by decantation, centrifugation, and drying. For a white, pharmaceutical grade quality, after centrifugation the lactose is redissolved and then treated with active carbon and filtration to remove impurities and color before recrystallization.

The kinetics and sensitivities of lactose crystallization regarding impurities have been investigated intensively by Visser (1983). He pointed at impurities that have great influence on crystal growth kinetics. These impurities might be intrinsic to lactose, like β-lactose, but related lactose impurities, such as lactose phosphate and minerals from cheese production or milk, also exert an influence on lactose crystallization. This is not relevant for process design and efficiency, as variations in crystal growth kinetics will also influence crystal shape and morphology. This might seem trivial, but particle shapes and surface properties are of great importance in determining the functionality of lactose in pharmaceutical applications by influencing flow characteristics and therefore blending properties.

A lower pH during crystallization was found to be an important factor in accelerating the rate of crystallization (Twieg & Nickerson, 1968). On the other hand, organic acids slowed down the crystallization rate. It was also found that impurities accelerate crystal growth in initial stages of crystallization, whereas in later stages they inhibit the growth. van Kreveld and Michaels (1965) found a number of different shapes as a function of supersaturation, with shapes ranging from the familiar tomahawk to triangular forms.

In conclusion, the initial crystallization process produces the base material that will be used for additional downstream processes, that is, processes that are performed after the initial crystallization like milling or sieving, and eventually for the excipient to be combined with APIs in a dosage form. The quality and functional attributes of these excipients originate in this material and to a great extent are defined by it. Controlling quality and consistency right through the crystallization process is therefore of high importance.

5.2.3 Production of lactose grades by milling and sieving

Traditionally, the milled and sieved lactose grades are the most widely used products in formulating dosage forms. These types of lactose were being used in pharmaceuticals from the early 19th century onward, as demonstrated by descriptions in pharmacopeias and textbooks from that period. One of the old texts refers to the use of lactose as an excipient in the homeopathic formulations. The first British Pharmacopoeia (1864) already contained a lactose (saccharum lactis) and the monograph *A Companion to the British Pharmacopoeia* (Squire, 1866) describes how saccharum lactis could be used by mixing it with children's food or cow's milk to provide a good substitute for mother's milk. The *Organon of Medicine* by Hahnemann describes use of milk sugar for compounding powdered materials. Lactose commercialization was started in the late 19th century as a result of the growing dairy processing industry in the Netherlands, and its need to convert by-products from the cheese industry into sources of additional value. In 1898 "Hollandse Melksuiker" (HMS) company in Uitgeest, the Netherlands, was founded. Lactose was produced from cheese whey, which made it commercially available.

Nowadays, the milled and sieved grades of lactose are the largest products utilized by the pharmaceutical industry in terms of volume. A wide variety of products with different PSD is available, varying from very fine types, in the 5 μm range, to coarse types with, for example, 200–300 μm average particle size.

Milled and sieved grades of lactose are used in many different dosage forms. They are suitable for use in filling sachets and capsules, and also as an excipient in WG and DG of formulations for tableting applications. A tableting formulation needs to be easy to process and form into tablets, and an excipient plays a vital role in achieving this.

Milled and sieved grades of lactose lack either flowability or compactibility for direct application in tablets and capsules. Generally speaking, milled lactose does not flow easily, while sieved lactose cannot be easily compacted. The main use of milled grades is there to be found in WG. The granulation process combines the ability of fine lactose to form tablets with the flow of a relative large granule.

Milling can be regarded as a straightforward process (Naik & Chaudhuri, 2015) in which all sorts of mechanical interactions break up large particles into smaller ones. Many other factors are involved in this, however, such as the input material, process settings, like airflow, and mechanical interactions *between* particles and *of* particles *with* equipment, all of which affect the constitution of the resulting powder. For example, in addition to size reduction, amorphization (Lerk et al., 1984) and formation of other crystal forms can occur (Della Bella, Müller, Soldati, Elviri, & Bettini, 2016; Shariare, de Matas, York, & Shao, 2011) during milling. Two distinct stages in the comminution process were found during ball milling (Pazesh, Gråsjö, Berggren, & Alderborn, 2017) of lactose: first, a comminution stage that was followed by a much slower amorphization stage. The authors of this study hypothesize that lactose fragmentation is the result of compression stress and amorphization is the result of shear stress. Size reduction of lactose under milling should therefore be performed via a process that minimizes shear stresses to prevent induction of amorphous phases.

Sieving is used to create products with specific particle sizes but lacks the comminution process. For sieving, a material is introduced on a mesh screen that is vibrated. This can be done with, for example, ultrasound or simply by shaking the sieves. Sieving is also employed in the analysis of lactose grades, and many milled and sieved products on the market are characterized by sieve fractions.

In general, grades of lactose with finer PSD are produced by milling and more coarse grades by separation with sieves. In DPI applications, however, it is not only the average particle size that plays a role in excipient functionality, but also the presence of a fraction of fines (Steckel, Markefka, TeWierik, & Kammelar, 2006). Sieving not only permits a coarse average particle size to be made, but also specifically removes fine fractions. Combinations of milling and sieving are quite common in lactose production. This allows for production of many grades with strong control on particle size by creating the required size of particle by milling and then collecting them by sieving.

5.2.4 Production of inhalation grade lactose

Another application of milled and sieved lactose is found in DPIs. In this process the functional requirements strongly depend on the type of inhalation and can vary from highly cohesive to free flowing. DPIs are commonly used to treat a wide range of respiratory diseases. Within DPI formulations, the excipient must achieve both quality and functional requirements. The most common types of lactose used in these formulations are actually milled and sieved types, but the required quality and functionality is different from standard types.

In 1864 Newton patented a DPI for delivering potassium chlorate to the lungs and in 1949, the first DPI with lactose as an excipient was brought to the market by Abbott Laboratories (US Patent 2,470,296, 1949). The Intal (sodium cromoglicate) Spinhaler, a DPI, was launched in the late 1960s to treat asthma. That was followed by a number of lactose-based drug formulations in (gelatin) capsule devices, like the Rotahaler and the Cyclohaler as an alternative to pMDIs, which rely on the use of fluorinated propellant, such as chlorofluorocarbons (CFCs).

Until the late 1980s, however, DPIs remained as niche products in an inhalation therapy landscape governed by pMDIs. With the Montreal Protocol of 1987, CFCs were banned and the use of propellants in pMDIs had to be replaced with less environment-unfriendly propellants like HFAs (hydrofluoroalkanes). From that moment on, DPIs gained popularity in the pharmaceutical industry. A huge range of different devices is currently being marketed. These devices differ in many respects: active or passive devices, type and dosage of drug, and way of filling. Lactose was the first excipient to be used in DPIs, and today about 80% of the DPIs on the market contain lactose as the excipient.[2] Lactose is one of the few commonly accepted excipients for inhalation.

Special grades of lactose were developed for this application, with a particular emphasis on quality and functionality requirements. APIs are delivered to the lungs of patients, and the quality requirements for excipients are more stringent than on excipients for oral dosage forms.

The excipient in DPI is needed to prepare and fill devices with blends, followed by delivery of the API to the lung. On the one hand, the lactose needs to form a proper blend with an active ingredient. On the other hand, the blend needs to be weak enough to separate the fine API particles, enabling them to enter the pulmonary tract. For this specific functionality, the surface properties of lactose need to be strictly controlled.

5.2.5 Production of lactose grades for direct compression

The preparation of tablets by DC requires specific functional characteristics from the excipient. It should provide flow and compactibility, but must also guarantee to meet the quality parameters of the dosage form, like content uniformity of the API. This implies effective blending properties to create a good blend and prevent separation. Several lactose grades are specifically recommended for use in DC.

[2] Data obtained from www.Rxlist.com (consulted in 2017), by counting the total amount of registered dry powder inhalation formulations with and without lactose as excipient.

5.2.5.1 Production of spray-dried lactose

Spray-dried lactose was one of the first excipients specifically developed for DC. All global pharmaceutical lactose producers market a number of grades of spray-dried lactose. Spray drying a suspension of fine α-lactose monohydrate particles in water will result in the formation of agglomerates composed of the fine particles derived from the suspension, bonded together by lactose originating from the solution. This product combines excellent flow and compactibility. In addition to that, spray-dried lactose has very good blending properties, while it is also possible to achieve excellent content uniformity. This is caused by the morphology of the product, which is a porous agglomerate that can hold other ingredients of a formulation, like APIs, in the pores.

Spray drying of lactose powders has a long history. It was recognized that spray drying a solution of lactose results in an unstable material that is highly hygroscopic in nature. Improved methods were developed that, for example, include posttreatments of the spray-dried product with moistened air (Peebles & Manning, 1933). The spray drying conditions were modified in such a way that lactose crystallized to the monohydrate form during the process (Peebles & Manning, 1937). This required postdrying of the material to remove excess water.

In the 1970s, Foremost patented (Hutton & Palmer, 1972) a method to produce a spray-dried lactose product by atomizing a suspension of finely divided crystalline lactose. The result was a powder with excellent flow properties consisting of finely divided lactose particles that are cemented or bonded together by a noncrystalline or glassy form of solidified lactose. In the 1980s, DMV, Veghel, the Netherlands, filed a patent (Kussendrager, van den Bigglaar, & Vromans, 1989) which claims control on the amount of amorphous content. Control on amorphous content is necessary to produce a consistent product. The functionality of the amorphous fraction in spray-dried lactose was proven in tableting (Vromans, Bolhuis, Lerk, Van Den Biggelaar, & Bosch, 1987). They prepared a range of lactose samples with an amorphous fraction ranging from 0% to 75% and primary particle size ranging from 1–8 to 32–45 μm.

An amorphous content between 15% and 50% enhanced compactibility by up to a factor of four when compared with either a zero or very high (>50%) level of amorphous content. The effect of particle size within the finely divided lactose on the compaction was reported (Rassu, Eissens, & Bolhuis, 2006).

Two commercial types of spray-dried lactose were compared with each other. These were the standard DCL11 and an improved DCL14, DMV Veghel, the Netherlands, currently produced and marketed by DFE Pharma as SuperTab 11SD and SuperTab 14SD, respectively. The major difference in these two products is the particle size of the original lactose powder. It was shown that granule properties of both materials did not differ significantly, but that tableting provided significantly stronger tablets with DCL14 over DCL11, proving the beneficial effect of a finer particle size.

Spray-dried lactose from different suppliers, Fast Flo 316 (Foremost), FlowLac 100 (Meggle), and SuperTab 11SD (DFE Pharma), were also used in a low-dosage WG application because of their superior blending properties. It was shown that spray-dried lactose gave better content uniformity in the initial blends (Huang et al., 2013) than other types. Due to the versatile functionality of spray-dried lactose, it can be found in many pharmaceutical applications. Spray-dried lactose, Fast Flo 316, Foremost, was used for DG application (Wu & Sun, 2007). They showed that tabletability of dry granulated spray-dried lactose was insensitive to granule size. The main usage of spray-dried lactose is in DC due to its excellent flow, good blending properties, and ability to be tableted.

5.2.5.2 Production of anhydrous lactose grades

Anhydrous lactose has a good flow and excellent tableting and recompaction properties, making it useful in direct compression DG applications. In addition, due to the absence of free and crystal water, anhydrous lactose is traditionally utilized in water-sensitive formulations.

Lactose in solution consists of both anomeric forms: α- and β-lactose. Above 93.5°C, β-lactose has the lowest solubility and will crystallize when the right supersaturation conditions have been met (Walstra et al., 2005). For lactose, these conditions can be achieved in a double drum dryer, also referred to as a roller dryer, with nip feeding as described in a patent by The Dry Milk Company Inc., New York, NY (Supplee & Flanigan, 1934). The general principle of drum drying (Mujumdar, 1995, pp. 203–213) is that two of the interior steam-heated drums are rotated in contrary motion, while feeding-in the solution between the two drums. Evaporation is achieved by contact drying on the drum surface, and the dried product can be scraped from the drums using blades pressed onto the drum surface. Crystallization is achieved rapidly and the resulting product contains typically more than 70% of β-lactose in combination with other forms like anhydrous α-lactose and anhydrous mixed crystal forms. The water content of the product is typically $<0.5\%$ and it is regarded as an anhydrous product. This process produces particles that comprise aggregates of very fine primary crystals, and these are responsible for excellent tableting functionality. In a comparison of anhydrous lactose grades of different suppliers, DMV-Fonterra Excipients, Friesland Foods Domo, and Kerry Bioscience (Gamble et al., 2010), it was found that the greater level of fines in lactose anhydrous NF DT, Kerry Bioscience, did lead to stronger compacts, that is, tablets.

Anhydrous lactose is used in all types of pharmaceutical applications, though primarily for DC and recompaction applications. Anhydrous β-lactose is regarded as the preferred form of lactose for RC (Hein, Picker-Freyer, & Langridge, 2008) because precompaction does not alter the second compaction, that is the actual tableting, as is proven by the unchanged tablet hardness.

5.2.5.3 Production of agglomerated grades of lactose

Agglomerated products were originally developed to obtain a free-flowing material with good tableting properties. Agglomerated lactose is produced using finely milled α-lactose monohydrate, with water or an aqueous solution of lactose as a binder. The result is an agglomerate of finely divided lactose particles, but with a more open and porous structure than the product obtained by spray drying, leading to good flow properties. Several techniques exist for agglomeration, including fluid bed agglomeration and high-shear agglomeration. It has been shown that a low shear process, such as fluid bed agglomeration, helps in delivering good tableting properties (Zuurman, Riepma, Bolhuis, Vromans, & Lerk, 1994).

Meggle (Wasserburg, Germany) was the first company to market an agglomerated form of α-lactose monohydrate, Tablettose (Meggle), and was followed by a product from DMV, DCL15 (Bolhuis & de Waard, 2011). It was observed that these products behave differently in tableting applications (Bolhuis & Zuurman, 1995). Tablets were made from α-lactose monohydrate 100 Mesh, Tablettose, and Pharmatose DCL15. Tablet crushing strength at the same compaction force for DCL15 was twice that of 100 Mesh. Tablet crushing strength for Tablettose lay in-between those two. It was not directly clear which material property caused these differences. It was hypothesized that this was caused by different manufacturing methods for Tablettose versus Pharmatose DCL15. The higher compactability of DCL15 was attributed to the presence of more β-lactose on the surface of this product compared with Tablettose.

More recently, an agglomerated β-lactose was introduced to the market (Kussendrager & Walsma, 2005). SuperTab 24AN (DFE Pharma) was shown to be the best performing for tableting from a series of anhydrous β-lactose types (Velázquez-gonzález, Ramírez-flores, & Villafuerte-robles, 2015). In contrast to spray-dried lactose, no amorphous lactose is found in agglomerated products, as conditions during agglomeration induce crystallization of potential amorphous phases. The use of agglomerated products is found in all types of pharmaceutical dosage forms.

5.2.5.4 Coprocessing of lactose with other excipients

In recent years a number of products that combine excipients, the so-called coprocessed excipients, have come to market (Gohel & Jogani, 2005; Mirani, Patankar, Borole, Pawar, & Kadam, 2011). A coprocessed excipient constitutes of at least two individual components that are processed together (Bolhuis & Armstrong, 2006; Bolhuis & de Waard, 2011) to combine properties of the individual components in ways that cannot be achieved simply by physically mixing them. Essential functional characteristics of excipients are flow and compactability, which are needed for efficient tablet manufacture. Other attributes are drug loading capacity for high dosage drugs or ensuring excellent content uniformity in low-dosage drug formulations. An example of this can be found in a coprocessed formulation consisting of lactose, HPMC (hydroxylpropyl methylcellulose), and PVPP (polyvinylpolypyrrolidone) as disintegrant (Wang et al., 2015). It was found that product performance is not driven only by the individual constituents, the overall production method also had a significant effect on performance. Due to the use of spray drying, up to 30% of amorphous lactose was formed, and this has a major effect on determining the improved properties.

Combinations of lactose with soluble components like sugar alcohols such as xylitol or lactitol, for example SuperTab 40LL (DFE Pharma), have enhanced properties with the improved flow typical for agglomerated products. Other types involve coprocessing lactose with other standard excipients, such as cellulose and cellulose derivatives, like, for example, Microcellac, Cellactose, and Retalac from Meggle. In general, these products are produced by spray drying or other agglomeration techniques like spray fluidization. The purpose of particle engineering and creation of composite materials is to enhance the functionality, like, for example, tabletability, powder flow, or dispersibility, of the individual components (Li, Lin, Shen, Hong, & Feng, 2017; Mangal, Meiser, Morton, & Larson, 2015).

Table 5.3 provides a summary of currently marketed coprocessed excipients that contain lactose as one of the components, together with the claimed functionality benefit.

5.3 Functionality of lactose in pharmaceutical applications

For a long time, excipients were regarded as being inert fillers. A paradigm shift is now taking place, with excipients increasingly seen as functional ingredients that have uses far beyond simply being a filler. Excipients provide functionality to a formulation to process it and deliver the drug to where it is needed. Lot-to-lot excipient variation, and understanding this effect on excipient behavior, is essential for application development (Gamble et al., 2010; Kushner et al., 2014). Most pharmaceutical applications are for oral solid dosage forms, like tablets and capsules, and in dry powder inhalation. In addition to these, lactose monohydrate is used in small amounts for ointments and intravenous therapy solutions (Strickley, 2004). Lactose is normally used because most drugs (API) cannot be compressed directly into tablets as they (1) are too low-dosed (2) and/or lack the proper characteristics to create a compact, (3) and/or do not inhibit lubrication or disintegration properties required for tableting.

5.3.1 Pharmaceutical oral dosage forms: tablets, capsules, and sachets

Different manufacturing processes can be used by the pharmaceutical formulator to manufacture tablets, capsules, and sachets. Oral dosage forms are normally chosen because they are user-friendly, in other words, swallowing a tablet is easier and more pleasant than invasive alternatives like injections, and because oral dosages are very efficient in bringing a medicine (drug) into the patient with the highest efficacy. Currently, the tablet is considered the preferred dosage form for administration and manufacturing.

Table 5.3 Overview of commercially available coprocessed excipients containing lactose.

Manufacturer	Product	Production method	Components	Functional property
BASF	Ludipress		α-Lactose monohydrate, povidone, cross-povidone	Tableting, disintegration
DFE Pharma	SuperTab 40LL	Agglomeration	Anhydrous lactose, lactitol	Tableting, flowability, fully soluble
Kerry	Disintequick	Spray drying	75% α-lactose monohydrate, 25% microcrystalline cellulose	tableting
Kerry	Disintequick	Spray drying	50% α-lactose monohydrate, 50% microcrystalline cellulose	Tableting
Kerry	Lubritose	Spray drying	Lactose 96% Glyceryl monostearate 4%	Eliminates need for external lubricant and time-sensitive blending
Meggle	MicroceLac 100	Spray drying	75% α-lactose monohydrate, 25% microcrystalline cellulose	Tableting
Meggle	Cellactose 80	Spray drying	α-Lactose monohydrate, 25% powdered cellulose	Tableting
Meggle/roquette	Starlac	Spray drying	α-Lactose monohydrate, 15% maize starch	Tableting
Meggle	CombiLac	Spray drying	70% α-lactose monohydrate, 20% microcrystalline cellulose, 10% white maize starch	Tableting
Meggle	RetaLac	Spray-agglomeration	50% α-lactose monohydrate, 50% HPMC	Dissolving in water

As partly shown in Fig. 5.1 lactose and MCC are the most frequently used excipients in oral dosage forms, with about 45% of drug products formulated with combinations of lactose and MCC.

Lactose and MCC are normally used in combination as the inherent property differences complement one another well in enabling robust manufacture of oral

Table 5.4 Schematic of process steps to make tablets (Gohel & Jogani, 2005).

Step	DC	DG/RC	WG
1	Mixing ↓	Mixing ↓	Mixing ↓
2	*Compression*	Slugging ↓	Binder solution ↓
3		Size reduction ↓	Massing ↓
4		Mixing ↓	Wet screening ↓
5		*Compression*	Drying ↓
6			Sieving ↓
7			Mixing ↓
8			*Compression*

solid dosage forms. The three most common manufacturing processes used to make oral solid dosages like tablets are DC, DG or RC, and WG (Table 5.4). Leane et al. (2015) published a proposed manufacturing classification system for oral solid dosage forms, enabling the selection of the "best" process for a given type of drug. These same processing steps (DG/RC and WG) are also common for capsule filling. In this case, however, the final (tablet) compression step is replaced by either a capsule fill or sachet fill step. The lactose (or formulation) functionalities that enable these manufacturing processes are commonly the flow, density/size, and compactability.

5.3.1.1 *Functionality-related characteristics of lactose in relation to tablet preparation*

To understand why lactose functionality is key for tablet manufacturing, it is necessary to describe the tableting process. In general, this consists of two parts: (1) the flow of the powder from a storage container with or without mechanical agitation and transport, the hopper, into the tablet dies and (2) compaction of the powder inside the die, between two punches into a tablet. The flow of lactose enables a formulation to enter the tablet die rapidly and in a homogeneous form, to enable a constant volume/mass. Hancock and Garcia-Munoz (2013) described that for tableting formulations, the die fill density should be above 0.5 g/mL to ensure good processing. Others (e.g., Sun, 2010) state that the key excipient should have a powder flow function, determined with a Schulze ring shear tester (Tan, Morton, & Larson, 2015), of above 6.7 to enable high-speed tableting.

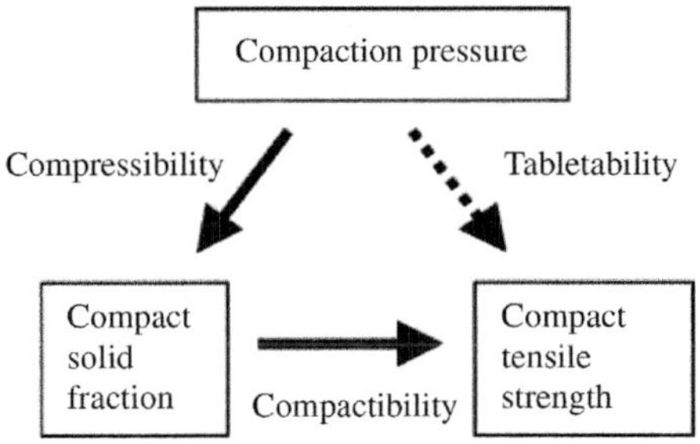

FIGURE 5.4

Illustration to show relationship between compaction pressure, solid fraction, and tensile strength of a given powder (Tye, Sun, & Amidon, 2005). Compressibility is the effect of application of pressure on a powder on the compact density, compactibility is the effect of compaction a powder on the strength of the compact, and tabletability is the effect of application of pressure on a powder on the strength of the compact.

As lactose is available in many grades, a pharmaceutical formulator is able to choose the correct grade, having both a high bulk density and good flow properties (enabling constant tablet weight).

The tableting process (densification part) can be described by the relationship between compaction pressure, compact solid fraction, and compact (tablet) tensile strength (Picker-Freyer, 2008) (Fig. 5.4). Hancock and Garcia-Munoz (2013) stated that the theoretical ideal (lactose) excipient properties for tableting are to achieve a compact, for example, a tablet, with a tensile strength larger than 1.0 MPa and with a solid fraction larger than 0.85, that is, a porosity smaller than 0.15, by compressing a powder with a stress between 20 and 125 MPa. DC grades of lactose are ideal for tableting, as they exhibit good flow and density, while having good compaction properties. Other excipients, such as MCC, do not exhibit all these parameters (e.g., they lack flow and density). Once tablets are made, they need to be strong enough not to break or create dust during the packaging process (low friability).

5.3.1.2 Tableting by direct compression

DC is the simplest way to manufacture tablets for oral use, and is commonly used for medium- to high-dose formulations. Successful DC is facilitated by a mixing process that enables the excipient(s) and drug powders to be uniformly distributed and not segregated in a flowing formulation, with high compaction properties. In DC, the formulation is fed into tableting equipment, with the powders flowing from a hopper via a tablet fill shoe into a tablet die. This is precompressed (using tablet punches), followed by a main compression step into a tablet. After the compression step the tablets are removed from the tablet die.

DC lactose grades are designed to maximize the good compaction characteristics of fine grade lactose powders (which are brittle and ductile). They are either spray dried (spray-dried lactose), granulated (granulated lactose), or roller dried (anhydrous lactose). The amorphous part of spray-dried lactose is ideally suited to creating hard tablets, due to its ductile behavior and good flow properties. Spray-dried lactose has a morphology ideally suited for low- to medium-dose formulations. Due to the large surface area of spray-dried lactose, however, it can be prone to overlubrication. Granulated lactose manufactured via fluid bed granulation or high-shear granulation will behave in very different ways when it comes to compaction and disintegration properties. Both granulated products will be suitable for medium- to high-dose formulations, due to the storage capacity of its cavities. Granulated lactose has few negative interactions during the lubrication step. Anhydrous lactose, due to its very small β-lactose microcrystals, leading to excellent compacting properties, can create the hardest tablets.

Different types and grades (size) of lactose behave differently in the tableting process due to their size, shape, and morphology (structure). DC lactose such as spray-dried lactose, anhydrous lactose, granulated lactose, granulated anhydrous lactose, and milled and sieved lactose all have different tableting behaviors. Whiteman and Yarwood (1988) showed that from the six types tested, anhydrous lactose has the best tableting properties, followed by spray-dried lactose. In addition, they discovered that the properties of spray-dried lactose produced by different manufacturers are so different that they should be regarded as different products, and therefore not interchangeable. The latter is in line with research carried out by Gamble et al. (2010), which showed that on anhydrous lactose products with apparently similar specifications behaved very differently when put through the same DC process. Sieved lactose is coarse in size and can be brittle during compaction. These are therefore often used in combination with highly compactable drug powders with poor flow properties, for which the coarse fast-flowing sieved lactose compensates. Milled lactose is sometimes used for formulation in combination with smooth-flowing medium compactable drug powders.

The DC process has an economy advantage as all materials needed can be purchased to specifications that allow for simple blending and tableting (Shangraw, 1989). Lactose is commonly used in DC due to its safety and costs. The disadvantage of a DC route to manufacture is the longer formulation development time to make a robust formulation.

5.3.1.3 Tableting by wet granulation

For low-dose formulations ($<5\%$ w/w), a granulation process is typically the preferred manufacturing approach. During WG, fine drug particles are mixed and gathered with excipients, such as fine particles of lactose to create larger permanent agglomerates. This process of agglomeration can be carried out with water or with the addition of binder excipients into the liquid media. The granules created are large and free-flowing, which facilitates flow and compaction. The small primary particles are drivers for the compaction (Zuurman, Bolhuis, & Vromans, 1995).

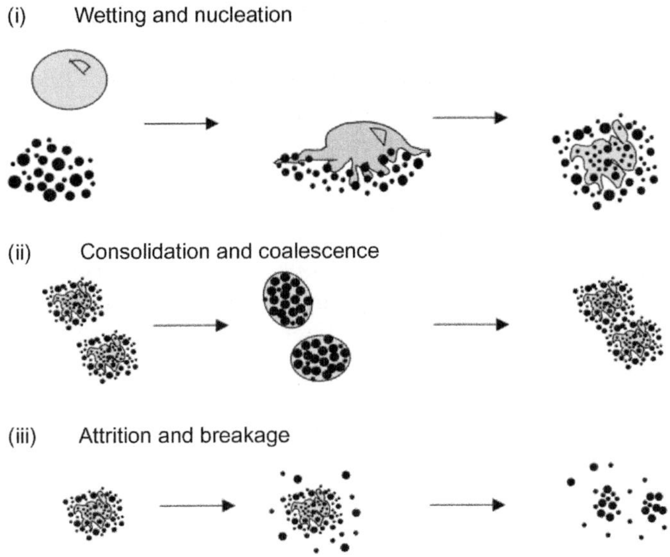

FIGURE 5.5

Three stages of wet granulation (Cantor, Augsburger, & Gerhardt, 2008).

WG (Fig. 5.5) can be performed in a fluid bed, high-shear granulator, or through a continuous process using a specialist device, such as a twin-screw extruder. The drug particles are mixed with the excipient (e.g., fine grade lactose) and moistened. The liquid will enable particle growth by coalescence, with groups of particles consolidating into larger agglomerates. Agglomerates that are too large or too weak will break up into smaller ones and the process is repeated until a stable agglomerate distribution is formed. The moist agglomerates are then dried and crystalline bridges are formed by the liquid and/or binders (agglomerates harden). A detailed granulation process description, with effects of parameters, are well described in the literature (Ban, Goodwin, van den Ban, & Goodwin, 2017; Suresh, Sreedhar, Vaidhiswaran, & Venugopal, 2017).

Generally, with an increase in bulk density, the compactibility of a granule fraction decreases. Little variation is observed between the intergranular porosities of the granule fractions. Both α-lactose monohydrate and anhydrous (β-lactose) are commonly used (Bolhuis & Zuurman, 1995). Compactibility of granule fractions from one lactose type, however, is mainly determined by the total porosity of the granule powder bed. This means that (Fig. 5.5) the compactability of the total granule is determined by the primary lactose size in the granule (Zuurman et al., 1995). Mercury porosimetry determinations on tablets compacted from the granule fractions show a relationship between the tablet pore system and the strength of the compacted material. Compressing granulations with a low bulk density results in tablets with a small average pore diameter and a high crushing strength (Bolhuis & Zuurman, 1995). Many papers have been written to explain the WG kinetics between added liquid (with or without binder) and excipient (lactose) particle size on compactability and compressibility (Ilić, Kása, Dreu, Pintye-Hódi, & Srčič, 2009; Šantl, Ilić, Vrečer, & Baumgartner, 2011). It is very important to understand the factors involved in creating wetted mass "dough" (Keleb, Vermeire, Vervaet, & Remon, 2004; Shah, Hussain, Hubert, & Farag Badawy, 2008). For lactose as excipients in WG these factors are the rate and solubility of the lactose grade used, the size and shape of the lactose grade used, and the anomer used (α- or β-lactose). The effect of the different excipients (e.g., form conversion; Shah et al., 2008) and synergy of lactose versus other components in WG tableting (Saha & Shahiwala, 2009) are well described. Lot-to-lot drug and excipient (lactose) variation can lead to differences in granule size and densities.

The main advantage of WG is the ability to create powders which flow and compress while fixing the combination of formulation components in a granular composition. During compaction the granules are fractured, exposing new surfaces and improving compressibility. Drugs having low doses and poor flow properties (high cohesion) can be wet-granulated to enable formulation flow and distribution (homogeneity and uniformity). In addition, a wide variety of drugs and functional excipient powders can be processed together in a single batch, while low-density and dusty drug powders can be handled without significant dust contamination.

The main disadvantage of WG is the cost. In fact lactose grades used in WG are typically cheaper than those in DC and RC. There are more processing steps needed to make tablets, however, resulting in higher overall costs due to the need for the extra manufacturing space, time, energy consumption, labor costs, and equipment required. Although the WG process is expensive, there is a widespread (incorrect) belief that it cannot be replaced by DC (Shangraw, 1989).

5.3.1.4 Tableting by dry granulation/roller compaction

DG/RC is an agglomeration process that has been in use for more than 50 years (Kleinebudde, 2004). DG is a controlled crushing of precompacted powders, which have been densified by either slugging or passing between two counterrotating rolls. Many different companies supply roller compactors, and their specific behavior is described in many papers (Rowe, Charlton, & McCann, 2017; Saarinen, Antikainen, & Yliruusi, 2017). The principle of DG/RC is to feed powder through a roller gap, where the powder is compacted by the pressure between the rolls. The configuration of the rolls (smooth, fluted, knurled, or pocket rolls) will determine if the material is compacted into dense ribbons (flakes, sheets, or strips) or briquettes (Fig. 5.6).

Lactose monohydrate and anhydrous lactose are ideal functional excipients for DG/RC as the materials are brittle (low amount of plastic/elastic deformation like MCC or starch) and deaeration is good. Irrespective of the lactose type and size, understanding and controlling equipment is critical. Inghelbrecht and Remon (1998) showed that the best granules were obtained by managing the pressure, roll speed, and powder feeding (screw speed). Compaction pressure is the most

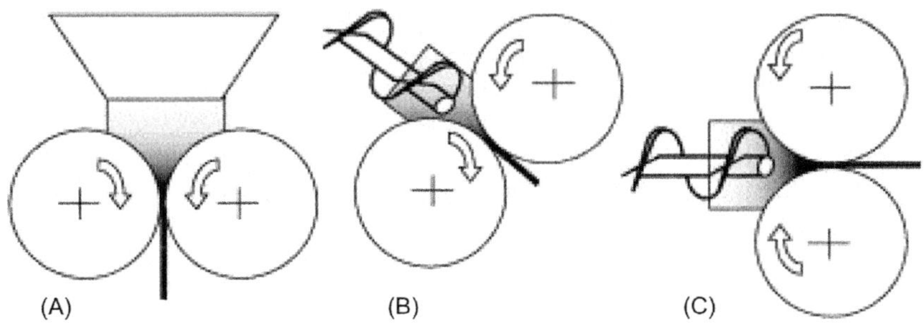

FIGURE 5.6

Configuration of roll compactors, with rolls in horizontal (A), inclined (B), or vertical (C) positions.

Taken from Gulgon, P., & Simon, O. (2003). Roll press design—Influence of force feed systems on compaction. Powder Technology, 130*(1–3), 41–48 [Elsevier].*

significant of the parameters that affect ribbons and granules during this process (Gupta, Peck, Miller, & Morris, 2005; Omar et al., 2015). Some researchers (Inghelbrecht & Remon, 1998) found that spray-dried lactose was not ideal for DG/RC, while others (Omar, Dhenge, Palzer, Hounslow, & Salman, 2016) actually found it to work better. When the output of DG/RC is milled and further used in tableting, anhydrous and lactose monohydrate 200 M are preferred more than MCC due to their recompactability behavior (Beten, Yüksel, & Baykara, 1994). For crystalline lactose De Boer et al. (1986) and Vromans et al. (1985) showed that fragmentation is the key mechanism. For anhydrous lactose, which consists of very small particles aggregated together, it is believed that the consolidation process is mostly fragmentation, while for lactose monohydrate it is fragmentation and plastic deformation (Vromans, 1987).

Advantages associated with RC are operational speed and the ability to link the process to a continuous operation. The process minimizes the creation of fines and dust, and has the potential to shorten disintegration times of a given formulation. Disadvantages are the potential to overcompact and thus lengthening the tablet dissolution times. Additional equipment like a mill and/or sieves may be required before tableting is possible.

5.3.1.5 Continuous manufacture of tablets

Lactose is commonly used in continuous manufacturing processes, due to its versatility and access to highly consistent grades. In a continuous manufacturing process, either a WG or DG process step (in reduced scale) is coupled to a blender and tableting equipment, or alternatively a continuous mixer is coupled to a tableting equipment.

Key requirements for a successful process are process control, drug and excipient powder consistency, and knowledge. This new way of manufacturing is favored and to some extent initiated by the US Food and Drug Administration (FDA) (Framework for innovative pharmaceutical development, manufacturing and quality assurance, 2004), and key success factors in enabling this method include utilizing process analytical technology (PAT) for successful in-process measurement and control, and multivariate analysis (MVA). Researchers, including Kushner et al. (2014), Haware, Bauer-Brandl, and Tho (2010), and Ferreira and Tobyn (2015) have shown that MVA (an advanced data analysis methodology) understanding of excipients such as lactose are key to the success of continuous manufacturing (by starting in batch). As lactose (and MCC) is well described in terms of physical and chemical properties in literature they are the preferred excipients to use for continuous manufacturing.

Continuous manufacturing in the pharmaceutical industry is relatively new. Predicted advantages are reduction of footprint and increased formulation and production flexibility. Disadvantages lie in the expensive PAT needs as well as high-end statistical process control required.

5.3.2 Application of lactose for inhalation

Inhalation has been used to deliver medications to the respiratory system for thousands of years (Dickhoff, de Boer, Lambregts, & Frijlink, 2006; Sanders, 2007), for example, via smoking herbal products. Delivery to the lungs is an attractive route for targeting lung-related diseases like asthma or COPD, as the API is directly deposited to the exact point where it is needed. In addition to local delivery, this route circumvents the first-pass metabolic pathway of the active ingredient that inevitably occurs after taking medication via the gastrointestinal route (Fig. 5.7).

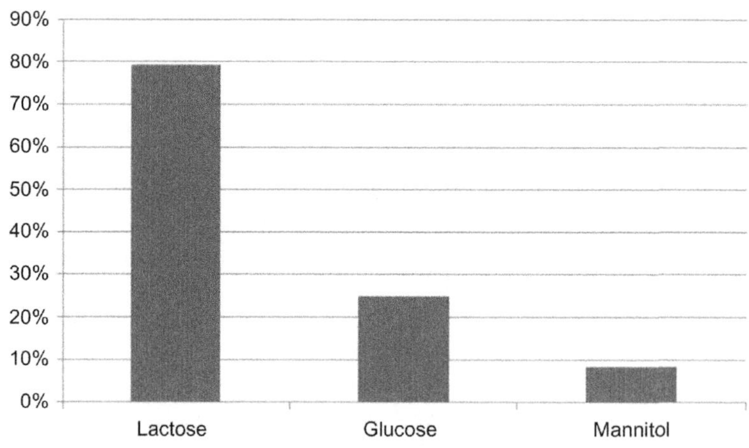

FIGURE 5.7

Excipient usage in DPI formulations with excipients, according RXList (2017). Some formulation contains both lactose and glucose. There are two formulation with mannitol of which one is Aridol, which is a purely mannitol to be used for bronchial challenge testing as mannitol is known to cause bronchial spasms.

Lactose is the most prevalent excipient in DPIs, covering more than 80% of all dry powder inhalation formulations.[3]

There are several reasons to incorporate an excipient in DPIs (Peters & Hebbink, 2016). First, this approach enables the manufacturer to fill an inhalation device with the generally low-dosed and highly cohesive API. In addition, the excipient should help the delivery of the API dosage to the place where it is needed in a predictable and reliable way (de Boer, Chan, & Price, 2012; Jones & Price, 2006; Jones et al., 2008; Shur, Harris, Jones, Kaerger, & Price, 2008; Shur et al., 2016). The required functional parameters of the excipients, therefore, are powder flow for device (blister, capsule, container) filling and to enable the API to be deagglomerated from the carrier particles during inhalation. This implies that the right balance between cohesiveness and deagglomeration needs to be established in the pharmaceutical blend. For inhalation applications, very fine API powders are used (1–5 μm) (Dickhoff, 2006). Such fine powders are cohesive and tend to stick to each other or to the walls of devices and filling equipment. To ensure the powder flows as required, a relative coarse excipient is needed, as this ensures effective powder flow to facilitate handling and filling of the formulation.

The lactose should be capable of forming a stable blend with the fine API material and no segregation of the powder blend should occur during pharmaceutical production. This requires that interaction should exist between carrier material and drug to form a stable ordered mixture (Hersey, 1975). At the same time, this interaction should not be too strong, as that would prevent the release of the API particles from the excipient during inhalation. In addition to achieving the right balance between binding and release of API particles to lactose particles, the inhalation device plays an important role in the manufacture and performance of a DPI. There are many different types of devices (Berkenfeld, Lamprecht, & McConville, 2015; de Boer et al., 2017; Daniher & Zhu, 2008; Grant, Walker, Hamilton, & Garrill, 2015). Due to different operational methods and different structures, each device poses different requirements on the powder formulation and hence on excipient functionality.

The three main types of DPI devices that are utilized are capsule devices, reservoir devices, and blister devices. As all formulations contain a relative high amount of lactose, lactose drives the processing steps. Filling of capsules can be performed in many ways and each process requires a specific flow, or density profile (Loidolt, Madlmeir, & Khinast, 2017). As lactose ranges from free-flowing to cohesive grades, there are plenty of avenues to target. The filling and emptying of reservoir devices requires a free-flowing powder blend, which is typically a coarse lactose grade. For filling of blister devices, a cohesive type is required, which in general is a fine grade of lactose or a type of lactose that contains a high amount of very fine lactose particles.

[3] Source: www.RXlist.com.

The correlation of excipient, API, and blend properties with DPI performance has been extensively researched. Generally the PSD, and in particular the amount of fines in the carrier present, determines the amount of drug that can be deposited in the lungs (Dickhoff, 2006; Kinnunen, Hebbink, Peters, Shur, & Price, 2014). A number of theories have been devised and investigated to explain this effect (de Boer, 2005; Grasmeijer et al., 2014; Shur et al., 2008), with four major mechanisms researched extensively.

The active sites theory states that active sites on the carrier are preferentially populated by API particles. The API particles are thereby rendered passive and are no longer available. If enough lactose fines are included, however, the active sites on the carrier will be populated with lactose, leaving the API particles available for delivery to the lungs. Another model is the API–lactose fines agglomeration theory (Lucas, Anderson, & Staniforth, 1998). It was found that performance improvement can be attributed to a redistribution of fine drug particles on the coarse surface. Larger agglomerates experience more drag force than a single small particle and are more easily detached from a coarse carrier surface. The increased cohesion theory was most recently developed. According to this theory, a critical airflow velocity is needed to start fluidization of a powder (Shur et al., 2008). More cohesive forces in the powder bed result in a higher critical airflow. The higher kinetic energy in the agglomerate, the more likely the deagglomeration will take place. It was shown that more cohesive power in a powder bed improves performance (Thalberg, Lindholm, & Axelsson, 2004). A fourth theory notes that lactose fines are larger than drug particles and therefore act as a buffer to prevent the drug particles from press-on forces (Dickhoff et al., 2006). During mixing, the larger lactose fines prevent drugs from being pressed onto the coarse surface, thereby enhancing performance. For every theory there are experimental results that both support and contradict them. The best explanation is that API delivery from a DPI is an event for which no single mechanism is solely responsible, but that many factors and interactions play a role.

Airflow and inhaler design, in combination with the formulation, also have an effect on DPI performance (Berkenfeld et al., 2015; De Boer et al., 2003a,b; de Boer, Hagedoorn, Gjaltema, Goede, & Frijlink, 2006; Pilcer & Amighi, 2010; Pilcer, Wauthoz, & Amighi, 2012; Telko & Hickey, 2005; Zeng, 2001). It was found that smoothing the lactose surface or changing the lactose shape improved the delivery of salbutamol sulfate (Zeng, Martin, Marriott, & Pritchard, 2000). Surface roughening of lactose was also shown to have a positive effect on performance (Dickhoff, 2006; Tan, Chan, & Heng, 2016).

It can be concluded that linking the performance of DPIs to individual lactose attributes is not a straightforward matter. It is clear that tight control on lactose quality and lactose functional attributes, such as particle size and surface properties, is an important requirement for controlling API delivery. This can lead to a need for customization of lactose carrier per end product.

5.4 Determination of formulation relevant attributes of lactose

In many applications, including pharmaceutical dosage forms, powder properties play an important role. There are a vast number of analysis techniques available for powders that can be used to describe the properties of that powder (Jawad, Martin, & Royall, 2015). Many of these techniques have been applied to lactose in order to establish relationships between powder properties and functionality of the material. This section will describe a number of chemical and powder physics techniques that have been developed for pharmaceutical applications and lactose in particular. This section will give an overview of the various properties possessed by all the different forms of lactose, together with ways of measuring them in order to achieve optimum functionality of the dosage forms.

5.4.1 Lactose attributes that are key to pharmaceutical applications

A number of functionality-related attributes have been identified for lactose in different applications. For tableting, the flow and density of the powders are key. For processing purposes, lactose powder flow is important. For filling of, for example, tablet machines and capsules, the material should flow very easily. For other applications, as for DPIs, different flow requirements are needed. Flow of powders, however, is a complex phenomenon and many attributes of the powder play a role. There are a number of different ways to quantify the flow, though indirect attributes like bulk and tapped density together with PSD are widely recognized as providing acceptable descriptors. The bulk density of lactose is a related functional characteristic as it affects filling in tablet dies, to give one example. There have been many attempts to find other attributes that relate to lactose functionality, which are either still under investigation or have failed to demonstrate a correlation with functionality.

5.4.2 Control methods

Next to functionality-related properties, lactose quality is most important, as each product must be proven safe in usage. The methods used to quantify this are physical, chemical, and biological. Many are described in regulatory documents, such as pharmacopoeias. A number of these methods deal with the identity of the product: is it really lactose, for example? Other methods relate to the purity: How pure is the lactose, are impurities present, and in what quantities? These impurities come from different sources: related sugars, like galactose or glucose, protein residues remaining from the purification processes, and ionic substances like salts. With regard to microbiology, lactose is controlled on several aspects, ranging from fungi to bacteria to prove the product is safe in usage and of no harm to a patient.

5.4.3 Description of methods

In this section, a number of product attributes and measurement methods will be described, together with potential relationships they have with functional properties or quality control, then how a method works and what future developments are likely to take place.

5.4.3.1 Determination of solid-state properties

A number of techniques can be used to investigate a solid material. In general these methods are centered around quantification of the polymorph constitution of lactose. Accurate measurement of the amorphous fraction in lactose samples is an important factor here (Lehto et al., 2006). Many of these are quality parameters and measured on regular basis, like anomer content and amorphous content. Most solid-state properties are related to production of lactose and to the types of crystals that are formed.

5.4.3.1.1 Lactose crystal structure determination

The lactose crystal of α-lactose monohydrate is well described (Fox, 2009; Roos, 2009). It is depicted in Fig. 5.8 and in general it is referred to as being tomahawk shape (Walstra et al., 2005; Visser, 1983).

This specific tomahawk shape of α-lactose monohydrate crystals is determined by the growth rates of the individual faces during crystallization. The shape and

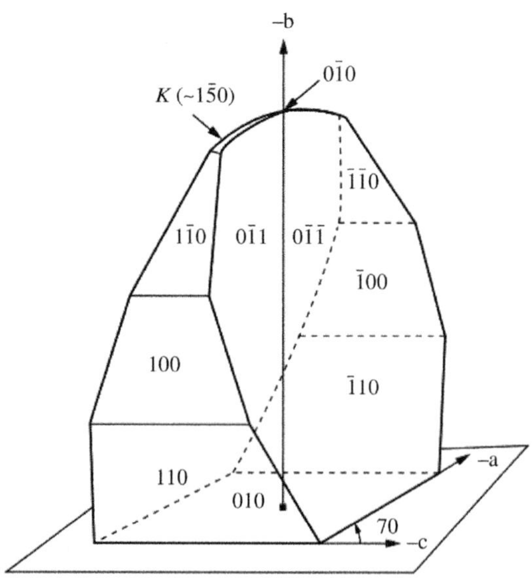

FIGURE 5.8

Crystal form of α-lactose monohydrate (Walstra et al., 2005).

morphology of the α-lactose crystal is well described, whereas for other crystalline forms these descriptions hardly exist. Kirk et al. (2007) described the crystal properties of all other known polymorphs. For pharmaceutical applications, after the α-lactose monohydrate, only the anhydrous β-lactose is of significance, with its crystal structure described (Buma & Wiegers, 1967). Due to the production process by fast evaporation of a lactose solution at high temperature, however (Section 5.2.5.2), the primary structure is of less importance, and properties are governed by the formed agglomerates.

5.4.3.1.2 Differential scanning calorimetry

With the aid of differential scanning calorimetry (DSC), phase transitions of materials can be measured as a function of temperature. This method delivers information on phase transitions and therefore on the type and purity of lactose, and also on water content.

To execute this method, two small pans, one empty and one loaded with the material under investigation, are heated simultaneously and kept at the same temperature throughout the experiment. The energy difference required to keep the two pans at the same temperature is recorded as a function of temperature. Phase transitions of the material can be measured in this way with high precision. These DSC thermograms have been reported for the different types of lactose (Garnier, Petit, & Coquerel, 2002; Jawad et al., 2015; Listiohadi, Hourigan, Sleigh, & Steele, 2009). In a typical DSC thermogram of α-lactose monohydrate crystal (Fig. 5.9) (Gombás, Szabó-Révész, Kata, Regdon, & Eros, 2002), dehydration of

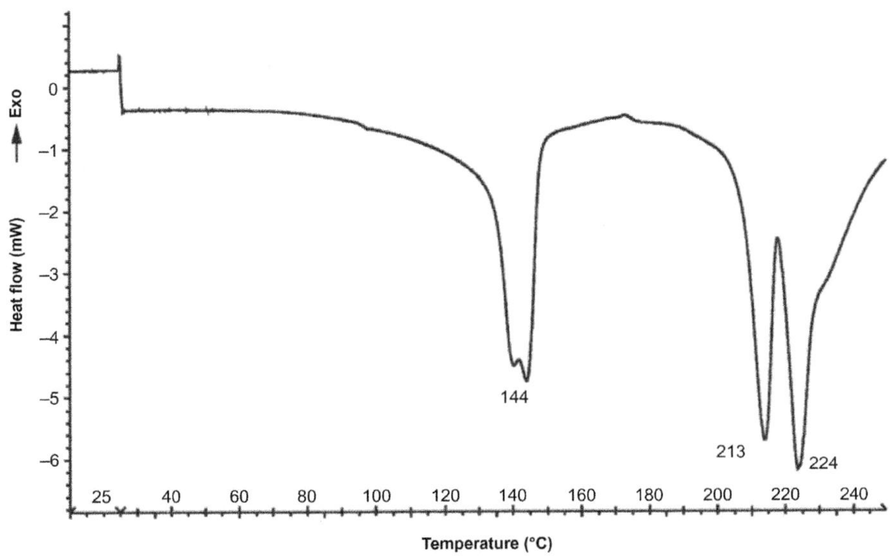

FIGURE 5.9

Typical DSC thermogram of crystalline α-lactose monohydrate (Gombás et al., 2002).

the crystal water around 130°C–150°C and a melting peak at 215°C can be found (Fanni & Parmentier, 1999). Anhydrous β-lactose has a melting peak at 225°C. Amorphous lactose was shown to have an exothermic peak around 160°C, which indicates crystallization of the amorphous phase to a crystalline form followed by melting peaks above 210°C. The exact position of the melting peaks depends on the history of amorphous phase preparation. It can be assumed that this phase consists of an α- and a β-lactose mixture in ratios that are determined in preparation. This ratio is also responsible for reported variations in the extent of crystallization at 160°C, together with the location and number of melting peaks. DSC is currently not used as a standard method for quality and functionality of lactose products.

5.4.3.1.3 Anomeric purity

In α-lactose monohydrate products, the ratio of α- and β-lactose is related to the history of the production process, so control of the anomeric ratio is a quality parameter. This also applies to anhydrous β-lactose, where the extent and control of anomeric purity is also seen as a quality parameter.

There are several ways to measure the anomeric purity of lactose. Pharmacopeia describes a method based on gas chromatography. First, lactose is reacted with trimethylsilylimidazole to prevent mutarotation and to make the compounds volatile enough for gas chromatography. Anomeric purity via this method can be determined by integrating the response peaks for both derivatized anomers. An alternative method via liquid nuclear magnetic resonance (NMR) (Jawad et al., 2012) was also reported. A sample of lactose is dissolved in a suitable anhydrous solvent, like DMSO (dimethyl sulfoxide), to slow down mutarotation. After rapid acquisition of the ^{1}H-NMR spectrum, the anomeric composition can be determined from the integral of the diagnostic α- or β-protons, giving separate signals in the 6–7 ppm region with a high precision.

For a number of lactose products the anomeric purity is determined on a regular basis. Direct relations with functionalities have not been established. Due to the relation with control on the lactose production process it is a stability measure.

5.4.3.1.4 Hygroscopicity of lactose

Interaction of pharmaceutical materials with moisture is often unwanted, as it might cause changes in the stability of the final formulation. The commercially available crystalline forms of lactose are not hygroscopic, as is shown by the published dynamic vapor sorption analysis of that material (Vollenbroek, Hebbink, Ziffels, & Steckel, 2010). Not more than about 0.05% of water is absorbed at 80% relative humidity by crystalline α-lactose monohydrate. The absorption of water by anhydrous β-lactose at relative humidity below about 80% is even lower (Pitchayajittipong, Price, Shur, Kaerger, & Edge, 2010), but at higher relative humidity, hysteresis is found in the dynamic vapour sorption (DVS) isotherms, caused by liquefaction of crystalline β-lactose followed by conversion and crystallization to the more favorable α-lactose monohydrate crystal. This conversion was found to be slow, requiring a period of 4 weeks at 98% relative humidity required for complete conversion. Amorphous lactose, on the other hand, is hygroscopic and at high enough relative humidity, around 50%, form conversion will take place.

This method is widely used in the pharmaceutical industry, but not in product release and product control, mainly because lactose is not hygroscopic and hardly absorbs any water. The method is also used to determine amorphous content, as described in the next section.

5.4.3.1.5 Amorphous content determination

During production processes an amorphous fraction can be formed (Della Bella et al., 2016). This may lead to caking, which is a disadvantage (Carpin et al., 2016, 2017), but it can also lead to advantages by enhancing the tabletability of the material, which plays an important role in the functionality of spray-dried lactose grades. It also plays a role in the adhesive properties of lactose, which is important for preparation of blends and in functionality for DPI applications. Amorphous lactose fractions originate from the processing and are present at the surface of the lactose crystals. Surface effects play an important role in DPI applications and control on the amorphous phase is of high importance.

The amorphous phase is capable of absorbing a substantial amount of water, which influences its properties (Roos, 2009; Timmermann, Steckel, & Trunk, 2006). With more water, the molecules in the amorphous become more mobile and at a certain mobility the material will crystallize (Vollenbroek, Hebbink, Ziffels, & Steckel, 2010). Because the formation of α-lactose monohydrate will result in a net mass increase (Buckton & Darcy, 1995, 1999), this can be used to quantify the amorphous content in a lactose sample. This approach uses the assumption that all amorphous lactose is converted to the α-lactose monohydrate, which is not necessarily true, because anomeric purity of the amorphous phase will be a mixture of the α- and the β-forms.

Several other techniques have been described to assess the amorphous content (Vollenbroek et al., 2010; Vollenbroek, Hebbink, Ziffels, & Steckel, 2010), including X-ray diffraction (Chen, Bates, & Morris, 2001), Raman (Katainen, Niemela, Harjunen, Suhonen, & Jarvinen, 2005), and solid-state NMR (SS-NMR) (Shah, Kakumanu, & Bansal, 2006), but most of these methods lack sensitivity at low levels of amorphous content. Spray-dried lactose has a higher amorphous content and methods to measure that have been described (Lehto et al., 2006). The presence of amorphous fraction is not necessarily negative as is shown in spray-dried lactose (Section 5.2.5.1).

In summary, amorphous content testing gives formulators the opportunity to control the level of the amorphous fraction. But all current methods lack sensitivity and further development would be needed.

5.4.3.1.6 X-ray diffraction

Polymorphism has an effect on how particles interact with each other and with other compounds, so it therefore exerts an influence on functionality. The morphology of a powder can be determined through X-ray powder diffraction (Einfal, Planinšek, & Hrovat, 2013). This is a common technique used to study crystallinity of samples, although the sensitivity of the test is limited. It is effective only for about 10% of a specific phase (Lehto et al., 2006) which limits the usefulness of this method in quantifying minor constituents. The effect of temperature and dehydration on different lactose forms was studied, using DSC and powder X-ray diffraction (PXRD) (Garnier et al., 2002). After heating at various temperatures, PXRD was measured, showing that after dehydration of the monohydrate crystal the hygroscopic form of anhydrous α-lactose was formed. Further heating resulted in the formation of the stable anhydrous α-lactose (Kirk et al., 2007).

Due to limitations in sensitivity, use of this method in release and control is limited.

5.4.3.1.7 Solid-state nuclear magnetic resonance

Use of SS-NMR makes it possible to study the different crystalline and amorphous forms present in a sample. NMR provides information not only on the chemical position of a nucleus in a molecule but also on the chemical environment. Different polymorphs of the same molecule, therefore, have different NMR characteristics, so for lactose, this provides the opportunity not only to study the different anomers but also to study the solid-state form of those anomers (Kirk et al., 2007; Pisklak, Zielińska-Pisklak, Szeleszczuk, & Wawer, 2016a,b). This method has been recently employed to confirm the type of produced polymorph (Carpin et al., 2017; Della Bella et al., 2016) used in studies on the effect of those specific forms on, for example, caking. The sensitivity of this method is similar to X-ray diffraction, that is, in the order of 10% of a specific phase.

5.4.3.2 Particle sizing

PSD is an important characteristic of a powder. The European Pharmacopoeia recognizes the PSD of lactose as a functionality-related characteristic (FRC), and many studies have shown that it is at least correlated with functionality. PSD also has a correlation with flow properties. There are several ways to measure PSD but the most common techniques are sieving and laser diffraction, which are also used as control parameters for product manufacture and product quality. In this section a number of techniques will be described.

5.4.3.2.1 Sieving

Sieving is an old technique, but it is still applied in both lactose product release and for research purposes. Sieving of lactose samples was applied as the main model output (Rosenboom, Antonyuk, Heinrich, & Kraft, 2015) to evaluate and optimize the WG step of α-lactose monohydrate. In another study, lactose granules were prepared using a camera system to the granule sizes and compared that with sieve analyses (Kumar et al., 2015) to predict granule size during the granulation process. It was found that some calculation adjustments were needed to compare the two different ways. Wet sieving was applied to generate a fine grade lactose (Adi, Larson, & Stewart, 2007) with a narrow PSD. For achieving particle sizes below 20 μm with a narrow distribution, wet sieving was preferred over wet milling. A disadvantage of this approach, however, is that the sieve fraction is always limited to a relatively low number. It is also important to put robust protocols in place to determine the endpoint of a sieving action in order to ensure stability.

For sieving, the sieves and powders need to be brought into motion. This can be achieved with several methods like mechanical or ultrasonic vibrations, air jets, or by tapping. The results of the obtained particle size are dependent on the applied method. The methods are relatively easy to use and give results as an actual mass fraction of a certain sieve size. This makes it a common technique for quality control and release of lactose products.

5.4.3.2.2 Laser diffraction

Nowadays, the most common technique for determination of PSD is laser diffraction. Powders are dispersed in a laser beam within dispersing liquids or as a dry dispersion in air or gas. Diffracted light is collected under several angles, while the use of models like the Mie or Fraunhofer theory (Ma, Merkus, de Smet, Heffels, & Scarlett, 2000; Telko & Hickey, 2005) makes it possible to convert the diffraction pattern into a PSD. It is assumed that the diffraction is caused by particles with perfect sphere geometry to create the PSD from the actual diffraction pattern. Care must be taken during preparation of the dispersion to prevent particle size changes due to (partial) solubilization or effects like Ostwald ripening (Watling et al., 2010). In dry dispersion, in air or gas, it is important to be careful over method development as, for example, milling of lactose with the dispersion line was observed (Bonakdar, Ali, Dogbe, Ghadiri, & Tinke, 2016).

Despite these concerns on usage, this method is widely recognized and used as the method of choice for particle sizing. Technology (hardware) and optical models (software) continue to evolve and should lead to even better results in the future.

5.4.3.2.3 Other particle sizing techniques

Of course any other particle sizing technique available in the market can be utilized, such as electrozone sensing (Shekunov, Chattopadhyay, Tong, & Chow, 2007). Application of new and different techniques, however, requires validation to be accepted by the pharmaceutical industry, regulatory authorities, and pharmacopoeias.

Apart from the techniques that require a large number of particles, single particle techniques like microscopy have extensively been used in pharmaceutical research and applied to lactose. These techniques make it possible to assess not just a one-dimensional (1D) parameter of a particle but also a second or a third (Gamble, Tobyn, & Hamey, 2015). The disadvantage is that these techniques are in general laborious and the number of analyzed particles is limited. As automated and computer-controlled system develop, however, these built-in disadvantages as being steadily reduced. An automated microscopy system was used to analyze the deposited agglomerates of fine lactose and budesonide particles (Kinnunen et al., 2014). By combining microscopy with compound identification using Raman spectroscopy, particle size and system performance could be determined.

At the moment these alternative sizing technologies are not commonly used as methods for product quality control.

5.4.3.3 Particle surface analysis

The surface of a product is strongly related to functional properties. It plays a role in formation of blends and in delivery of drugs: a larger relative surface will enable more interaction with additional compounds, like, for example, APIs. A number of techniques exist to assess the surface area of a powder instead of its bulk properties. These techniques are generally not performed for quality control of lactose products, though attempts have been made to link the results of these techniques to functional parameters.

5.4.3.3.1 Specific surface area

The multilayer adsorption technique BET (Brunauer, Emmett, & Teller, 1938) makes evaluation of the specific surface area possible. In general, lactose has a relatively low surface area in the order of 0.1–1 m^2/g, making it necessary to use special and sensitive techniques.

5.4.3.3.2 Inverse gas chromatography

The development of inverse gas chromatography (IGC) enables measurement of surface energy (Jones, Young, & Traini, 2012; Mohammadi-Jam & Waters, 2014). In contrast to gas chromatography, where an unknown gas sample is analyzed with the aid of a known and standard column, IGC is an unknown column analyzed with known gas probes. Total surface energy is the sum of dispersive (van der Waals) and polar energy contribution. The dispersive factor can be measured by utilizing nonpolar probes (a series of alkanes) and the polar contribution with polar probes [such as THF (tetrahydrofuran) and ethyl acetate] (Das, Tucker, & Stewart, 2015). IGC can be utilized in an infinite dilution or in finite dilution mode. With the first method, the concentration of probe molecules is so low that they do not interfere with each other on the surface of the analyte, while in finite dilution experiments, the concentrations are much higher, and multilayer adsorption will play a role.

As such, not just the size but also the shape of the detected peak will change as concentrations increase. In theory, much more of the surface area is analyzed and heterogeneity of a surface can be assessed (Ho et al., 2010; Jefferson, Williams, & Heng, 2011; Shah et al., 2017). It was found that unmilled samples of lactose showed much less heterogeneity than milled samples, while the surface heterogeneity of blends of milled and unmilled types of lactose could be attributed to the milled fraction. Recent investigations have looked into converting dispersive energy as function of surface coverage in a model that shows surface heterogeneity (Jefferson et al., 2011).

IGC has the potential to become relevant technique for quality control of products. At the moment, however, the technology still faces some challenges in the translation of acquired data to functionality. Further developments are needed.

5.4.3.3.3 Atomic force microscopy, cohesive-adhesive balance

Atomic force microscopy (AFM) has also been applied to lactose surfaces to assess the roughness of the surface. By applying environmentally controlled AFM, crystallization of amorphous lactose by increasing humidity levels (Price & Young, 2004) was visualized. In multicomponent systems, such as pharmaceutical formulations, adhesive and cohesive forces are extremely importance as they determine the stability of a blend. The interaction between two particles, either the same material or the different materials, was measured with an AFM by modifying the tip with a small particle of a material, followed by scanning of a surface with this tip (Begat, Morton, Staniforth, & Price, 2004; Price, Young, Edge, & Staniforth, 2002).

5.4.4 Physical and chemical properties of α-lactose monohydrate and anhydrous β-lactose

Different properties of α-lactose monohydrate and the anhydrous β-lactose are described in Table 5.5.

Table 5.5 Selected chemical properties of lactose.

	α-Lactose monohydrate	Anhydrous β-lactose	
Solubility at 20°C	7.4	48	g/100 g water[1]
True density	1.545	1.59	g/cm^3
Melting point	202	252	°C[2]
Specific optical rotation	+91.1	+33.5	degree[3]
Specific heat	1,251	1,193	J/g
Molecular weight	360.34	342.31	g/mol
Heat of solution	−50.24	−9.62	J/g
Heat of combustion	16,106	16,465	kJ/g[4]

1: 19.1 g/100 g water at mutarotational equilibrium.
2: disintegration.
3: +55.5 degrees at mutarotational equilibrium.
4: Westhoff et al. (2014); Rajah and Blenford (1988); Clarke & Stegeman (1939);
Rajah and Blenford (1988); Westhoff et al. (2014); Fanni and Parmentier (1999); Rajah & Blenford, 1988; Fox, 2009.

The major differences between the two forms are solubility at room temperature and specific optical rotation. With regard to properties in solutions, care should be taken in the determination as, in solution, lactose will mutarotate toward a thermodynamic equilibrium, so values for properties like optical rotation will change over time until equilibrium is reached.

5.5 Concerns with the use of lactose in pharmaceutical dosage forms

A number of concerns have been raised concerning the use of lactose in pharmaceutical dosage forms. These concerns relate to interaction of lactose with the dosage form in ways that might potentially affect product quality and stability. Other concerns relate to lactose itself, as most of the world's adult population produces insufficient levels of the enzyme lactase required to hydrolyze lactose, while residual bovine proteins may be present. Many potentially unstable API combinations have been successfully formulated with lactose. The amount of lactose swallowed in medical treatment is at such a low level that digestion is in general not affected. Pharmaceutical grades of lactose are highly refined and studies with allergic patients did not reveal an issue in swallowing these grades of lactose.

5.5.1 Lactase persistence

Lactose intolerance is a concern raised regularly concerning medical use of lactose. This is because lactose-intolerant individuals cannot digest the sugar in milk (lactose) because they have a deficiency of lactase, an enzyme produced by cells in the lining of the small intestine. Lactase is required to metabolize lactose. Reduced levels of this enzyme, which can sometimes be temporary, due to infection, cause symptoms such as abdominal gas, diarrhea, or abdominal cramps.

The ability to digest lactose, the sugar found in the milk of nearly all mammals, depends on the presence of lactase, an enzyme which occurs naturally in babies and young children but tends to reduce sometime after weaning. Approximately 65% of the world's population has reduced lactase levels in their gut by late childhood, and many others develop lactase deficiency as they reach middle and late middle age (Gerbault et al., 2011; Hertzler, Huynh, & Savaiano, 1996; Heyman, 2006; Itan, Powell, Beaumont, Burger, & Thomas, 2009). Yet low or even no lactase production does not lead to complete intolerance of lactose, particularly in the low levels of lactose in pharmaceutical preparations. It should also not be confused with milk allergy, which is a different condition caused by an adverse reaction to the proteins in cow's milk (see Chapter 3: Lactose intolerance and other related food sensitivities for further discussions of lactose intolerance and milk allergies).

So what are the practical implications for pharmaceutical use of lactose, even for patients that have a very low level of lactase? The scientific evidence suggests that between 6 and 12 g of lactose can be taken by almost any patient with no ill effects. As the dosage increases, an increasing number of lactase deficient patients may experience the classic symptoms that accompany fermentation of lactose in the large intestine: flatulence, stomach cramps, and potential diarrhea (Bril, Shoham, & Marcus, 2011).

Reactions to lactose will vary, depending on whether food is eaten at the same time as medication is taken, as this normally reduces the likelihood of symptoms, and are also related to the frequency and number of tablets taken (Montalto et al., 2008; Zarbock, Magnuson, Hoskins, Record, & Smith, 2007). To put this into perspective, if a glass of milk (250 mL) delivers the total acceptable dose of lactose (12 g), two tablets that use lactose as an excipient deliver between 0.2 and 1.4 g of lactose. That is at most 12% of the acceptable limit and normally much less (Silanikove et al., 2015). The delivered dose of lactose by inhalation is even lower, as dosage forms are in general not exceeding 25 mg of lactose.

It is now clear that 12 g of pharmaceutical lactose in a single dosage, and 18 g in a day, can be tolerated by virtually any human being, regardless of whether they are lactase persistent (i.e., they continue to produce lactase into adulthood) or not (i.e., the majority of the population where lactase levels decrease sometime after weaning). (Corgneau, Scher, Ritie-pertusa, Petit, & Nikolova, 2017; Perino, Cabras, Obinu, & Cavalli Sforza, 2009) (see Prof. Dr. Martin Smollich, 2016, FortbildungsKongress Schladming). To go beyond this limit it would be necessary to take a very large number of tablets with a lactose excipient. This is rarely necessary and, in these unusual cases, clinicians will exercise caution in the formulation of the drugs they prescribe.

Lactose has been one of the most widely used and successful excipients in pharmaceutical history. Recent research shows that even patients producing very low levels of lactase, or none at all, will normally experience no ill effects from taking medicines that use lactose as an excipient. The scientific evidence increasingly confirms safety and viability of lactose for the future.

5.5.2 Bovine protein allergy

All lactose used in pharmaceutical industry is of bovine milk origin, so concerns may be raised related to bovine protein allergy, which might be caused by residual protein in the lactose product. In contrast to lactose intolerance, which is caused by partial or total absence of the lactase enzyme that metabolizes the disaccharide, protein allergy is caused by an immune reaction of the body on contact with the allergen (Rangel et al., 2016). For intolerance to lactose, the amount taken is the major factor in determining an adverse reaction, while for allergies even small amounts of the offending agent might result in allergic reactions (Audicana Berasategui et al., 2011; Robles & Motheral, 2014). Bovine milk protein allergy is reported for 2%–5% of infants and children, but milk protein allergy usually disappears in childhood. In ingredients with origins in dairy products, like lactose, the allergen is cow's milk protein. Reported cases of allergic reactions with pharmaceutical formulations can be traced to defective purification systems of the lactose (Fiocchi et al., 2003). It was shown that even children with severe cow's milk allergy can safely consume preparations containing lactose (Wiecek, Wos, Horowska-Ziaja, Flak-Wancerz, & Grzybowska-Chlebowczyk, 2016). See Chapter 3, Lactose intolerance and other related food sensitivities, for more information on further discussions on lactose intolerance and milk allergies.

5.5.3 Incompatibility and instability with active pharmaceutical ingredients

The concern is about the compatibility of lactose with certain APIs. Lactose is a reducing sugar and as such potentially reactive with amine compounds. A number of products have been successfully formulated with lactose and have been shown to be stable. For instance, trandolapril, a secondary amine, is thermally stable with a number of excipients (including lactose) (Roumeli et al., 2013). No interactions between trandolapril and lactose were found below 100°C. Another example of a product that is successfully formulated with lactose is pregabalin, a primary amine, administered in hard gelatin capsules with, among others lactose monohydrate as inactive ingredient (Lovdahl et al., 2002).

Although lactose is a reducing sugar, it is important to take note of other ingredients or related impurities, in a formulation such as glucose. Glucose is known to be a related impurity of lactose but also of MCC (Hoaglund Hyzer et al., 2017), and it is more reactive in Maillard reactions (Chapter 6: Lactose in the dairy production chain) than lactose. A study with bisoprolol fumarate showed that accelerated conditions at 40°C and 75% relative humidity were needed to reduce the amount of bisoprolol by about 0.3% after 6 months' storage (Szalka, Lubczak, Naróg, Laskowski, & Kaczmarski, 2014), no matter whether anhydrous lactose or lactose monohydrate was used as the excipient.

Stability of aspirin and niacinamide formulations made with lactose monohydrate or anhydrous lactose was studied (Du & Hoag, 2001). It was found that moisture and heat were required for degradation reaction and that lactose monohydrate and anhydrous lactose showed the same stability, indicating that bound, crystal water does not influence the reaction rate.

It can be concluded that instability of amine-containing drugs in the presence of lactose cannot be excluded, but that there are many cases in which amine products are successfully formulated with lactose. Additional factors required for instability are heat and moisture. Excluding those will in general result in a stable formulation. Tableting by DC or DG and preparation of inhalation formulations avoid the usage of water, resulting in a very low risk.

5.6 Regulatory

The pharmaceutical industry is highly regulated by authorities both concerning production methods (GMP guidelines) and on product tests and specifications (pharmacopeias). For pharmaceutical products and APIs, official GMP guidelines are issued to regulate production (ICH guidelines). Because excipients are very diverse products, which are also used in numerous other applications, no mandatory governmental GMP standard applies to these products. The GMP guideline is managed by the International Pharmaceutical Excipients Council (IPEC), a global nonprofit organization, consisting of five regional councils (IPEC-Americas, IPEC Europe, IPEC Japan, IPEC China, and IPEC India). The councils comprise excipient manufacturers, distributors, and pharmaceutical companies that use these excipients. The Federation has as its key objectives harmonization of compendial standards and GMP guidelines. In addition, the council provides a source of advice and expertise for other stakeholders on excipients.

5.6.1 Pharmacopoeias

A pharmacopoeia is published by authorities or government-related institutions and contains directions for identifying preparations of pharmaceutical dosage forms. The pharmacopeia describes, for each product, the tests that have to be performed, including the method and the specification. Next to mandatory characteristics, the European Pharmacopoeia also defines FRCs (Functionality Related Characteristics). FRCs are controllable physical or chemical characteristics of an excipient that impact on its functionality. For lactose those are defined as the PSD and Hausner ratio, ratio of poured bulk density and tapped bulk density of the powder, as indicator of powder flow.

The three major pharmacopoeias (United States Pharmacopeia-National Formulary (USP-NF), Japanese Pharmacopeia (JP), and European Pharmacopeia (EP)) have separate harmonized monographs for α-lactose monohydrate and for (anhydrous) lactose, and these are depicted in Tables 5.6 and 5.7.

5.7 Future of lactose in pharmaceuticals

5.7.1 Final considerations

Lactose is the most frequently cited diluent in oral products, and probably maintains its high usage rate because positive commercial and pharmaceutical factors outweigh potential negatives, while its long history of use means formulators and manufacturers are well aware of its capabilities.

An excipient needs to be chemically compatible with a specific API and contain minimal reactive impurities that could also react with the API. Chemical incompatibility between an excipient and an API, whether real or perceived, rules

Table 5.6 α-Lactose monohydrate.

Test	Description/notes	Specification	EP	USP-NF	JP
Identification		Complies	✓	✓	✓
Appearance of solution	Clear		✓	✓	✓
	Nearly colorless		×	✓	✓
	Not more colored than BY7		✓	×	×
Acidity/alkalinity	Titration with sodium hydroxide	nmt 0.4 mL	✓	✓	✓
Specific optical rotation		+54.4 to +55.9	✓	✓	✓
Absorbance	UV at 400 nm	nmt 0.04	✓	✓	✓
	UV at 210–220 nm	nmt 0.25	✓	✓	✓
	UV at 270–300 nm	nmt 0.07	✓	✓	✓
Heavy metals	As lead	nmt 5 ppm	✓	✓	✓
Loss on drying	80°C/2 h (Note 2)	nmt 0.5%	✓	✓	✓
Water content	Karl Fischer	4.5 to 5.5%	×	✓	✓
Sulfated ash	Residue on Ignition	nmt 0.1%	✓	✓	✓
Anaerobic bacteria		nmt 100 cfu/g	✓	✓	✓
Fungi and yeast		nmt 50 cfu/g	×	✓	✓
Escherichia coli		Absent	✓	✓	✓
Salmonella		Absent	×	×	✓
Particle size distribution	FRC in EP		✓	✓	×
Hausner ratio	FRC in EP		✓	×	×

nmt, not more than.

Table 5.7 Anhydrous lactose.

Test	Description/notes	Specification	EP	USP-NF	JP
Identification		Complies	✓	✓	✓
Appearance of solution	Clear		✓	✓	✓
	Nearly colorless		×	✓	✓
	Not more colored than BY7		✓	×	×
Acidity/alkalinity	Titration with sodium hydroxide	nmt 0.4 mL	✓	✓	✓
Specific optical rotation		+54.4 to +55.9	✓	✓	✓
Absorbance	UV at 400 nm	nmt 0.04	✓	✓	✓
	UV at 210–220 nm	nmt 0.25	✓	✓	✓
	UV at 270–300 nm	nmt 0.07	✓	✓	✓
Heavy metals	As lead	nmt 5 ppm	✓	✓	✓
Loss on drying	80°C/2 h	nmt 0.5%	✓	✓	✓
Water content	Karl Fischer	nmt 1.0%	×	✓	✓
Sulfated ash	Residue on ignition	nmt 0.1%	✓	✓	✓
Anaerobic bacteria		nmt 100 cfu/g	✓	✓	✓
Fungi and yeast		nmt 50 cfu/g	×	✓	✓
Escherichia coli		Absent	✓	✓	✓
Salmonella		Absent	×	×	✓
Particle size distribution	FRC in EP		✓	✓	×
Hausner ratio	FRC in EP		✓	×	×
Anomer ratio	FRC in EP		✓	✓	✓

out use of that excipient. As discussed in Section 5.5.3, lactose is seen as undergoing Maillard reaction with primary amines, secondary amines, peptides, and proteins, and this may be cited as a reason not to use lactose. Lactose is not necessarily ruled out, however, and there are plenty of potentially incompatible drugs that are successfully formulated with lactose as long as the correct processing conditions are used. Examples include pregabalin and sodium alendronate, which are both primary amines.

Water is the medium in which many of the reactions take place, and therefore excipients containing a high amount of free water (MCC and starches) may allow reaction of drugs with excipients or impurities. Lactose is considered a dry excipient with minimal amounts of free water (Section 5.4.3.1.4), minimizing side reactions in this medium.

Water solubility is a desirable characteristic for an excipient, so that tablets will dissolve after swallowing. Lactose is one of the few water-soluble excipients. Lactose is not hygroscopic (Section 5.4.3.2), which ensures that formulation with lactose will not attract water.

Excipients should not exert an adverse effect either on the patient directly or on drug absorption. Lactose intolerance might be perceived as an adverse effect, though the quantities of lactose ingested via tablets or inhalation are far below the generally accepted quantities that cause unwanted gastrointestinal effects (Section 5.5.1). There is some potential for allergic reactions to residual bovine proteins (Section 5.5.2), but it has been shown that allergic reactions are highly unlikely after taking oral products because of the denaturing effect of gastrointestinal conditions. Even in inhalation products there are no reports of allergic reactions (Fiocchi et al., 2003; Thoren, Wallin, Whitehead, & Sandström, 2001).

With regard to tableting, an excipient should have good tabletability and flow properties. Powder flow can always be influenced by particle size, whereas tableting properties are less easy to influence. Of all the common diluents lactose probably has the best balance of flow and tableting properties and can be used as the sole diluent in DC formulations. For inhalation, the flow and PSD should be controllable, and both can be achieved with lactose.

Commercial factors favor lactose because a multiplicity of grades is available, and similar products can be obtained from different suppliers, allowing for multiple sourcing.

The production of lactose has a long history, and many different grades can be made in multiple ways. New processing techniques are being developed, which offer better control of lactose for pharmaceutically relevant parameters. In addition, new techniques are in development to allow production of new shapes and forms of lactose. This is achieved, for example, by spray drying in the presence of a template (Ebrahimi, Saffari, & Langrish, 2015). After removal of the template a lactose framework remains with unique properties. Spherical crystals can be made by an antisolvent method (Lamešić, Planinšek, Lavrič, & Ilić, 2016; Muhammad, Tang, Chan, & Dehghani, 2012). The agglomerates formed are highly spherical with a high specific surface area. Compaction was found to be enhanced when compared to the commercial lactose grade for DC. Lactose can also play a role in new ways of preparing formulations, for example, by three-dimensional (3D) printing (Jonathan & Karim, 2016; Khaled, Burley, Alexander, & Roberts, 2014; Norman, Madurawe, Moore, Khan, & Khairuzzaman, 2017; Ursan, Chiu, & Pierce, 2013).

5.7.2 New developments in pharmaceutics and the role of lactose

The pharmaceutical industry in the future is expected to focus on a specific range of areas. There is a clear trend for chronic deceases, such as asthma, to increase, resulting in formulations specifically for these, with DPIs as a good example. There is a worldwide trend to focus on specific groups, like geriatrics and pediatrics. This will give rise to more specific and personalized medication, for example, orally disintegrating medications or 3D printing of formulations. Others relate to regulatory requirements, as regulatory bodies become increasingly more cautious, resulting in a higher need for control in the formulation process, in production of pharmaceuticals and in use of excipients. Policy makers and funding bodies for medications are also becoming more demanding, while demand for medication in emerging economies is in many cases faster growing than that in the industrialized economies. Both trends lead to demand for cost-effective medications. Finally, a growing trend for medical research is for medication that focuses on prevention rather than reaction treatment, leading to the need for developing preventive medications.[4]

As is described in this chapter, lactose is a versatile excipient that is safe to use, relatively cheap, and widely available in many forms. Development of new forms of lactose continues to make this excipient very much ready for the future.

5.8 Conclusion

This chapter describes the application of lactose as excipient in the pharmaceutical industry. About 60%–70% of pharmaceutical dosage forms contain lactose, and in volume it is one of the biggest pharmaceutical excipients. Lactose can have several functions in a dosage form: as a filler to provide bulk to for instance tablets, as a binder to provide the strength to a dosage form to keep it together, and to provide the flow to a formulation to be capable of producing it. Next to that, the excipient can assist in delivering the drug to the place of action like in DPIs, where lactose is used as a carrier to give bulk to the very low-dose drug, assists in filling of the drug into inhalation devices, and provides the necessary dispersion of the drug compound in the inhalation airstream to make it reach the lungs. A wide range of commercial pharmaceutical grades of lactose is available with specific properties for each specific formulation challenge. Pharmaceutical grade lactose is produced by double crystallization of several whey streams in the dairy industry. The product obtained consists mainly of α-lactose monohydrate, which after milling and sieving processes is used as a basic excipient in many oral solid dosage forms and in inhalation formulations. Further processing of α-lactose monohydrate by, for instance, spray drying, agglomeration, and roller drying yields materials that possess enhanced functionality on specific parameters like powder flow and tabletability. There are many different production processes in the pharmaceutical industry that all require specific grades of lactose. Oral solid dosage forms that are prepared via WG need fine grades of lactose, like milled grades. For DG processes lactose grades that are recompactible, like anhydrous lactose, are used and for DC processes, lactose with good flow and good compactibility, such as anhydrous, agglomerated, or spray-dried grades, are required. There are a couple of concerns in relation to usage of lactose in pharmacy like lactose intolerance. It is generally recognized and accepted that a daily

[4] https://www.pwc.com/gx/en/industries/pharmaceuticals-life-sciences/pharma-2020/industry-strategies-trends-analysis.html (assessed August 2017).

intake of 10–12 g does not result in physical problems and it is hard to reach that amount of lactose by pharmaceutical intake. Regulatory documents describe the requirements to guarantee the safe usage of drug products and pharmaceutical grade lactose needs to comply with those. In conclusion, lactose is a versatile excipient that is safe to use, relatively cheap, and widely available in many forms.

References

Adi, H., Larson, I., & Stewart, P. (2007). Use of milling and wet sieving to produce narrow particle size distributions of lactose monohydrate in the sub-sieve range. *Powder Technology*, *179*(1–2), 95–99.

Audicana Berasategui, M. T., Barasona Villarejo, M. J., Corominas Sánchez, M., de Barrio Fernández, M., García Avilés, M. C., García Robaina, J. C., Gastaminza Lasarte, G., et al. (2011). Potential hypersensitivity due to the food or food additive content of medicinal products in Spain. *Journal of Investigational Allergology and Clinical Immunology*, *21*(7), 496–506.

van den Ban, S., Goodwin, D. J., van den Ban, S., & Goodwin, D. J. (2017). The impact of granule density on tabletting and pharmaceutical product performance. *Pharmaceutical Research*, *34*(5), 1002–1011.

Begat, P., Morton, D. A. V., Staniforth, J. N., & Price, R. (2004). The cohesive-adhesive balances in dry powder inhaler formulations I: Direct quantification by atomic force microscopy. *Pharmaceutical Research*, *21*(9), 1591–1597.

Berkenfeld, K., Lamprecht, A., & McConville, J. T. (2015). Devices for dry powder drug delivery to the lung. *AAPS PharmSciTech*, *16*(3), 479–490.

Beten, D. B., Yüksel, N., & Baykara, T. (1994). The changes in the mechanic properties of a direct tableting agent microcrystalline cellulose by precompression. *Drug Development and Industrial Pharmacy*, *20*(14), 2323–2331.

Bolhuis, G. K., & Armstrong, N. A. (2006). Excipients for direct compaction—an update. *Pharmaceutical Development and Technology*, *11*(1), 111–124.

Bolhuis, G. K., & de Waard, H. (2011). In M. Çelik (Ed.), *Compaction properties of directly compressible materials* (2nd ed.). CRC Press.

Bolhuis, G. K., & Zuurman, K. (1995). Tableting properties of experimental and commercially available lactose granulations for direct compression. *Drug Development and Industrial Pharmacy*, *21*(18), 2057–2071.

Bonakdar, T., Ali, M., Dogbe, S., Ghadiri, M., & Tinke, A. (2016). A method for grindability testing using the Scirocco disperser. *International Journal of Pharmaceutics*, *501*(1–2), 65–74, [Elsevier].

Bril, S., Shoham, Y., & Marcus, J. (2011). The "mystery" of opioid-induced diarrhea. *Pain Research Management*, *16*(3), 197–199.

Brunauer, S., Emmett, P. H., & Teller, E. (1938). Adsorption of gases in multimolecular layers. *Journal of the American Chemical Society*, *60*(2), 309–319.

Buckton, G., & Darcy, P. (1995). The use of gravimetric studies to assess the degree of crystallinity of predominantly crystalline powders. *International Journal of Pharmaceutics*, *123*(2), 265–271.

Buckton, G., & Darcy, P. (1999). Assessment of disorder in crystalline powders—a review of analytical techniques and their application. *International Journal of Pharmaceutics, 179*(2), 141−158.

Buma, T. J., & Wiegers, G. A. (1967). X-ray powder patterns of lactose and unit cell dimensions of beta-lactose. *Netherlands Milk and Dairy Journal-Nederlands, 21*(3−4), 208.

Cantor, S., Augsburger, L., & Gerhardt, A. (2008). Pharmaceutical granulation processes, mechanism and the use of binders. In S. W. Hoag (Ed.), *Pharmaceutical dosage forms—tablets* (3rd ed., pp. 261−301), CRC Press.

Carpin, M., Bertelsen, H., Bech, J. K., Jeantet, R., Risbo, J., & Schuck, P. (2016). Caking of lactose: A critical review. *Trends in Food Science & Technology, 53*, 1−12.

Carpin, M., Bertelsen, H., Dalberg, A., Roiland, C., Risbo, J., Schuck, P., & Jeantet, R. (2017). Impurities enhance caking in lactose powder. *Journal of Food Engineering, 198*, 91−97, [Elsevier Ltd.].

Chen, X., Bates, S., & Morris, K. (2001). Quantifying amorphous content of lactose using parallel beam X-ray powder diffraction and whole pattern fitting. *Journal of Pharmaceutical and Biomedical Analysis, 26*(1), 63−72.

Clarke, T. H., & Stegeman, G. (1939). Heats of combustion of some mono- and disaccharides. *Journal of the American Chemical Society, 61*(7), 1726−1730.

Corgneau, M., Scher, J., Ritie-pertusa, L., Petit, J., & Nikolova, Y. (2017). Recent advances on lactose intolerance : Tolerance thresholds and currently available answers. *Critical Reviews in Food Science and Nutrition, 57*(15), 3344−3356.

Daniher, D. I., & Zhu, J. (2008). Dry powder platform for pulmonary drug delivery. *Particuology, 6*(4), 225−238.

Das, S., Tucker, I., & Stewart, P. (2015). Surface energy determined by inverse gas chromatography as a tool to investigate particulate interactions in dry powder inhalers. *Current Pharmaceutical Design, 21*(27), 3932−3944.

De Boer, A., Hagedoorn, P., Gjaltema, D., Goede, J., Kussendrager, K., & Frijlink, H. (2003a). Air classifier technology (ACT) in dry powder inhalation Part 2. The effect of lactose carrier surface properties on the drug-to-carrier interaction in adhesive mixtures for inhalation. *International Journal of Pharmaceutics, 260*(2), 201−216.

De Boer, A., Hagedoorn, P., Gjaltema, D., Goede, J., Kussendrager, K., & Frijlink, H. (2003b). Air classifier technology (ACT) in dry powder inhalation Part 1. Introduction of a novel force distribution concept (FDC) explaining the performance of a basic air classifier on adhesive mixtures. *International Journal of Pharmaceutics, 260*(2), 187−200.

de Boer, A. H., Chan, H. K., & Price, R. (2012). A critical view on lactose-based drug formulation and device studies for dry powder inhalation: Which are relevant and what interactions to expect? *Advanced Drug Delivery Reviews, 64*(3), 257−274.

de Boer, A. H., Hagedoorn, P., Gjaltema, D., Goede, J., & Frijlink, H. W. (2006). Air classifier technology (ACT) in dry powder inhalation. *International Journal of Pharmaceutics, 310*(1−2), 81−89.

de Boer, A. H., Hagedoorn, P., Hoppentocht, M., Buttini, F., Grasmeijer, F., & Frijlink, H. W. (2017). Dry powder inhalation: Past, present and future. *Expert Opinion on Drug Delivery, 14*(4), 499−512, [Taylor & Francis].

de Boer, A.H. (2005). *Optimisation of dry powder inhalation: the application of air classifier and laser diffraction technology for the generation and characterisation of aerosols from adhesive mixtures*. University of Groningen.

de Boer, A. H., Vromans, H., Leur, C. F., Bolhuis, G. K., Kussendrager, K. D., & Bosch, H. (1986). Studies on tableting properties of lactose. Part III. The consolidation behaviour of sieve fractions of crystalline α-lactose monohydrate. *Pharmaceutisch Weekblad*, *8*(2), 145–150.

Della Bella, A., Müller, M., Soldati, L., Elviri, L., & Bettini, R. (2016). Quantitative determination of micronization-induced changes in the solid state of lactose. *International Journal of Pharmaceutics*, *505*(1–2), 383–393.

Dickhoff, B. H. J. (2006). Adhesive mixtures for powder inhalation. *Groningen, The Netherlands: University of Groningen*.

Dickhoff, B. H. J., de Boer, A. H., Lambregts, D., & Frijlink, H. W. (2006). The effect of carrier surface treatment on drug particle detachment from crystalline carriers in adhesive mixtures for inhalation. *International Journal of Pharmaceutics*, *327*(1–2), 17–25.

Du, J., & Hoag, S. W. (2001). The influence of excipients on the stability of the moisture sensitive drugs aspirin and niacinamide: Comparison of tablets containing lactose monohydrate with tablets containing anhydrous lactose. *Pharmaceutical Development and Technology*, *6*(2), 159–166.

Durham, R. J. (2009). Modern approaches to lactose production. In M. Corredig (Ed.), Dairy-Derived Ingredients: Food and Nutraceutical Uses (pp. 103–144). Elsevier.

Ebrahimi, A., Saffari, M., & Langrish, T. (2015). Spray drying and post-processing production of highly-porous lactose particles using sugars as templating agents. *Powder Technology*, *283*, 171–177.

Einfal, T., Planinšek, O., & Hrovat, K. (2013). Methods of amorphization and investigation of the amorphous state. *Acta Pharmaceutica (Zagreb, Croatia)*, *63*(3), 305–334.

Fanni, J., & Parmentier, M. (1999). Physical and chemical properties of molecular compounds of lactose. *Journal of Dairy Science*, *82*(12), 2558–2563, [Elsevier].

FDA. https://www.fda.gov/downloads/Drugs/GuidanceComplianceRegulatoryInformation/Guidances/UCM070305.pdf. (consulted 21-01-2019).

Ferreira, A. P., & Tobyn, M. (2015). Multivariate analysis in the pharmaceutical industry: Enabling process understanding and improvement in the PAT and QbD era. *Pharmaceutical Development and Technology*, *20*(5), 513–527.

Fiocchi, A., Restani, P., Leo, G., Martelli, A., Bouygue, G. R., Terracciano, L., Ballabio, C., et al. (2003). Clinical tolerance to lactose in children with cow's milk allergy. *Pediatrics*, *112*(2), 359–362.

Fox, P. F. (2009). Lactose: Chemistry and properties. In P. McSweeney, & P. F. Fox (Eds.), *Advanced dairy chemistry: Volume 3: Lactose, water, salts and minor constituents* (pp. 1–15). New York, NY: Springer New York.

Gamble, J. F., Chiu, W.-S., Gray, V., Toale, H., Tobyn, M., & Wu, Y. (2010). Investigation into the degree of variability in the solid-state properties of common pharmaceutical excipients—anhydrous lactose. *AAPS PharmSciTech*, *11*(4), 1552–1557.

Gamble, J. F., Tobyn, M., & Hamey, R. (2015). Application of image-based particle size and shape characterization systems in the development of small molecule pharmaceuticals. *Journal of Pharmaceutical Sciences*, *104*(5), 1563–1574.

Garnier, S., Petit, S., & Coquerel, G. (2002). Dehydration mechanism and crystallisation behaviour of lactose. *Journal of Thermal Analysis and Calorimetry*, *68*(2), 489–502.

Gerbault, P., Liebert, A., Itan, Y., Powell, A., Currat, M., Burger, J., Swallow, D. M., et al. (2011). Evolution of lactase persistence: An example of human niche construction. *Philosophical Transactions of the Royal Society B: Biological Sciences*, *366*(1566), 863–877.

Gohel, M. C., & Jogani, P. D. (2005). A review of co-processed directly compressible excipients. *Journal of Pharmacy and Pharmaceutical Sciences*, *8*(1), 76–93.

Gombás, Á., Szabó-Révész, P., Kata, M., Regdon, G., & Eros, I. (2002). Quantitative determination of crystallinity of α-lactose monohydrate by DSC. *Journal of Thermal Analysis and Calorimetry*, *68*(2), 503–510.

Grant, A. C., Walker, R., Hamilton, M., & Garrill, K. (2015). The ELLIPTA® dry powder inhaler: Design, functionality, in vitro dosing performance and critical task compliance by patients and caregivers. *Journal of Aerosol Medicine and Pulmonary Drug Delivery*, *28*(6), 474–485.

Grasmeijer, F., Lexmond, A. J., van den Noort, M., Hagedoorn, P., Hickey, A. J., Frijlink, H. W., & de Boer, A. H. (2014). New mechanisms to explain the effects of added lactose fines on the dispersion performance of adhesive mixtures for inhalation. *PloS One*, *9*(1), e87825, [Public Library of Science].

Guigon, P., & Simon, O. (2003). Roll press design—influence of force feed systems on compaction. *Powder Technology*, *130*(1–3), 41–48, [Elsevier].

Gupta, A., Peck, G. E., Miller, R. W., & Morris, K. R. (2005). Effect of the variation in the ambient moisture on the compaction behavior of powder undergoing roller-compaction and on the characteristics of tablets produced from the post-milled granules. *Journal of Pharmaceutical Sciences*, *94*(10), 2314–2326.

Hancock, B. C., & Garcia-Munoz, S. (2013). How do formulation and process parameters impact blend and unit dose uniformity? Further analysis of the product quality research institute blend uniformity working group industry survey. *Journal of Pharmaceutical Sciences*, *102*(3), 982–986.

Haware, R. V., Bauer-Brandl, A., & Tho, I. (2010). Comparative evaluation of the powder and compression properties of various grades and brands of microcrystalline cellulose by multivariate methods. *Pharmaceutical Development and Technology*, *15*(4), 394–404.

Heikonen, M.K., & Lallukka, Y. (1985). European Patent Application No. 86302329. A process for preparing anhydrous lactose. Finland.

Hein, S., Picker-Freyer, K. M., & Langridge, J. (2008). Simulation of roller compaction with subsequent tableting and characterization of lactose and microcrystalline cellulose. *Pharmaceutical Development and Technology*, *13*(6), 523–532.

Hersey, J. A. (1975). Ordered mixing: A new concept in powder mixing practice. *Powder Technology*, *11*(1), 41–44.

Hertzler, S. R., Huynh, B. C., & Savaiano, D. A. (1996). How much lactose is low lactose? *Journal of the American Dietetic Association*, *96*(3), 243–246.

Heyman, M. B. (2006). Lactose intolerance in infants, children, and adolescents. *Pediatrics*, *118*(3), 1279–1286.

Ho, R., Muresan, A. S., Hebbink, G. A., Heng, J. Y. Y., Ho, R., Muresan, A. S., Hebbink, G. A., et al. (2010). Influence of fines on the surface energy heterogeneity of lactose for pulmonary drug delivery. *International Journal of Pharmaceutics*, *388*(1–2), 88–94.

Hoaglund Hyzer, C. S., Williamson, M. L., Jansen, P. J., Kopach, M. E., Scherer, R. B., & Baertschi, S. W. (2017). Mechanistic studies of the *N*-formylation of Edivoxetine, a secondary amine-containing drug, in a solid oral dosage form. *Journal of Pharmaceutical Sciences*, *106*(5), 1218–1238.

Hourigan, J. A., Lifran, E. V., Vu, L. T. T., Listiohadi, Y., & Sleigh, R. W. (2013). Lactose: Chemistry, Processing, and Utilization. In G. W. Smithers, & M. A. Augustin (Eds.), *Advances in Dairy Ingredients* (pp. 21–41). John Wiley & Sons, Inc.

Huang, W., Shi, Y., Wang, C., Yu, K., Sun, F., & Li, Y. (2013). Using spray-dried lactose monohydrate in wet granulation method for a low-dose oral formulation of a paliperidone derivative. *Powder Technology*, *246*, 379–394.

Hutton, J. T., & Palmer, G. M. (1972). Lactose product and method. US Patent 3,639,170.

Ilić, I., Kása, P., Jr., Dreu, R., Pintye-Hódi, K., & Srčič, S. (2009). The compressibility and compactibility of different types of lactose. *Drug Development and Industrial Pharmacy*, *35*(10), 1271–1280, [Informa Healthcare].

Inghelbrecht, S., & Remon, J. P. (1998). The roller compaction of different types of lactose. *International Journal of Pharmaceutics*, *166*, 135–144.

Itan, Y., Powell, A., Beaumont, M. A., Burger, J., & Thomas, M. G. (2009). The origins of lactase Persistence in Europe. (M. M. Tanaka, Ed.) *PLoS Computational Biology*, *5*(8), e1000491.

Jawad, R., Elleman, C., Vermeer, L., Drake, A. F., Woodhead, B., Martin, G. P., & Royall, P. G. (2012). The measurement of the β/α anomer composition within amorphous lactose prepared by spray and freeze drying using a simple 1H-NMR method. *Pharmaceutical Research*, *29*(2), 511–524.

Jawad, R., Martin, G. P., & Royall, P. G. (2015). *Chemical and compositional characterisation of lactose as a carrier in dry powder inhalers. Pulmonary Drug Delivery* (pp. 143–170). Chichester, UK: John Wiley & Sons, Ltd.

Jefferson, A. E., Williams, D. R., & Heng, J. Y. Y. (2011). Computing the surface energy distributions of heterogeneous crystalline powders. *Journal of Adhesion Science and Technology*, *25*(4–5), 339–355.

Jonathan, G., & Karim, A. (2016). 3D printing in pharmaceutics: A new tool for designing customized drug delivery systems. *International Journal of Pharmaceutics*, *499*(1), 376–394.

Jones, M. D., & Price, R. (2006). The influence of fine excipient particles on the performance of carrier-based dry powder inhalation formulations. *Pharmaceutical Research*, *23*(8), 1665–1674.

Jones, M. D., Harris, H., Hooton, J. C., Shur, J., King, G. S., Mathoulin, C. A., Nichol, K., et al. (2008). An investigation into the relationship between carrier-based dry powder inhalation performance and formulation cohesive-adhesive force balances. *European Journal of Pharmaceutics and Biopharmaceutics: Official Journal of Arbeitsgemeinschaft für Pharmazeutische Verfahrenstechnik e.V*, *69*(2), 496–507.

Jones, M. D., Young, P., & Traini, D. (2012). The use of inverse gas chromatography for the study of lactose and pharmaceutical materials used in dry powder inhalers. *Advanced Drug Delivery Reviews*, *64*(3), 285–293, [Elsevier B.V].

Katainen, E., Niemela, P., Harjunen, P., Suhonen, J., & Jarvinen, K. (2005). Evaluation of the amorphous content of lactose by solution calorimetry and Raman spectroscopy. *Talanta*, *68*(1), 1–5, [Elsevier].

Keleb, E. I., Vermeire, A., Vervaet, C., & Remon, J. P. (2004). Single-step granulation/tabletting of different grades of lactose: A comparison with high shear granulation and compression. *European Journal of Pharmaceutics and Biopharmaceutics*, *58*(1), 77–82.

Khaled, S. A., Burley, J. C., Alexander, M. R., & Roberts, C. J. (2014). Desktop 3D printing of controlled release pharmaceutical bilayer tablets. *International Journal of Pharmaceutics*, *461*(1–2), 105–111, [Elsevier B.V].

Kinnunen, H., Hebbink, G., Peters, H., Huck, D., Makein, L., & Price, R. (2014). Extrinsic lactose fines improve dry powder inhaler formulation performance of a cohesive batch of budesonide via agglomerate formation and consequential co-deposition. *International Journal of Pharmaceutics*, *478*(1), 53–59, [Elsevier B.V].

Kinnunen, H., Hebbink, G., Peters, H., Shur, J., & Price, R. (2014). Defining the critical material attributes of lactose monohydrate in carrier based dry powder inhaler formulations using artificial neural networks. *AAPS PharmSciTech*, *15*(4), 1009–1020.

Kirk, J. H., Dann, S. E., & Blatchford, C. G. (2007). Lactose: A definitive guide to polymorph determination. *International Journal of Pharmaceutics*, *334*(1–2), 103–114. Retrieved February 11, 2015, from http://www.ncbi.nlm.nih.gov/pubmed/17110062.

Kleinebudde, P. (2004). Roll compaction/dry granulation: Pharmaceutical applications. *European Journal of Pharmaceutics and Biopharmaceutics*, *58*, 317–326.

Kumar, A., Dhondt, J., De Leersnyder, F., Vercruysse, J., Vanhoorne, V., Vervaet, C., Remon, J. P., et al. (2015). Evaluation of an in-line particle imaging tool for monitoring twin-screw granulation performance. *Powder Technology*, *285*, 80–87, [Elsevier B.V].

Kushner, J., Langdon, B. A., Hicks, I., Song, D., Li, F., Kathiria, L., Kane, A., et al. (2014). A quality-by-design study for an immediate-release tablet platform: Examining the relative impact of active pharmaceutical ingredient properties, processing methods, and excipient variability on drug product quality attributes. *Journal of Pharmaceutical Sciences*, *103*(2), 527–538.

Kussendrager, K. D., & Andreae, A. C. (1984). Process for preparing lactose products. US Patent 4,594,110.

Kussendrager, K. D., & Walsma, B. (2005). Anhydrous lactose agglomerates and the preparation thereof. US Patent 8,173,173

Kussendrager, K. D., van den Bigglaar, H. A. M., & Vromans, H. (1989). Spray dried lactose and process for preparing the same. US Patent 4,802,926.

Lamešić, D., Planinšek, O., Lavrič, Z., & Ilić, I. (2016). Spherical agglomerates of lactose with enhanced mechanical properties. *International Journal of Pharmaceutics*, *516*(1–2), 247–257.

Leane, M., Pitt, K., Reynolds, G., Anwar, J., Charlton, S., Crean, A., Creekmore, R., et al. (2015). A proposal for a drug product Manufacturing Classification System (MCS) for oral solid dosage forms. *Pharmaceutical Development and Technology*, *20*(1), 12–21.

Lehto, V.-P., Tenho, M., Vähä-Heikkilä, K., Harjunen, P., Päällysaho, M., Välisaari, J., Niemelä, P., et al. (2006). The comparison of seven different methods to quantify the amorphous content of spray dried lactose. *Powder Technology*, *167*(2), 85–93.

Lerk, C. F., Andreae, A. C., de Boer, A. H., De Hoog, P., Kussendrager, K., & Van Leverink, J. (1984). Transitions of lactoses by mechanical and thermal treatment. *Journal of Pharmaceutical Sciences*, *73*(6), 857–859.

Li, Z., Lin, X., Shen, L., Hong, Y., & Feng, Y. (2017). Composite particles based on particle engineering for direct compaction. *International Journal of Pharmaceutics*, *519*(1), 272–286, [Elsevier B.V].

Listiohadi, Y., Hourigan, J. A., Sleigh, R. W., & Steele, R. J. (2009). Thermal analysis of amorphous lactose and alpha-lactose monohydrate. *Dairy Science and Technology*, *89*(1), 43–67.

Loidolt, P., Madlmeir, S., & Khinast, J. G. (2017). Mechanistic modeling of capsule filling processes. *International Journal of Pharmaceutics*, *532*(1), 47–54, [Elsevier].

Lovdahl, M. J., Hurley, T. R., Tobias, B., & Priebe, S. R. (2002). Synthesis and characterization of pregabalin lactose conjugate degradation products. *Journal of Pharmaceutical and Biomedical Analysis*, *28*(5), 917–924.

Lucas, P., Anderson, K., & Staniforth, J. N. (1998). Protein deposition from dry powder inhalers: Fine particle multiplets as performance modifiers. *Pharmaceutical Research*, *15*(4), 562–569.

Ma, Z., Merkus, H. G., de Smet, J. G. A., Heffels, C., & Scarlett, B. (2000). New developments in particle characterization by laser diffraction: Size and shape. *Powder Technology*, *111*(1–2), 66–78.

Mangal, S., Meiser, F., Morton, D., & Larson, I. (2015). Particle engineering of excipients for direct compression: Understanding the role of material properties. *Current Pharmaceutical Design*, *21*(40), 5877–5889.

Mirani, A. G., Patankar, S. P., Borole, V. S., Pawar, A. S., & Kadam, V. J. (2011). Direct compression high functionality excipient using coprocessing technique: A brief review. *Current Drug Delivery*, *8*(4), 426–435.

Mohammadi-Jam, S., & Waters, K. E. (2014). Inverse gas chromatography applications: A review. *Advances in Colloid and Interface Science*, *212*, 21–44, [Elsevier B.V].

Montalto, M., Gallo, A., Santoro, L., D'Onofrio, F., Curigliano, V., Covino, M., Cammarota, G., et al. (2008). Low-dose lactose in drugs neither increases breath hydrogen excretion nor causes gastrointestinal symptoms. *Alimentary Pharmacology and Therapeutics*, *28*(8), 1003–1012.

Muhammad, S. A. F. S., Tang, P., Chan, H.-K., & Dehghani, F. (2012). The effect of lactose micro-spherical crystals prepared by conditioning with supercritical fluid on salbutamol sulphate inhalation performance. *The Journal of Supercritical Fluids*, *71*, 92–101.

Mujumdar, A. S. (Ed.), (1995). *Handbook of industrial drying systems* (3rd ed.). Taylor & Francis Ltd.

Naik, S., & Chaudhuri, B. (2015). Quantifying dry milling in pharmaceutical processing: A review on experimental and modeling approaches. *Journal of Pharmaceutical Sciences*, *104*(8), 2401–2413.

Norman, J., Madurawe, R. D., Moore, C. M. V., Khan, M. A., & Khairuzzaman, A. (2017). A new chapter in pharmaceutical manufacturing: 3D-printed drug products. *Advanced Drug Delivery Reviews*, *108*, 39–50.

Omar, C. S., Dhenge, R. M., Osborne, J. D., Althaus, T. O., Palzer, S., Hounslow, M. J., & Salman, A. D. (2015). Roller compaction: Effect of morphology and amorphous content of lactose powder on product quality. *International Journal of Pharmaceutics*, *496*(1), 63–74.

Omar, C. S., Dhenge, R. M., Palzer, S., Hounslow, M. J., & Salman, A. D. (2016). Roller compaction: Effect of relative humidity of lactose powder. *European Journal of Pharmaceutics and Biopharmaceutics*, *106*, 26–37, [Elsevier B.V].

Pazesh, S., Gråsjö, J., Berggren, J., & Alderborn, G. (2017). Comminution-amorphisation relationships during ball milling of lactose at different milling conditions. *International Journal of Pharmaceutics*, *528*(1–2), 215–227, [Elsevier B.V].

Peebles, D. D., & Manning, P. D. V. (1933). Method for manufacture of lactose containing material. US Patent 1,928,135.

Peebles, D.D., & Manning, P. D. V. (1937). Manufacture of stable powdered products containing milk sugar. US Patent 2,835,586.

Perino, A., Cabras, S., Obinu, D., & Cavalli Sforza, L. (2009). Lactose intolerance: A non-allergic disorder often managed by allergologists. *European Annals of Allergy and Clinical Immunology*, *41*(1), 3–16.

Peters, H. J. W., & Hebbink, G. A. (2016). Selection of excipients for dry powder inhalers. *ONdrugDelivery*, *2016*(72), 187–189.

Picker-Freyer, K. M. (2008). Tablet production systems. In S. C. Gad (Ed.), *Pharmaceutical manufacturing handbook: Production and processes* (pp. 1053–1098). Wiley.

Pifferi, G., & Restani, P. (2003). The safety of pharmaceutical excipients. *Il Farmaco*, *58*(8), 541–550.

Pilcer, G., Wauthoz, N., & Amighi, K. (2012). Lactose characteristics and the generation of the aerosol. *Advanced Drug Delivery Reviews*, *64*(3), 233–256, [Elsevier B.V].

Pilcer, G., & Amighi, K. (2010). Formulation strategy and use of excipients in pulmonary drug delivery. *International Journal of Pharmaceutics*, *392*(1–2), 1–19.

Pisklak, D. M., Zielińska-Pisklak, M. A., Szeleszczuk, Ł., & Wawer, I. (2016a). 13C solid-state NMR analysis of the most common pharmaceutical excipients used in solid drug formulations Part II: CP kinetics and relaxation analysis. *Journal of Pharmaceutical and Biomedical Analysis*, *122*, 29–34.

Pisklak, D. M., Zielińska-Pisklak, M. A., Szeleszczuk, Ł., & Wawer, I. (2016b). 13C solid-state NMR analysis of the most common pharmaceutical excipients used in solid drug formulations, Part I: Chemical shifts assignment. *Journal of Pharmaceutical and Biomedical Analysis*, *122*, 81–89.

Pitchayajittipong, C., Price, R., Shur, J., Kaerger, J. S., & Edge, S. (2010). Characterisation and functionality of inhalation anhydrous lactose. *International Journal of Pharmaceutics*, *390*(2), 134–141.

Price, R., & Young, P. (2004). Visualization of the crystallization of lactose from the amorphous state. *Journal of Pharmaceutical Sciences*, *93*(1), 155–164.

Price, R., Young, P., Edge, S., & Staniforth, J. (2002). The influence of relative humidity on particulate interactions in carrier-based dry powder inhaler formulations. *International Journal of Pharmaceutics*, *246*(1–2), 47–59.

Rajah, K. K., & Blenford, D. E. (Eds.), (1988). *The ALM guide to lactose properties and uses*. The Hague: Association of Lactose Manufacturers.

Rangel, A. H., do, N., Sales, D. C., Urbano, S. A., Galvão Júnior, J. G. B., Andrade Neto, J. C. d, & Macêdo, C. dS. (2016). Lactose intolerance and cow's milk protein allergy. *Food Science and Technology (Campinas)*, *36*(2), 179–187, [SBCTA].

Rassu, G., Eissens, A. C., & Bolhuis, G. K. (2006). Tableting properties of an improved spray-dried lactose. *Journal of Drug Delivery Science and Technology*, *16*(6), 455–459, [Elsevier Masson SAS].

Robles, J., & Motheral, L. (2014). Hypersensitivity reaction after inhalation of a lactose-containing dry powder inhaler—case report. *Journal of Pediatric Pharmacology and Therapeutics*, *19*(3), 206–211.

Roos, Y. H. (2009). In P. McSweeney, & P. F. Fox (Eds.), *Advanced dairy chemistry* (Vol. 3). New York, NY: Springer New York.

Rosenboom, J.-G. J.-G., Antonyuk, S., Heinrich, S., & Kraft, M. (2015). Characterisation of lactose powder and granules for multivariate wet granulation modelling. *Chemical Engineering Science*, *123*, 395–405.

Roumeli, E., Tsiapranta, A., Pavlidou, E., Vourlias, G., Kachrimanis, K., Bikiaris, D., & Chrissafis, K. (2013). Compatibility study between trandolapril and natural excipients used in solid dosage forms. *Journal of Thermal Analysis and Calorimetry*, *111*(3), 2117–2123.

Rowe, J.M., Charlton, S.T., & McCann, R.J. (2017). Development, scale-up, and optimization of process parameters: Roller compaction theory and practice. *Developing solid oral dosage forms: Pharmaceutical theory and practice* (2nd ed.), 869–915. Academic Press.

Saarinen, T., Antikainen, O., & Yliruusi, J. (2017). Simultaneous comparison of two roller compaction techniques and two particle size analysis methods. *AAPS PharmSciTech*, *18*(8), 3198–3207.

Saha, S., & Shahiwala, A. F. (2009). Multifunctional coprocessed excipients for improved tabeltting performance. *Expert Opinion on Drug Delivery*, *6*, 197–208.

Sanders, M. (2007). Inhalation therapy: An historical review. *Primary Care Respiratory Journal*, *16*(2), 71–81, [General Practice Airways Group].

Šantl, M., Ilić, I., Vrečer, F., & Baumgartner, S. (2011). A compressibility and compactibility study of real tableting mixtures: The impact of wet and dry granulation versus a direct tableting mixture. *International Journal of Pharmaceutics*, *414*(1–2), 131–139.

Shah, B., Kakumanu, V. K., & Bansal, A. K. (2006). Analytical techniques for quantification of amorphous/crystalline phases in pharmaceutical solids. *Journal of Pharmaceutical Sciences*, *95*(8), 1641–1665.

Shah, K. R., Hussain, M. A., Hubert, M., & Farag Badawy, S. I. (2008). Form conversion of anhydrous lactose during wet granulation and its effect on compactibility. *International Journal of Pharmaceutics*, *357*(1–2), 228–234.

Shah, U. V., Karde, V., Ghoroi, C., & Heng, J. Y. Y. (2017). Influence of particle properties on powder bulk behaviour and processability. *International Journal of Pharmaceutics*, *518*(1–2), 138–154.

Shangraw, R. F. (1989). Compressed tablets by direct compression. *Pharmaceutical dosage forms: Tablets*, *2*(1), 195–246.

Shariare, M. H., de Matas, M., York, P., & Shao, Q. (2011). The impact of material attributes and process parameters on the micronisation of lactose monohydrate. *International Journal of Pharmaceutics*, *408*(1–2), 58–66.

Sharp, P.F. (1940). US patent application No. 2319562. Stable crystalline anhydrous alpha lactose product and process. US.

Shekunov, B. Y., Chattopadhyay, P., Tong, H. H. Y., & Chow, A. H. L. (2007). Particle size analysis in pharmaceutics: Principles, methods and applications. *Pharmaceutical Research*, *24*(2), 203–227.

Shur, J., Harris, H., Jones, M. D., Kaerger, J. S., & Price, R. (2008). The role of fines in the modification of the fluidization and dispersion mechanism within dry powder inhaler formulations. *Pharmaceutical Research*, *25*(7), 1631–1640.

Shur, J., Price, R., Lewis, D., Young, P. M., Woollam, G., Singh, D., & Edge, S. (2016). From single excipients to dual excipient platforms in dry powder inhaler products. *International Journal of Pharmaceutics*, *514*(2), 374–383.

Silanikove, N., Leitner, G., & Merin, U. (2015). The interrelationships between lactose intolerance and the modern dairy industry: Global perspectives in evolutional and historical backgrounds. *Nutrients*, *7*(9), 7312–7331.

Smollich, 2016. http://pharmacon.de/2016/laktose-kein-problem-in-arzneimitteln/ (accessed 23-01-2019).

Squire, P. (1866). *A Companion to the British Pharmacopoeia.* London: John Churchill and Sons.

Steckel, H., Markefka, P., TeWierik, H., & Kammelar, R. (2006). Effect of milling and sieving on functionality of dry powder inhalation products. *International Journal of Pharmaceutics*, *309*(1–2), 51–59.

Strickley, R. G. (2004). Solubilizing excipients in oral and injectable formulations. *Pharmaceutical Research*, *21*(2), 201–230.

Sun, C. C. (2010). Setting the bar for powder flow properties in successful high speed tableting. *Powder Technology*, *201*(1), 106–108, [Elsevier B.V].

Supplee, G.C., & Flanigan, G.E. (1934). Process of producing beta lactose. US Patent 1,954,602.

Suresh, P., Sreedhar, I., Vaidhiswaran, R., & Venugopal, A. (2017). A comprehensive review on process and engineering aspects of pharmaceutical wet granulation. *Chemical Engineering Journal*, *328*, 785–815, [Elsevier B.V].

Szalka, M., Lubczak, J., Naróg, D., Laskowski, M., & Kaczmarski, K. (2014). The Maillard reaction of bisoprolol fumarate with various reducing carbohydrates. *European Journal of Pharmaceutical Sciences*, *59*(1), 1–11, [Elsevier B.V].

Tan, B. M. J., Chan, L. W., & Heng, P. W. S. (2016). Improving dry powder inhaler performance by surface roughening of lactose carrier particles. *Pharmaceutical Research*, *33*, 1–13.

Tan, G., Morton, D., & Larson, I. (2015). On the methods to measure powder flow. *Current Pharmaceutical Design*, *21*(40), 5751–5765.

Telko, M. J., & Hickey, A. J. (2005). Dry powder inhaler formulation. *Respiratory Care*, *50*(9), 1209–1227.

Thalberg, K., Lindholm, D., & Axelsson, A. (2004). Comparison of different flowability tests for powders for inhalation. *Powder Technology*, *146*(3), 206–213.

Thoren, P., Wallin, A., Whitehead, P. J., & Sandström, T. (2001). The effect of different concentrations of lactose powder on the airway function of adult asthmatics. *Respiratory Medicine*, *95*(11), 870–875.

Timmermann, I.-L., Steckel, H., & Trunk, M. (2006). Assessing the re-crystallization behaviour of amorphous lactose using the RH-perfusion cell. *European Journal of Pharmaceutics and Biopharmaceutics: Official Journal of Arbeitsgemeinschaft für Pharmazeutische Verfahrenstechnik e.V*, *64*(1), 107–114.

Twieg, W., & Nickerson, T. (1968). Kinetics of lactose crystallization. *Journal of Dairy Science*, *51*(11), 1720–1724.

Tye, C. K., Sun, C., & Amidon, G. E. (2005). Evaluation of the effects of tableting speed on the relationships between compaction pressure, tablet tensile strength, and tablet solid fraction. *Journal of Pharmaceutical Sciences*, *94*(3), 465–472.

Ursan, I. D., Chiu, L., & Pierce, A. (2013). Three-dimensional drug printing: A structured review. *Journal of the American Pharmacists Association*, *53*(2), 136–144.

van Kreveld, A., & Michaels, A. S. (1965). Measurement of crystal growth of α-lactose. *Journal of Dairy Science*, *48*(3), 259–265.

Velázquez-gonzález, K., Ramírez-flores, E., & Villafuerte-robles, L. (2015). Influence of different types of lactose on powder flow and tablets dissolution—article—pharma excipients. *Journal of Applied Pharmaceutical Science*, *5*(9), 8.

Visser, R. A. (1983). *Crystal growth kinetics of alpha-lactose hydrate*. Catholic University of Nijmegen.

Vollenbroek, J., Hebbink, G. A., Ziffels, S., Steckel, H., Vollenbroek, J., Hebbink, G. A., Ziffels, S., et al. (2010). Determination of low levels of amorphous content in inhalation grade lactose by moisture sorption isotherms. *International Journal of Pharmaceutics*, *395*(1–2), 62–70.

Vollenbroek, J., Hebbink, G. A., Ziffels, S., & Steckel, H. (2010). Determination of low levels of amorphous content in inhalation grade lactose by moisture sorption isotherms. *International Journal of Pharmaceutics*, *395*(1–2), 62–70.

Vromans, H. Studies on Consolidation and Compaction Properties of Lactose (1987) Thesis. University of Groningen, The Netherlands.

Vromans, H., Bolhuis, G., Lerk, C., Van Den Biggelaar, H., & Bosch, H. (1987). Studies on tableting properties of lactose. VII. The effect of variations in primary particle size and percentage of amorphous lactose in spray dried lactose products. *International Journal of Pharmaceutics*, *35*(1–2), 29–37.

Vromans, H., De Boer, A. H., Bolhuis, G. K., Lerk, C. F., & Kussendrager, K. D. (1986). Studies on tableting properties of lactose: The effect of initial particle size on binding properties and dehydration characteristics of α-lactose monohydrate. *Drug Development and Industrial Pharmacy*, *12*(11–13), 1715–1730.

Vromans, H., De Boer, A. H., Bolhuis, G. K., Lerk, C. F., Kussendrager, K. D., & Bosch, H. (1985). Studies on tableting properties of lactose. Part 2. Consolidation and compaction of different types of crystalline lactose. *Pharmaceutisch Weekblad*, *7*(5), 186–193.

Walstra, P., Wouters, J. T. M., & Geurts, Tom J. (2005). *Dairy science & technology* (2nd ed.). Abingdon, Oxford, UK: CRC Press, Taylor & Francis Group, 782 pp.

Wang, S., Li, J., Lin, X., Feng, Y., Kou, X., Babu, S., & Panicucci, R. (2015). Novel coprocessed excipients composed of lactose, HPMC, and PVPP for tableting and its application. *International Journal of Pharmaceutics*, *486*(1–2), 370–379.

Watling, C. P., Elliott, J. A., Scruton, C., Cameron, R. E., Watling, C. P., Elliott, J. A., Scruton, C., et al. (2010). Surface modification of lactose inhalation blends by moisture. *International Journal of Pharmaceutics*, *391*(1–2), 29–37.

Westhoff, G. M., Kuster, B. F., Heslinga, M. C., Pluim, H., & Verhage, M. (2014). Lactose and derivatives. *Ullmann's Encyclopedia of Industrial Chemistry*. (pp. 1–9). Weinheim: Wiley-VCH Verlag GmbH & Co KGaA.

Whiteman, M., & Yarwood, R. J. (1988). The evaluation of six lactose-based materials as direct compression tablet excipients. *Drug Development and Industrial Pharmacy*, *14* (8), 1023–1040.

Wiecek, S., Wos, H., Horowska-Ziaja, S., Flak-Wancerz, A., & Grzybowska-Chlebowczyk, U. (2016). Lactose intolerance in children with IgE-dependent allergy to milk proteins. *Pediatric, Allergy, Immunology, and Pulmonology*, *29*(2), 86–90.

Wu, S.-J., & Sun, C. (Calvin) (2007). Insensitivity of compaction properties of brittle granules to size enlargement by roller compaction. *Journal of Pharmaceutical Sciences*, *96*(5), 1445–1450, [Elsevier].

Zarbock, S. D., Magnuson, B., Hoskins, L., Record, K. E., & Smith, K. M. (2007). Lactose: The hidden culprit in medication intolerance? *Orthopedics*, *30*(8), 615–617.

Zeng, X. M. (2001). Particulate interactions in dry powder formulations for inhalation. Igarss 2014. Abingdon, UK: Taylor & Francis.

Zeng, X. M., Martin, G. P., Marriott, C., & Pritchard, J. (2000). The influence of carrier morphology on drug delivery by dry powder inhalers. *International Journal of Pharmaceutics*, *200*(1), 93–106.

Zuurman, K., Bolhuis, G. K., & Vromans, H. (1995). Effect of binder on the relationship between bulk density and compactibility of lactose granulations. *International Journal of Pharmaceutics*, *119*(1), 65–69.

Zuurman, K., Riepma, K. A., Bolhuis, G. K., Vromans, H., & Lerk, C. F. (1994). The relationship between bulk density and compactibility of lactose granulations. *International Journal of Pharmaceutics*, *102*(1–3), 1–9.

CHAPTER

Lactose in the dairy production chain

6

Kasper A. Hettinga
Dairy Science & Technology, Food Quality & Design Group, Wageningen University & Research, Wageningen, The Netherlands

6.1 Introduction

Lactose, the main carbohydrate in milk, is a unique sugar produced in the mammary gland of mammals. In this chapter, lactose is discussed from a chain perspective: from mammary gland, through dairy processing, to the consumer and customer of ingredients. A graphical representation of this is shown in Fig. 6.1, which depicts the lactose from its initial production to its fate along the dairy production chain, including references to the different sections of this chapter in which the respective topics are discussed.

First, lactose is introduced from the perspective of primary production. Lactose synthesis in the mammary gland will be explained, as well as its role in driving milk yield. This section will also deal with the importance of lactose for the newborn. Second, an overview of the physical and chemical properties of lactose, as relevant for dairy consumer products and ingredients, will be given. Third, the role of lactose in dairy products is discussed, especially looking at the conversion of lactose during fermentation and its removal for lactose-free dairy products. Fourth, the specific role of lactose in dairy-based ingredient is reviewed. This will deal with both the role of lactose in a wide range of dairy-based ingredients and specific lactose-based ingredients, all of which can be used in many food, feed, and pharmaceutical applications. Fifth, a consumer perspective on lactose is given, looking at the intake levels of lactose by the average consumer, as well as the role of lactose in perception and quality of dairy products and food products to which dairy-based ingredients have been added. Finally, a perspective is given on future research needs with regard to lactose in the dairy chain.

6.2 Primary milk production

Lactose starts its journey along the dairy production chain on the farm (Fig. 6.1). In the mammary gland of mammals, lactose is synthesized from glucose.

Lactose. DOI: https://doi.org/10.1016/B978-0-12-811720-0.00006-4

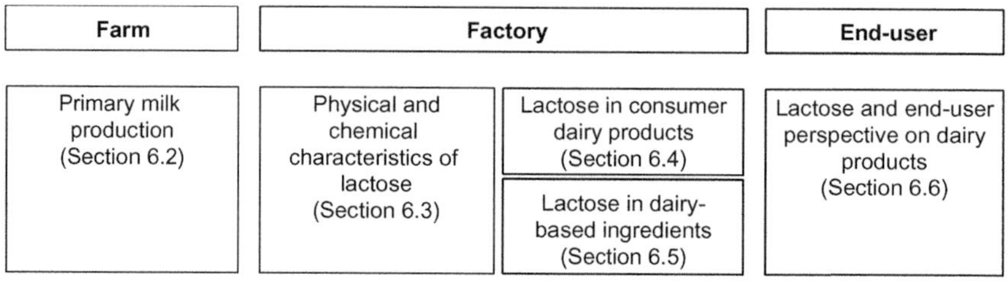

FIGURE 6.1

Lactose along the dairy chain, its production, characteristics, and the importance in processing and consumption, including references to the different sections of this chapter.

The amount of lactose synthesized at this stage has a major impact on the volume of milk the mammal produces. Lactose is also an important energy source in milk for the newborn. In this section, this primary production perspective on lactose will be given, with a focus on bovine milk production.

6.2.1 Role of glucose in milk synthesis

Milk is synthesized in the mammary gland. The gross anatomy of the mammary gland differs a lot between different mammalian species, with the number of glands and teats being very different among them. The microscopic anatomy, however, is very similar among mammalian species. Mammary glands are made up of a branching network of ducts that end in alveoli. Each alveolus consists of epithelial cells surrounding a lumen. The alveoli function by absorbing components from the blood and transforming them into the basic components of milk, followed by the secretion of these components into the lumen. Most of the milk components are synthesized locally in the mammary epithelial cells of the mammary gland, whereas few others (vitamins, minerals and some specific proteins) are transported unchanged from the blood to the milk. Fig. 6.2 gives an overview of the synthesis and secretion of all milk components.

Glucose is a major component absorbed from the blood by the mammary gland (Fig. 6.2). It is crucial for lactose synthesis, for which it is the sole precursor. However, the glucose taken up into the mammary epithelial cells has many roles besides lactose synthesis. The glucose requirements of lactating dairy cattle are thus of particular importance, as glucose is not only required to synthesize lactose. One of its other functions is being a major precursor of glycerol (to form milk triglycerides). Moreover, since ruminants require a large part of NADPH to be synthesized from glucose in the pentose phosphate pathway, the de novo synthesis of fatty acids, in which NADPH is used, also involves glucose utilization. The combination of all these processes imposes a large drain on

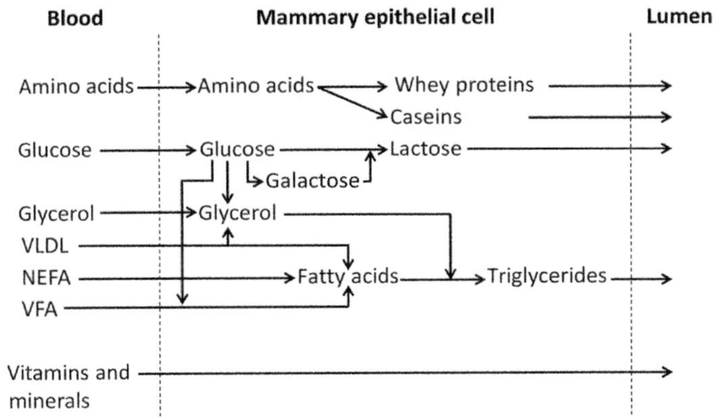

FIGURE 6.2

Principle pathways of synthesis and secretion of milk components, including the precursors for the synthesized milk components that are absorbed from the blood. *NEFA*, Nonesterified fatty acids; *VFA*, volatile fatty acids; *VLDL*, very-low-density lipoproteins.

available glucose. For each kg of bovine milk, 60–70 g of glucose is required, of which about two-thirds is used for lactose synthesis (Cant, Trout, Qiao, & Purdie, 2002). The production of 40 kg of milk per day thus requires about 2.5 kg of glucose per day. However, the direct absorption of glucose from the intestinal tract into blood is low. Carbohydrates in the diet of the cow are largely fermented in the rumen, so that only a small amount of carbohydrates may deliver some glucose to be absorbed. Dairy cattle therefore rely heavily on the process of gluconeogenesis, a metabolic pathway that results in the generation of glucose from noncarbohydrate carbon substrates such as pyruvate, lactate, glycerol, and glucogenic amino acids (Schutz, 2011). In cows, propionic acid alone accounts for more than 50% of all the glucose produced in the gluconeogenesis process. The glucogenic amino acids may contribute up to 30% of glucose production (Young, 1977). However, synthesis of glucose from amino acids gives rise to substantial excess of nitrogen that is then excreted as urea, which represents a loss of nitrogen to the cow (Jonker, Kohn, & Erdman, 1998). Lactate, pyruvate, and glycerol are minor contributors to gluconeogenesis in the cow (Young, 1977). Especially in early lactation, glucose and its precursors may be in short supply, giving rise to a negative energy balance. This negative energy balance may lead to cows mobilizing body fat, making them vulnerable to metabolic and reproductive disorders. Hence, careful balancing of the diet for glucogenic nutrients, by supplying more plants rich in carbohydrates, is essential for optimal production and health of high producing dairy cattle (Steeneveld, Schukken, van Knegsel, & Hogeveen, 2013).

6.2.2 Biosynthesis of lactose and related carbohydrates

The mammary gland uses glucose for several milk biosynthesis pathways as explained in Section 6.2.1. The main use of glucose, however, is for lactose synthesis, as further described in this section.

6.2.2.1 Lactose synthesis

Lactose synthesis, as shown in Fig. 6.2, is based on glucose derived fully from arterial blood supply. However, lactose is formed by condensation of one glucose and one galactose molecule. Therefore besides glucose, galactose is also required. This galactose has to be synthesized in the mammary gland from the absorbed glucose. Fig. 6.3 shows the biochemical process of both glucose activation and epimerization to uridine diphosphate (UDP)-galactose, which combined with glucose, and with the help of lactose synthase, leads to the final synthesis of lactose.

The lactose synthase system catalyzes the formation of lactose from glucose and galactose. This final stage of lactose synthesis takes place in the Golgi, which is one of the organelles in the mammary epithelial cells. After the lactose has been synthesized, it is transported by cytoplasmic vesicles to the lumen, where it is secreted.

As mentioned in Section 6.1, lactose is uniquely synthesized in the mammary gland of mammals. The reason why lactose is synthesized, and not another disaccharide, may be some of the specific properties of lactose. One major property that may underlie this is that lactose cannot diffuse out of the Golgi or the secretory vesicles. Therefore, the lactose can be easily secreted and also aid in drawing water into these organelles, thereby determining the volume of milk produced (Holt, 1983), as will be further explained in Section 6.2.3.

6.2.2.2 Regulation of lactose synthesis

The lactose synthase system is a complex of galactosyltransferase and α-lactalbumin. The α-lactalbumin is, besides part of the lactose synthase system, also one of the major whey proteins secreted by the mammary epithelial cells. Usually, the galactosyltransferase, that is part of the lactose synthase system, will add galactose to *N*-acetylglucosamine, which is not the catalytic activity that is needed for lactose synthesis in the mammary gland. The binding of α-lactalbumin

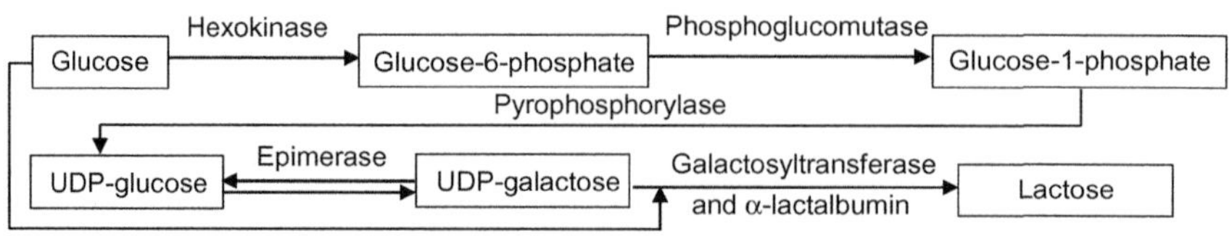

FIGURE 6.3

Mammary gland synthesis pathway of lactose.

changes the substrate specificity of the galactosyltransferase from *N*-acetylglucosamine to glucose, thereby making lactose synthesis in the mammary gland possible. Synthesis of lactose thus depends on the gene transcription of α-lactalbumin, so that the α-lactalbumin can be synthesized and become part of the lactose synthase system. This transcription of α-lactalbumin, that is thus essential to milk synthesis, is regulated by the hormone prolactin, indicating that the lactose synthase system is only active in the mammary gland during pregnancy and lactation (Akersk, Bauman, Capuco, Goodman, & Tucker, 1981). Several studies have been done to confirm this importance of α-lactalbumin for the lactose synthase complex, and thereby lactose synthesis in the mammary gland. First, it was shown that in the nonlactating mammary gland, no α-lactalbumin is formed and the lactose synthase system is therefore not active at this stage (Anderson, Rudolph, McManaman, & Neville, 2007), which is probably due to the absence of the hormone prolactin. Second, using gene knockouts of α-lactalbumin in mice, the requirement for α-lactalbumin in the synthesis of lactose has been confirmed (Stacey et al., 1995; Stinnakre, Vilotte, Soulier, & Mercier, 1994). The low-lactose milk produced in this way is too viscous to be removed from the mammary gland easily, which is related to the osmotic role of lactose in milk yield, as will be explained further in Section 6.2.3.

6.2.2.3 Synthesis of other carbohydrates

Besides lactose, also other carbohydrates are synthesized in the mammary gland. The number as well as amount of these other carbohydrates (mainly oligosaccharides) is highly variable between mammalian species. These oligosaccharides are generally produced at (much) lower levels than lactose. The reducing end of such milk oligosaccharides is composed of a lactose moiety. To this lactose unit, lacto-*N*-biose units (Gal-(β1,3)-GlcNAc) or lactosamine units (Gal-(β1,4)-GlcNAc) are attached. Next, a wide range of other carbohydrate units can be attached, resulting in a wide variety of different oligosaccharides (Urashima, Asakuma, & Messer, 2007). These different oligosaccharides have a wide range of functions that are highly structure-dependent, as reviewed by Bode (2012). Oligosaccharides, both isolated from milk and produced enzymatically, are frequently used as dairy-based ingredient, as will be further discussed in Section 6.5.3.

6.2.3 Variation in lactose synthesis and its effects on overall milk synthesis

The lactose that is secreted by the mammary gland, as explained in the previous section, has important consequences for the overall volume of milk that is produced by the mammary gland, due to its role in the osmolarity of milk. This lactose synthesis depends on the availability of glucose. However, glucose availability also has effects beyond the synthesis of lactose. These topics will be discussed in this section.

6.2.3.1 Role of lactose in milk yield

The effect of lactose on milk yield is mainly due to the important role of lactose in the osmotic pressure of milk. The osmotic pressure of milk is identical to the osmotic pressure of blood, which is very constant. The osmotic pressure of milk is directly proportional to the number of particles in solution; it is therefore obvious that it is primarily determined by the major constituents of low molecular weight, the lactose and the salts, and not by proteins and fat globules. In general, more than 50% of the osmotic pressure of milk is determined by the lactose that is secreted by the mammary epithelial cells. This thereby explains the role of lactose in determining the total milk yield. If more lactose is synthesized in the mammary gland, this will lead to a need for more water to be drawn into the lumen of the mammary gland from the blood, to keep the osmotic pressure of the milk in the lumen stable. To maintain a water balance and maximize milk production, cows need to have access to sufficient water for the high water metabolism required for lactation (Murphy, 1992). In conclusion, cows will thus produce more liters of milk, if the synthesis of lactose increases. At the same time, this also leads to a relative constant lactose content (% w/w) of milk. This is also reflected in the variation in lactose levels between cows, which is much smaller than the variation in the other two macronutrients, fat and protein (Stoop, Bovenhuis, & van Arendonk, 2007). Also seasonal variation, which is for a large part determined by variation in feed composition, is smaller for lactose than for the other main milk components (Heck, van Valenberg, Dijkstra, & van Hooijdonk, 2009).

6.2.3.2 Osmolarity of milk

Although lactose is important for the osmotic pressure of milk, it is not the sole contributor. Salts are the other major contributor to osmotic pressure. While lactose is synthesized in the mammary gland, minerals are both actively and passively transported from the blood capillary into the lumen of the mammary gland. This is the main reason why there is some variation in lactose content of milk (Wheelock, Rook, & Dodd, 1965). First, looking at different lactation stages, we see that lactose is slowly decreasing from approximately 4.6% at peak lactation to about 4.4% in late lactation. This is due to the slowly increasing rate of passive transfer of minerals into the lumen of the mammary gland over lactation, due to increased leaking over the tight junctions between the mammary epithelial cells. In early lactation, the tight junctions between mammary epithelial cells function well and minimize this passive diffusion. As lactation progresses, these tight junctions loose effectiveness, and will then allow more minerals to passively diffuse into the milk (Auldist, Coats, Rogers, & McDowell, 1995; Nguyen & Neville, 1998). This increased amount of minerals diffusing into milk will then contribute to the osmotic pressure, and therefore to have a stable osmotic pressure, a lower lactose concentration in the milk will be reached. This negative correlation between lactose content and minerals is a generic pattern that has been shown in the milk of many mammals (Sharp, 1938).

The same process also plays a role during inflammation of the mammary gland. Here, because of leakage between the damaged mammary epithelial cells, which is due to the inflammatory process, passive mineral diffusion is increased. Simultaneously, synthesis of milk components (including lactose) is reduced, due to damage to the mammary epithelial cells. These two effects together lead to a relatively low-lactose content, which can drop by about 0.5% point (Auldist et al., 1996), and relatively high mineral content of milk during a mammary inflammation.

6.2.3.3 Regulation of lactose in nonbovine mammals

The lactose synthesis and its regulation, as described earlier, are general processes that occur in all mammals, although the examples given were mainly focused on cows and bovine milk, which is the major industrial source of milk and lactose derived from it. It is important to realize that although the basic processes underlying lactose synthesis and its role in determining the osmotic pressure of milk are the same for all mammals, not all mammals have equal amounts of lactose in their milk. It ranges from 0.1% in seal milk to 7.5% in human milk, as given in Table 6.1. On the higher end of this range, the lactose concentration is limited by the osmotic pressure of the mammal's body. The variation in lactose content within a species is tightly regulated, similar to the cow, through its effect on

Table 6.1 Lactose content of a number of different mammals.

Mammal	Lactose content (%)
Seal	0.1
Black bear	0.4
Rabbit	3.0
Reindeer	3.9
Goat	4.3
Cow	4.6
Indian elephant	4.7
Sheep	4.8
Camel	5.0
Pig	5.5
Donkey	6.3
Human	7.5

Based on Sharp, P. F. (1938). Relation between lactose and ash content of the milk of different mammals. Journal of Dairy Science, 21, *127–128. doi:10.3168/jds.S0022-0302(38)95626-7; Jenness, R., Regehr, E. A., & Sloan, R. E. (1964). Comparative biochemical studies of milk. II. Dialyzable carbohydrates.* Comparative Biochemistry and Physiology, 13, *339–352; Warren, C. D., Chaturvedi, P., Newburg, A. R., Oftedal, O. T., Tilden, C. D., & Newburg, D. S. (2001). Comparison of oligosaccharides in milk specimens from humans and twelve other species.* Advances in Experimental Medicine and Biology, 501, *325–332; Osman Swar, M. (2011). Donkey milk-based formula: A substitute for patients with cow's milk protein allergy.* Sudanese Journal of Paediatrics, 11, *21–24. (Jenness, Regehr, & Sloan, 1964; Osman Swar, 2011; Warren et al., 2001).*

osmotic pressure, where more lactose synthesis leads to a larger milk volume due to its effect of drawing water into the lumen to keep osmotic pressure the same on both sides of the epithelial cells.

6.2.3.4 Effects beyond lactose synthesis

As mentioned in Section 6.2.1, lactose synthesis strongly depends on glucose availability, because this is the sole precursor for lactose synthesis. However, blood glucose is involved in milk synthesis, not only as a direct precursor for lactose synthesis, but also for other milk synthesis–related functions and as energy carrier, as mentioned before. When it comes to energy use, glucose can be reduced to pyruvate through glycolysis (Abraham & Chaikoff, 1959), which can then be further incorporated in the Krebs cycle for energy production in the form of ATP. These roles of glucose in the metabolism of the mammary gland have inspired researchers to study the effect of blood glucose levels and insulin concentration (as major regulator of glucose levels in the blood) on the mammary milk synthesis process. This research has shown that increasing glucose levels in blood after metabolic interventions such as infusion of insulin, or infusion of energy in the form of starch or glucose, have wide ranging effect on milk synthesis. Not only glucose metabolism itself was altered, but also seemingly unrelated processes such as milk protein synthesis (Mackle et al., 1999; Menzies, Lefevre, Macmillan, & Nicholas, 2009). Cows in negative energy balance (see Section 6.2.1) have been infused with glucose or insulin, but this did not necessarily result in an increased lactose synthesis. In early lactation, lactose production reduces with improved energy balance, consequently resulting in a lower milk yield, as explained earlier in this section, and further expanded upon in literature (Wheelock et al., 1965). In early lactation, lactose production thus depends on the metabolic status, resulting in an impact on overall milk yield depending on metabolic state (van Knegsel, Remmelink, Jorjong, Fievez, & Kemp, 2014). These studies on glucose and insulin infusion indicate that lactose synthesis is not a self-contained process in the mammary gland, but is intricately related to the synthesis of other milk components at the same time, probably due to the multiple roles of lactose in the milk synthesis process in the mammary gland.

6.2.4 Importance of milk carbohydrates for the newborn

Milk is the sole source of lactose in nature, and milk of all mammals contains lactose. There are several benefits of lactose as a carbohydrate in milk. First, per unit energy, it has a lower impact on osmolarity compared to monosaccharides. With osmotic pressure being tightly regulated, as described in Section 6.2.3, this means that milk can contain more energy at the same osmotic pressure. In addition, the viscosity of a lactose solution is lower, at a similar weight-percentage solution compared to monosaccharides. Both these effects on osmotic pressure and viscosity make lactose a suitable sugar for milk, although this would be true for many other disaccharides as well. However, an added benefit of lactose compared to other disaccharides is its low sweetness, its easy hydrolysis in the newborn digestive system, and especially its properties in the mammary gland (see Section 6.2.2). Compared to lactose, the milk oligosaccharides have an even larger difference to monosaccharides on both these two parameters.

A large variation of lactose content is present between species, although the variation within a species is rather limited, as explained in Section 6.2.3. Due to the variation between species, the relative contribution of lactose to energy intake is highly variable between mammals. Compared to the milk fat and milk protein, lactose has up till now been associated with a much smaller range of functions, and is often seen as a mere supply of energy. To release the energy from lactose, the lactose needs to be hydrolyzed by lactase. Lactase synthesis already starts in utero. In most mammals, lactase synthesis remains active until weaning, after which it disappears (Antonowicz & Lebenthal, 1977). The diminishing synthesis of lactase may be triggered by reduced lactose synthesis in the mammary gland (Chapter 2: Digestion, absorption, metabolism, and physiological effects of lactose). In humans, lactase synthesis postweaning does exist (Chapter 1: The evolution of lactose digestion and Chapter 2: Digestion, absorption, metabolism, and physiological effects of lactose), which is due to genetic polymorphisms that emerged around 7,000–10.000 years ago (Chapter 1: The evolution of lactose digestion).

Compared to lactose, oligosaccharides on the other hand have been shown to have a much broader range of functions. In the period after birth, the microbiota in the intestine needs to develop. The presence of oligosaccharides in milk is important for this development of the microbiota of newborns, by providing specific carbohydrates that are not digested in the small intestine and reach the colon intact, where they can be fermented by the microbiota. Similar effects on microbiota stimulation may exist for lactose, although its prebiotic effect is controversial (Chapter 4: Lactose—a conditional prebiotic?). The importance of oligosaccharides for the microbiota is one of the reasons why a lot of research is done on the isolation and synthesis of oligosaccharides for addition to infant formula. This will be further discussed in Section 6.5, in which lactose-based ingredients are further discussed.

6.3 Physical and chemical characteristics of lactose

As mentioned in Section 6.2, lactose synthesized by the cow affects milk production level. This effect, and especially the effects of lactose further down the chain (see Fig. 6.1), depends strongly on the physical and chemical characteristics of lactose. In this section, these characteristics are explained, allowing a better understanding of the role of lactose in dairy products and dairy-based ingredients (Sections 6.4–6.6).

As explained in Section 6.2.2, lactose is a disaccharide composed of a galactose and glucose unit. Both sugar units exist in a six-ring configuration, as shown in Fig. 6.4.

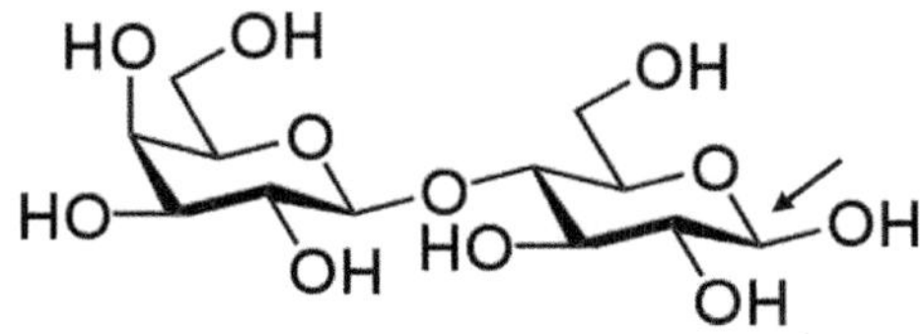

FIGURE 6.4

Chemical structure of lactose with the arrow indicating the C1-atom of glucose that determines the lactose configuration in α-lactose and β-lactose (further explained in Section 6.3.1).

The glucose unit of lactose can be hydrolyzed by water, leading to an open chain structure. Lactose in solution is in an equilibrium between the open and closed structure. This means that lactose is a reducing sugar, because it has a reactive carbonyl group when present as an open structure. Although only a small fraction ($<0.1\%$) of lactose is present in an open chain structure, this open chain structure is chemically important, because it is chemically reactive, for example, in the Maillard reaction (Section 6.3.4). Increasing temperature and increasing or decreasing pH all lead to an increase in the open chain structure (Walstra, Wouters, & Geurts, 2006). In the following subsections, both basic physical and chemical properties as well as some chemical changes that can occur to lactose during processing are described.

6.3.1 Mutarotation

Lactose occurs in two basic configurations: α-lactose and β-lactose. These two configurations differ in the position of the hydroxyl group on the C1-atom of the glucose (Fig. 6.4). There are marked differences between α- and β-lactose, especially in solubility and crystallization behavior; α-lactose is much less soluble and crystallizes as a monohydrate, whereas β-lactose is more soluble and forms anhydrous crystals. One could imagine that this could have an impact on digestion and metabolism of lactose, but previous research did not find an effect of the lactose configuration on its digestibility (Zunft & Schulze, 1990). This may be due to a process called mutarotation. Mutarotation is a process whereby, in solution, α- and β-lactose can convert into each other. As mentioned earlier, a small fraction of lactose exists as an open chain structure, which is an intermediate in the interconversion between α- and β-lactose (Walstra et al., 2006). The rate of mutarotation depends on temperature and pH, because both have a large influence on the fraction of lactose that is present in the open chain structure (Troy & Sharp, 1930). After a certain amount of time, depending on the mutarotation rate, there will be an equilibrium between α- and β-lactose, of which the ratio depends on the temperature (Haase & Nickerson, 1966).

6.3.2 Solubility

Compared to other disaccharides, the solubility of lactose is rather low. As an example, the solubility at 25°C is only 10% of that of sucrose. Fig. 6.5 shows that there is a large difference in the initial solubility of α- and β-lactose. But as mentioned in Section 6.3.1, mutarotation will cause an interconversion, leading to a different solubility of lactose after time (indicated with the solid lines in Fig. 6.5). So, initially, little α-lactose can be dissolved in water (e.g., 7 g/100 g water at 20°C), but after mutarotation that causes part of the α-lactose to be interconverted to β-lactose, more α-lactose will dissolve that will again be partly interconverted into β-lactose. The opposite occurs when dissolving β-lactose, which will lead to crystallization of α-lactose. At 20°C, the final lactose solubility of 19.2% is in the form of 7% α-lactose and 11.2% β-lactose (Walstra et al., 2006), whereas at body temperature (37°C) the final solubility is about 24% (Hunziker & Nissen, 1926).

6.3.3 Lactose crystallization

Although lactose has a low solubility, it does not crystallize easily. As discussed in Section 6.3.2, and indicated in Fig. 6.5, a high degree of supersaturation of lactose can occur. Only at a supersaturation degree over 2.1, lactose will spontaneously crystallize. At lower degrees of supersaturation, rapid spontaneous

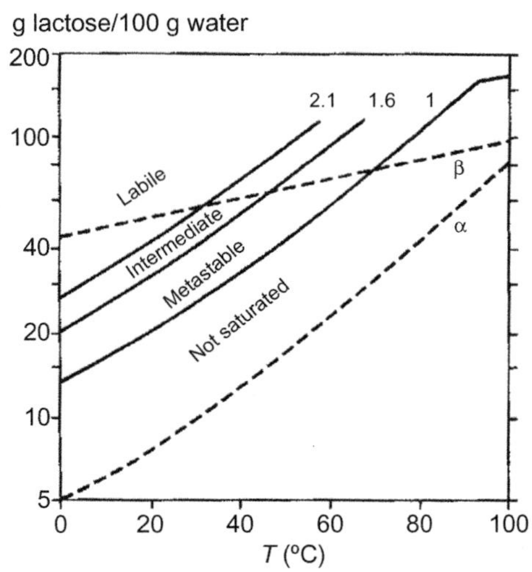

FIGURE 6.5

Solubility of α- and β-lactose. The initial solubility is indicated with the dotted line, with final solubility given by curve 1. Curves 1, 1.6, and 2.1 indicate the relative degree of supersaturation of lactose as a function of temperature.

Reprinted with permission from Walstra, P., & Jenness, R. (1984). Dairy chemistry and physics. *New York: John Wiley & Sons (Walstra & Jenness, 1984).*

crystallization will not occur, but can be induced by adding finely powdered lactose. Inducing lactose crystallization through seeding with finely powdered lactose is frequently used in dairy production and production of dairy-based ingredients (see Sections 6.4.3 and 6.5.1 and Chapter 5: Application of lactose in the pharmaceutical industry). In this section, the crystallization process will be further explained.

Lactose can crystallize in the form of both α- and β-lactose. As mentioned in Section 6.3.1, the α-lactose will crystallize as a monohydrate, whereas β-lactose will crystallize anhydrously. Besides crystallizing in these two forms, lactose can also exist in a noncrystalline state, which is called amorphous or glassy lactose. This amorphous state of lactose contains a mixture of α- and β-lactose, in the same ratio as they were present in the solution from which the amorphous lactose was created (Jawad et al., 2012).

The most common form of lactose in food products is α-lactose, and is formed when a supersaturated lactose solution crystallizes at temperatures below 93.5°C. The monohydrate crystals are very hard and poorly soluble. β-Lactose is formed when a supersaturated lactose solution crystallizes at temperatures above 93.5°C. These crystals of β-lactose are better water-soluble than those of α-lactose (Gänzle, Haase, & Jelen, 2008).

Where both crystalline forms of lactose are not very hygroscopic, amorphous lactose is highly hygroscopic. Amorphous lactose is mainly formed when a lactose-containing solution is very quickly dried, like with spray drying (Gänzle et al., 2008). Amorphous lactose is actually an extremely viscous liquid of which the behavior is very dependent on the temperature. The viscosity decreases with increasing temperature, as indicated in Fig. 6.6A. If the temperature is below a certain threshold (called glass transition temperature, T_g), the material behaves as a glass. At increasing temperatures, the viscosity drops, and the material becomes rubbery. This glass transition temperature strongly depends on a combination of water activity and temperature (Jouppila & Roos, 1994), as shown in Fig. 6.6B. Above the glass transition temperature, crystallization of the glass lactose can occur, which results in the expulsion of water that was embedded in the glassy lactose, which can further decrease the glass transition temperature.

Glass transition is generally occurring for all sugars, although the relation between temperature and water activity of the glass transition temperature is very different for different sugars (Foster, Bronlund, & Tony Paterson, 2006).

6.3.4 Chemical reactions involving lactose

Lactose can play a role in many different chemical reactions in dairy products. One of the major chemical reactions that impacts the consumer perception of many dairy products is the Maillard reaction, also called glycation. The Maillard reaction is the reaction between a carbonyl group of reducing sugars and the ε-amino group of a protein. In milk, the main reactants are lactose (in its open chain structure) and the lysine amino acids of the milk proteins, because those

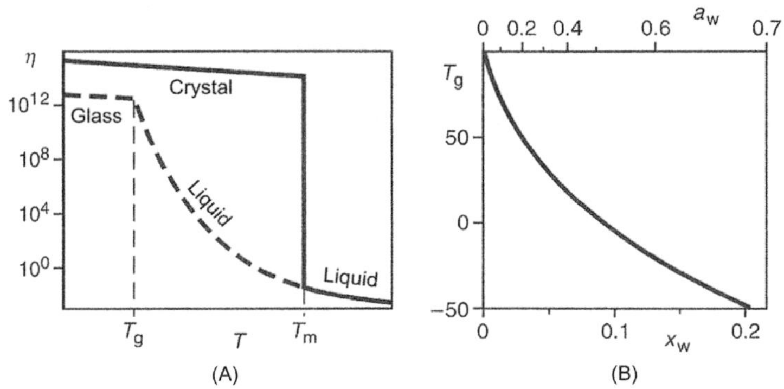

FIGURE 6.6

Glass transitions. (A) Approximate relation between the viscosity η (Pa·s) of a material and temperature (T); under crystalline conditions (solid line), and under glass conditions (broken line). T_g means glass transition and T_m melting point. (B) Relation between the glass transition temperature (T_g, °C), the mass fraction of water (x_w), and the water activity (a_w).

Reprinted with permission from Walstra, P., Wouters, J. T. M., & Geurts, T. J. (2006). Dairy science and technology *(2nd ed.). Taylor and Francis.*

have an ε-amino group. The Maillard reaction leads to changes in flavor, color (i.e., browning), texture, and nutritional value (i.e., loss of lysine, which is an essential amino acid). The Maillard reaction can be subdivided in three stages: the early, advanced, and final stage. In the early stage, the so-called Amadori products (mainly lactulosyllysine in milk) are formed via the formation of a Schiff's base and the Amadori rearrangement. Next, in the advanced stage, the Amadori products are broken down to many breakdown products of the sugar–amino compounds formed in the early stage, like carboxymethyllysine. The final stage of the Maillard reaction consists of the condensation of amino compounds and sugar fragments into polymerized protein and brown pigments, like melanoidins (Van Boekel, 1998).

The Maillard reaction has several consequences for heated dairy products. It both changes the sensory perception (taste and color), which can be either positive or negative (see Section 6.6.2), and reduces the nutritional quality of the milk product. The sugar involved in the Maillard reaction in dairy products is usually lactose. Due to the abundance of lactose in almost all dairy products, the Maillard reaction can occur in almost any dairy product. The reaction can occur during both heating processes in the dairy industry and the storage after the product has been produced. Compared to monosaccharides, lactose is slightly less reactive in the Maillard reaction (Martins, Jongen, & van Boekel, 2000). Dairy products containing monosaccharides, through either addition or lactose hydrolysis (see Section 6.4.2) are thus more sensitive to the Maillard reaction.

6.4 Lactose in consumer dairy products

Lactose, as present in the raw cow's milk, has major implications further down the dairy production chain (Fig. 6.1). It impacts both the production of consumer products (this section) and dairy-based ingredients (Section 6.5).

For consumer products, lactose plays several important roles during their production. In fermented products, it is the main carbohydrate (energy source) used by the bacteria involved in the fermentation. Next, there is an increase in demand for lactose-free dairy products, which has its own technological challenges. Finally, in concentrated and dried dairy products, lactose determines many of the properties of the final product. The technological aspects of lactose in the production of consumer products will be discussed in this section, whereas the impact on the perception of dairy products by the consumer is discussed in Section 6.6.

6.4.1 Fermentation

Fermentation is used for many dairy products, including yoghurt, butter, and cheese. The major reason for the popularity of fermenting dairy products is that it enhances its shelf life, and leads to flavor and texture changes in the final product. The two roles of lactose in this process are being energy source for the lactic acid bacteria that are used for fermenting dairy products and a precursor for specific flavor components. Only bacteria that are able to produce lactase are able to use lactose as energy source. In nature, many bacteria have lactase activity. Lactic acid bacteria used in dairy fermentations generally display high lactose utilization (de Vos & Vaughan, 1994; Rao & Dutta, 1978), although not all lactic acid bacterial strains are able to utilize lactose for lactic acid production (Kandler, 1983).

The lactic bacteria take up carbohydrates, like lactose, using a range of transporters (de Vos & Vaughan, 1994), after which they convert the carbohydrates to lactic acid (Kandler, 1983). This lactic acid formation is the main reason for the shelf life enhancement of fermented dairy products, as it inhibits the growth of other bacteria, both directly and through the lowered pH (O'Bryan, Crandall, Ricke, Ndahetuye, & Ndahetuye, 2015). The general pathways involved in the specific conversion of lactose by lactic acid bacteria are described in Fig. 6.7.

Two major categories of lactic acid bacteria are the homofermentative and heterofermentative lactic acid bacteria. The homofermentative lactic acid bacteria (e.g., *Lactococcus*, *Lactobacillus*) are used most frequently in food fermentation. These bacteria generally grow very fast and mainly convert sugars (like lactose) into lactic acid according to the EMP pathway (Fig. 6.7), without producing many other components. Their fast growth is mainly due to the simple metabolism of these homofermentative lactic acid bacteria. Simultaneously, these bacteria rely on many nutrients being available, besides lactose, because they can hardly produce other nutrients themselves. The heterofermentative lactic acid

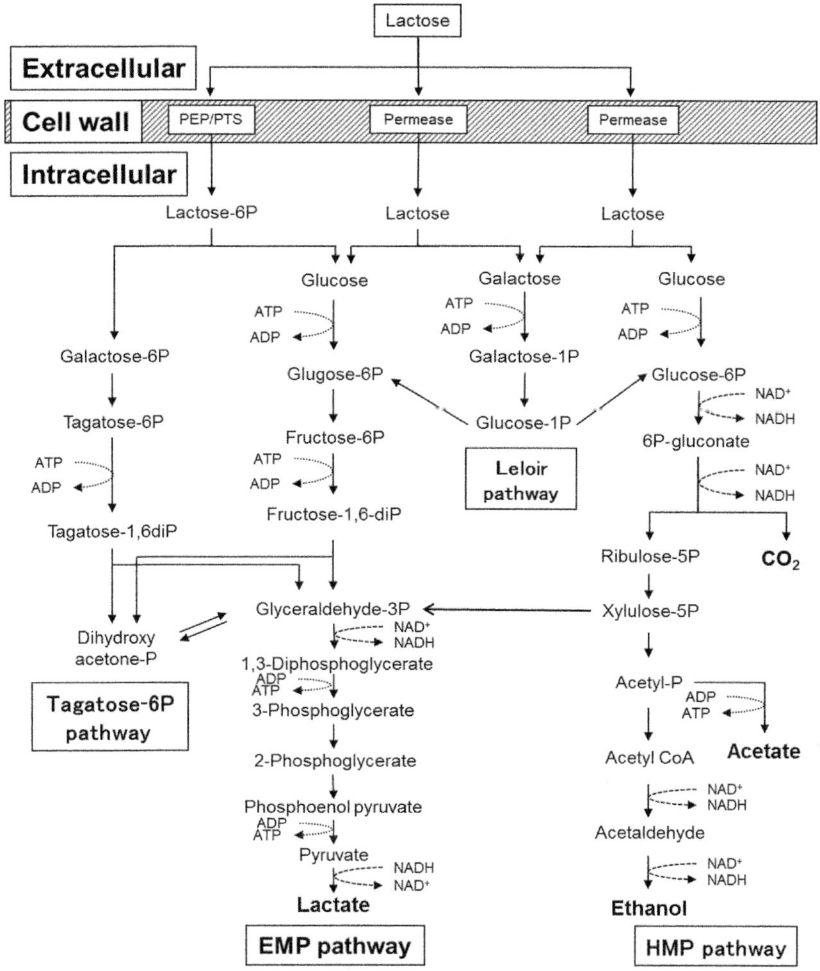

FIGURE 6.7

Uptake and breakdown reactions of lactose by lactic acid bacteria. EMP, Embden-Meyerhof-Parnas; HMP, Hexose Monophosphate.

Reprinted with permission from Harutoshi, T. (2013). Exopolysaccharides of lactic acid bacteria for food and colon health applications. In D. J. M. Kongo (Ed.), Lactic acid bacteria—R & D for food, health and livestock purposes. *Intech.*

bacteria (e.g., leuconostoc), on the other hand, produce a much wider range of components. These bacteria especially play a role in flavor and gas formation in fermented dairy products (Sharpe, 1979).

Two distinctive characteristics in the active biochemical pathways of lactose breakdown are the ability to break down galactose and the mutarotation of the lactic acid formed. All lactic acid bacteria that utilize lactose can use the glucose released from it for their metabolism. However, the majority of lactic acid bacteria do not use the galactose. Many lactic acid bacteria have a lactose/galactose antiporter system on their cell membrane to get rid of the galactose upon lactose absorption (de Vos & Vaughan, 1994). When comparing fermented dairy products, existence of galactose metabolism can differ. Yoghurt producing cultures generally do not use the galactose released from lactose, leading to free galactose in the final product (Ohlsson et al., 2017; Settachaimongkon et al., 2014). Some cheese starters, on the other hand, contain galactose-utilizing bacteria. The galactose use of these bacteria is very slow, thus leading to very slow disappearance of galactose during cheese ripening (Hettinga, 2014).

The other distinction that can be made between different lactic acid bacteria is with regard to their production of either of the two optical isomers of lactic acid, D- and L-lactic acid. Generally, *Lactococcus* spp. produce L-lactic acid, *Leuconostoc* spp. produce D-lactic acid, and *Lactobacillus* spp. can produce either of the lactic acid isomers. In our bodies, only L-lactic acid is produced, and this is also the only form our body can metabolize further when it is present in our diet (Ewaschuk, Naylor, & Zello, 2005).

Although these distinctions seem simple, all lactic acid bacteria use different biochemical pathways leading to a different range of metabolites. This leads to a much wider range of metabolites being produced by the different lactic acid bacteria, showing that in real life there is a much more complex distinction between these bacteria. This distinction is, however, often related to other metabolic pathways than lactose metabolization (Settachaimongkon et al., 2014).

The monosaccharides formed by the lactic acid bacteria when hydrolyzing lactose can be used for energy production, according to the pathways indicated in Fig. 6.7. These monosaccharides can, however, also be the building blocks for synthesis of exopolysaccharides by the lactic acid bacteria. These exopolysaccharides are long carbohydrate chains build from one or multiple different monosaccharides. Exopolysaccharides can result in ropiness and increased viscosity of fermented dairy products. Starters that produce exopolysaccharides are therefore used in many dairy products for improving the texture of the product (Harutoshi, 2013; also see Section 6.6.2).

Due to all these differences between individual lactic acid bacteria, most fermented dairy products are made by combining different types of lactic acid bacteria, to get the right balance between acidification, flavor formation, and exopolysaccharide synthesis.

6.4.2 Lactose-free dairy products

Where fermented dairy products are naturally low in lactose, there is also an increasing demand for products that are completely lactose-free. To produce such products that are completely free of lactose, different technological approaches exist. They can roughly be split into two approaches: (1) removing lactose from the milk and (2) splitting the lactose in the milk into glucose and galactose. In most cases, a combination of both methods is applied. Due to the relative low sweetness of lactose compared to glucose, splitting all lactose enzymatically would increase the sweetness of the product, which in most products is unwanted. Therefore first removing part of the lactose followed by splitting the remaining lactose gives a product with a similar sweetness as its lactose-containing counterpart (see also Section 6.6.2).

6.4.2.1 Removing lactose

Several methods exist to remove lactose from milk. Crystallization, chromatography, and membrane separation are the three main processes that can be used. Crystallization can be induced to remove lactose from concentrated solutions. This method is mainly used for lactose isolation from whey streams (see Chapter 5: Application of lactose in the pharmaceutical industry), but cannot be used for plain milk due to a too low concentration of lactose. Chromatography, on the other hand, can be used to separate components like lactose based on different residence times of components in chromatographic columns. By having a resin in the column with different affinities to different components, or that separate components by size, different fractions can be collected through this residence time difference (Harju, Kallioinen, & Tossavainen, 2012). For lactose, strong cation exchange columns can be used. The formation in such columns of lactose-cation complexes slows the flow of lactose, thereby increasing its residence time in the column relative to other milk components (Harju & Heikkila, 1990). This method can very specifically separate lactose from milk, compared to the other methods discussed in this subsection.

Another process that can be used to reduce or remove lactose from milk, and is currently the most popular process in dairy industry, is the membrane filtration process. Ultrafiltration can be used to separate the milk proteins from a lactose-rich stream, although in this case both lactose and salts are removed from the product. A combination of ultrafiltration and nanofiltration can be used to collect a lactose-rich and a lactose-poor stream, as depicted in Fig. 6.8. The ultrafiltration

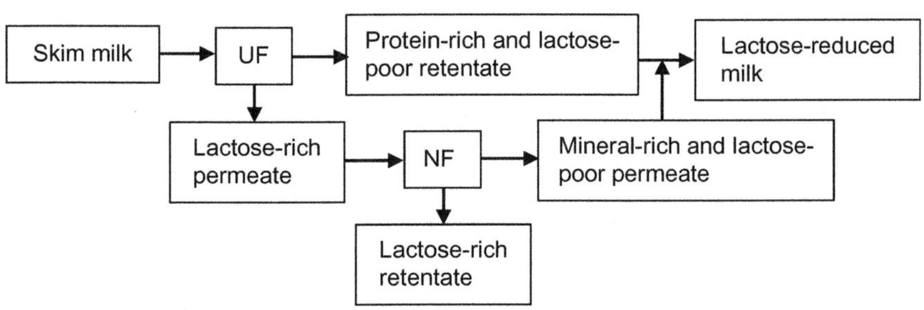

FIGURE 6.8

A combination of ultrafiltration (UF) and nanofiltration (NF) for creating a lactose-reduced milk.

retentate, containing the proteins, and the nanofiltration permeate, containing most minerals, can be combined to produce a lactose-reduced milk that still contains all proteins and most of the minerals from milk.

It is, however, difficult to obtain a fully lactose-free product with these membrane processes (Harju et al., 2012). Often, these membrane processes are done in such a way that about half of the lactose is removed (Harju et al., 2012; Troise et al., 2016), so that after splitting the remaining lactose, a product of similar sweetness, is obtained as regular lactose-containing milk (see Section 6.6.2).

6.4.2.2 Splitting lactose

After removal of part of the lactose, as described previously, the remaining lactose may be split enzymatically to produce products that are completely lactose-free. The enzyme which splits lactose in our body or in microbial cells, called lactase, can be isolated for that purpose from microbes. Especially the yeasts from the Kluyveromyces family are often used for industrial production of lactase. This isolated lactase can then be added to dairy products, to split the lactose. Most of this isolated lactase has an optimum pH of around 6 and an optimal temperature of around 37°C (Mahoney, 1997; Seyis & Aksoz, 2003).

To prevent lactase from ending up in the final product, where it can cause shelf life and off-flavor problems due to its ability to hydrolyze protein, the milk containing the lactase can be heat-treated at such a high temperature that the lactase is inactivated (Troise et al., 2016). An alternative to this heat treatment is the immobilization of lactase on a carrier. This carrier can easily be removed from the lactase-treated dairy product, thereby also removing the lactase itself (Greenberg & Mahoney, 1981).

Lactase treatment is a relatively low-tech solution to lactose removal, not requiring specialized equipment (Zadow, 1986). Although the optimal temperature of isolated lactase is usually around 37°C, keeping milk at this temperature costs energy and is a risk from a microbial point of view, as it may cause spoilage. Therefore lactose hydrolysis at fridge temperatures, but with longer incubation times to compensate for the slower rate of hydrolysis, can also be applied in practice.

6.4.2.3 Maillard reaction in lactose-free dairy products

One disadvantage of lactose-free dairy products in which lactase has been used is that the glucose formed during lactose hydrolysis is much more reactive in the Maillard reaction than lactose (Section 6.3.4). Therefore browning due to the Maillard reaction occurs much faster in products after lactase treatment.

A decision with regard to process design that influences the extent of browning is whether the heat treatment of lactose-free dairy products is done before or after lactase treatment. Usually, the milk is heated after lactase treatment, to inactivate lactase, but this may lead to extensive Maillard reaction. Heat treatment can also be done prior to lactase treatment to reduce the extent of the Maillard reaction (Mendoza, Olano, & Villamiel, 2005), but then the hygienic requirements during lactase treatment are higher, and remaining active lactase in the final product may lead to off-flavors due to protein breakdown, as described previously.

6.4.3 Lactose in concentrated and dried dairy products

In milk and most liquid dairy products, the level of lactose, or its hydrolysis products, is usually too low to influence the physical properties of these products. However, when concentrating milk and possibly subsequently drying milk, lactose starts to become an important component determining the physical properties, which will be further discussed in this section.

6.4.3.1 Effect of lactose in concentrated dairy products

Milk and dairy streams are being concentrated for multiple reasons. First, the end product may be a liquid concentrated dairy product, like (sweetened) evaporated/condensed milk. Concentration is also applied prior to drying, to reduce the energy use by the final drying process, improving the overall energy efficiency of powder production, for both milk powders and dairy-based ingredients. Finally, specific lactose concentration is also applied on whey-based dairy streams for the isolation of lactose, as described in Chapter 5, Application of lactose in the pharmaceutical industry.

When making concentrated dairy products, lactose solubility is one of the major limiting factors. As explained in Section 6.3.1, the solubility of lactose is rather limited in comparison to other sugars. Lactose crystallization can make concentrated dairy products unacceptable to the consumer, due to the perceived sandiness caused by lactose crystals in the product (Morrissey, 1985).

An important concentrated dairy product is sweetened evaporated milk. Usually, considerable amount of sugar (sucrose) is added to this product, which may lead to enhanced lactose crystallization. The finely powdered lactose crystals are used to induce lactose crystallization. By this seeding with powdered lactose crystals, more but smaller lactose crystals are formed, preventing the lactose crystals to cause perceivable sandiness.

Similar sandiness problems have also been described for ice cream, which essentially is also a concentrated dairy product. During freezing, more and more water becomes unavailable due to its crystallization, which causes the lactose to become supersaturated in the unfrozen part of the product. This supersaturation of lactose can induce its crystallization. In ice cream, like in liquid concentrated dairy products, seeding with finely powdered lactose crystals can prevent the lactose crystals of becoming too large, which can prevent perceived sandiness (Nickerson, 1962). Stabilizers used in ice-cream production also prevent sandiness, but through a different mechanism. The stabilizer prevents the nucleation of lactose, through their effect on the viscosity of the continuous phase, thereby preventing lactose crystals from being present in the final product (Nickerson, 1962).

Like with regular dairy products also concentrated dairy products can be made lactose-free, as described in Section 6.4.2. The processes described in that section for lowering or reducing lactose can also be applied for concentrated dairy products. Lactose-free sweetened condensed milk can, for example, be prepared by hydrolysis of lactose. For this specific product, the hydrolysis of lactose will also reduce the required sugar addition due to the increased sweetness after the hydrolysis process, because the monosaccharides formed during hydrolysis are sweeter than the original lactose (Kosikowski & Wierzbicki, 1972). However, the considerations for the Maillard reaction play an even larger role for this type of product, as the more reactive glucose is formed. This can lead to enhanced browning of the product during its shelf life. Therefore applying filtration to reduce lactose (Section 6.4.2) may be more appropriate. At the same time, Maillard reaction does not need to be completely prevented, because a certain browning and flavor development due to the Maillard reaction is often expected by consumers of this type of product (see also Section 6.6.2).

6.4.3.2 Lactose in milk powder

This part of Section 6.4.3 will deal with milk powder that is meant for direct use by consumers. Dried dairy components for use as dairy-based ingredient will be discussed in Section 6.5.

For most dairy powders, including milk powder, it is important to realize that lactose is the main component present in the powder particle, being the continuous phase in which the fat and protein are embedded. The lactose content of milk powders ranges from 37% in whole milk powder to 55% in skim milk powder. The state of the lactose (glassy or crystalline, see Section 6.3.3) is thus very important for the behavior of the powder particle, as it forms the continuous phase and most of the surface of the powder particle (Walstra et al., 2006).

In milk powder, the lactose is usually present, completely or partly, in a glassy state. As lactose is very hygroscopic in its glassy state, this causes the milk powder to be very hygroscopic as well. If water is allowed to be absorbed by the powder, for example, due to incorrect storage conditions, this first leads to liquid bridging (connecting powder particles by the absorbed water), causing a reduction in the glass transition temperature, which is in turn leading to a sticky surface. These processes can be followed by crystallization as part of the amorphous lactose. Such crystallization may lead to crystals across multiple milk powder particles, because the particles are already sticking together. This process is called caking and leads to irreversibly binding together of milk powder particles in the form of lumps. In the end, the complete milk powder batch can be bound together to one solid mass that cannot be separated anymore (Fitzpatrick et al., 2010).

6.5 Lactose in dairy-based ingredients

Milk is not only used to make dairy products, but its components are more and more applied in a wide range of dairy-based ingredients, often using whey as starting material. Lactose-enriched or lactose-reduced whey streams can also be applied in animal feed. Finally, lactose that is isolated and purified from specific dairy streams can also be further converted into specific lactose-based ingredients, applying a wide range of lactose conversions.

6.5.1 Lactose in dairy-based powdered ingredients

As mentioned earlier, most dairy-based ingredients use whey as starting material. The major dry material in whey is lactose. The majority of the whey-based ingredients are based on the whey proteins. These ingredients are usually made by filtration, especially ultrafiltration, of whey, to enrich the product in whey proteins. The permeate from this process is thus further enriched in lactose. This lactose can be concentrated and crystallized from such permeate streams to isolate lactose specifically, as explained in Chapter 5, Application of lactose in the pharmaceutical industry. As explained in that chapter, a major application of this isolated lactose is pharmaceutical applications. However, almost all dairy- and whey-based ingredients contain lactose, ranging from less than 1% in specific protein isolates to 85% in whey permeate. In all these ingredients, lactose plays a role in several properties of the powder. Two generally important processes that influence the quality of dairy-based ingredients, in which lactose plays an important role, are the Maillard reaction and powder caking.

Dairy powders are generally sensitive to the Maillard reaction, due to the presence of lactose and lysine. Also, the storage conditions are such that the Maillard reaction can continuously, but slowly, proceed. Depending on the water activity of the powder, the reaction rate of the Maillard reaction is usually higher than in liquid dairy products. In addition, dairy powders are usually stored for prolonged times at room temperatures, that depending on the country may be relatively high. The combination of these factors can lead to considerable Maillard reaction in dairy-based ingredients, causing a discoloration and loss of nutritive value, as explained in Section 6.3.4. Generally, the higher the water activity of such powdered ingredients, the faster the Maillard reaction (optimal water activity for the Maillard reaction is about 0.60, which translates to about approximately 12% moisture in dairy-based ingredient powders). If powdered ingredients are thus allowed to absorb moisture due to incorrect storage, this will increase the extent of the Maillard reaction (Thomas, Scher, Desobry-Banon, & Desobry, 2004).

In addition to inducing the Maillard reaction, additional moisture absorbed by dairy-based ingredients may also lead to caking of the powder, as explained for milk powder in Section 6.4.3. To deal with the hygroscopicity and the consequential caking of powdered dairy-based ingredients, but especially whey streams with a high lactose content, precrystallization of lactose may be performed (Silalai & Roos, 2010). This process is very similar to the lactose seeding described for concentrated dairy products in Section 6.4.3. After the product has been concentrated by evaporation, prior to drying, extremely small lactose crystals, as a fine powder, may be added to the concentrated liquid to induce crystallization. Due to this seeding with many small crystals, crystallization will lead to many small crystals in the concentrated products. The final product will still consist of an amorphous lactose phase in which the other components, including the lactose crystals, are embedded. Because of the many small crystals present in the powder particle, water drawn into the powder may still lead to crystallization of the amorphous lactose, but this crystallization will lead to extension of the many small existing crystals embedded in the powder particle, thereby reducing the chance of developing large lactose crystals spanning multiple particles, and thereby reducing the tendency of caking of the dairy-based ingredients.

6.5.2 Lactose in animal feed

In the first days after birth, the young dairy animals (cows, goats, etc.) are fed with the colostrum produced by the mother animal. After a few days, the milk is started being sold, so the young will be fed with milk replacers after these first days. Also for other mammals (e.g., pigs), weaning occurs shortly after birth. This weaning practice is mainly driven by the economic pressure to increase the production efficiency. In the early weaning period, dairy-based ingredients are the basis of milk replacers given to such young animals (Zijlstra, Whang, Easter, & Odle, 1996). The main advantages of using dairy-based ingredients are the high digestibility and the favorable palatability (Terré, Devant, & Bach, 2007). For economic reasons, these milk replacers usually do not contain casein, but are based on whey. Using whey proteins, a maximized growth rate of the young animal can be reached compared to using plant proteins (Yun et al., 2005). But milk replacers are not only about whey protein. As explained in Section 6.2, milk of different mammals differs in composition. Protein, minerals, and lactose content all vary considerably between animals. For this reason, lactose-enriched or lactose-reduced whey can be used as starting material for producing animal feed, to adjust the feed composition to best reflect the needs of the young animal. The lactose in animal feed has been shown to have many beneficial effects for the young animal, especially on the microbiota in the intestinal tract (Atkinson, Kratzer, & Stewart, 1957).

Mostly, the membrane processes explained in Section 6.3.2 are used to adjust the lactose content of whey to create a starting material with the correct protein–lactose–mineral ratio (Zydney, 1998).

6.5.3 Dairy-based ingredients

Lactose that has been isolated from whey can be used as ingredient in the pharmaceutical industry (Chapter 5: Application of lactose in the pharmaceutical industry). However, it can also be used as starting material for the creation of lactose-based ingredients. Using different processes, lactose is converted to components that have a wide range of functions, from pharmaceutical (lactulose), to nonnutritive sweetener (lactitol), for enhancing the gut microbiota (galacto-oligosaccharides). This section will give an overview of the processes used for making these lactose-based ingredients, as well as their application.

Many of the lactose-based ingredients are based on chemical conversion of lactose. Lactose, like any other sugar, can be chemically converted based on a range of different reactions from basic carbohydrate chemistry. The main chemical reactions are hydrolysis to monosaccharides (see Section 6.4.2), oxidation to lactobionic acid, isomerization to lactulose, and hydrogenation to lactitol.

Lactobionic acid, a sugar acid, which is formed by oxidation from lactose, is not normally found in nature. Many processes exist to convert lactose into lactobionic acid. All these processes oxidize a free aldehyde group of the lactose, to create an acid with a carboxy group (Morrissey, 1985). Lactobionic acid can bind to several minerals, thereby having a chelating effect. Besides food applications, lactobionic acid is also used as an antioxidant in cosmetic products, but can also serve as an excipient in pharmaceutical applications.

Lactulose can be formed by isomerization of lactose. This reaction happens spontaneously upon extensive heating of milk. Lactulose is therefore sometimes used as marker for the heating intensity the milk was subjected to. It can also be produced in industry by inducing the lactose isomerization, which is for example done for pharmaceutical applications. Lactulose can be isolated from the reaction mixture using ion exchange chromatography. Lactulose is neither broken down nor absorbed in the digestive tract and therefore ends up in the large intestine, where it is fermented to organic acids. This can lead to retention of water, and thereby help in case of constipation (Aider & Halleux, 2007), which is the main reason for its industrial pharmaceutical synthesis.

Another use of chemically converted lactose is as sugar replacer. Lactose can be hydrogenated to the sugar alcohol lactitol, which is a sweetener that has a similar sweetness to sucrose. The hydrogenation of lactose to form lactitol is done by heating lactose under pressure in the presence of a specific nickel catalyst. Lactitol is no longer a reducing sugar and is chemically very stable. Lactitol, like lactulose, is not digested, and therefore has laxating properties, due to its fermentation in the large intestine. But the main reason for the use of lactitol is that it does not provide calories to the products sweetened by it. The solubility of lactitol is comparable to that of sucrose, for which it is frequently used as a replacement. Lactitol cannot be used by bacteria in the oral cavity, and therefore is noncariogenic (Morrissey, 1985).

Lactose, which is a disaccharide, can be elongated to oligosaccharides by adding galactose units to form galacto-oligosaccharides, although technically lactose is also a galacto-oligosaccharide. The resulting oligosaccharides are not digested and have a prebiotic function in the large intestine, stimulating specific bacteria in the microbiota (Fanaro et al., 2007; van Hoffen et al., 2009; Walton et al., 2012). In infants, galacto-oligosaccharides are often combined with fructo-oligosaccharides for optimal effect on the microbiota development (Fanaro et al., 2007). The process of synthesizing galacto-oligosaccharides is based upon the addition of these galactose units to the galactose end of lactose, creating a linear chain with a length of up to about 8 sugar units. The process is usually based upon the use of the β-galactosidase enzyme. This enzyme catalyzes both the hydrolysis of lactose and the transgalactosylation, adding galactose units to the lactose or galacto-oligosaccharide molecules. The extent of hydrolysis and transgalactosylation occurring depends on the source of the enzyme. For optimal production of galacto-oligosaccharides, the enzyme needs to be added to a very concentrated lactose solution. As the transgalactosylation takes several hours at 37°C, the hygienic design of the process is very important. The final product after the enzyme reaction contains, besides galacto-oligosaccharides, also lactose and glucose units that were not polymerized by the enzyme. From this mixture, the galacto-oligosaccharides need to be isolated (Torres, do, Gonçalves, Teixeira, & Rodrigues, 2010).

6.6 Lactose and end-user perspective on dairy products

At the end of the dairy production chain (Fig. 6.1), the dairy products reach the end user. The lactose in liquid and dried dairy products affects the quality of dairy products, which can be perceived by both the industrial end user and the final human consumer. When considering lactose intake by the consumer, this can be through both the consumption of dairy products and as ingredient in many other food products. Lactose intake can also come from pharmaceutical products (Chapter 5: Application of lactose in the pharmaceutical industry).

6.6.1 Role of lactose in quality of liquid dairy products

Lactose plays an important role in the quality perception of liquid dairy products by the end user, mainly through its role in the sensory characteristics. Lactose has both a direct effect on sensory perception, through its inherent sweetness, and an indirect effect, by being reactant in the Maillard reaction. Lactose also is the energy source for the lactic acid bacteria that are used for the production of fermented dairy products.

6.6.1.1 Sweetness

Compared to other sugars, lactose is a relatively mild sweet component; this role of lactose on sweetness perception has been further discussed in Chapter 2, Digestion, absorption, metabolism, and physiological effects of lactose. The monosaccharides forming lactose has a higher sweetness than the lactose itself. This is the reason, as already mentioned in Section 6.4.2, that for lactose-free products usually a combination of removal and hydrolysis of lactose is used. This way, a product with a similar sweetness to the lactose-containing version can be made. At the same time, in products that are sweetened, using lactase can increase the sweetness of the milk itself, therefore requiring less sugar to be added to further increase the sweetness of the final product. For example, splitting 70% of the lactose in milk leads to a sweetness increase that is comparable to the addition of 2% sucrose to that milk (Zadow, 1986).

6.6.1.2 Maillard reaction

Besides a direct effect on sweetness, lactose can also affect the perception of dairy products in multiple ways through being a reactant in the Maillard reaction (see Section 6.3.4). First, it can cause the formation of specific flavor components, which can both be unwanted, for example, in plain dairy products, and also contribute positively, for example, in condensed milk or ghee. These flavor components can originate from different pathways within the Maillard reaction (Newton, Fairbanks, Golding, Andrewes, & Gerrard, 2012; van Boekel, 2006). In addition, the Maillard reaction also has an impact on the color, through the formation of melanoidins in the final stage of the Maillard reaction (Van Boekel, 2001). Just like for flavors, for some products it is seen as detrimental, whereas for others it is part of what is expected by consumers. As mentioned in Section 6.4.2, lactose hydrolysis leads to enhanced Maillard reaction through the formation of the more reactive glucose. As consumers expect a similar perception of lactose-free dairy products compared to their lactose-containing counterparts, this leads to new, specific challenges for these products (Adhikari, Dooley, Chambers, & Bhumiratana, 2010; Harju et al., 2012).

6.6.1.3 Fermented dairy products

Another category of dairy products that are influenced by lactose are fermented dairy products. Lactose supplies the energy for growth of the lactic acid bacteria used for the fermentation, and is also the precursor of lactic acid and some flavor components. By converting lactose to lactic acid, lactic acid bacteria cause a pH decrease, which leads to the typical sour taste of fermented dairy products. The lactic acid also prolongs the shelf life of fermented dairy products. In addition, lactose can be converted to several flavor compounds, for example, diacetyl, that play an important role in the flavor of many fermented dairy products (Urbach, 1995).

Besides an effect on flavor perception, fermentation can also influence the texture of dairy products. Lactic acid, due to its pH reducing effect, causes the aggregation of casein micelles in liquid milk that is being fermented, for example, to yoghurt. This increases the viscosity of the milk and can eventually lead to complete gelation (Lee & Lucey, 2004). Besides the direct effect of pH on the texture of dairy products, many bacteria used in dairy fermentation (especially in yoghurt fermentation) produce exopolysaccharides that can enhance the viscosity of dairy products (Harutoshi, 2013), as already explained in Section 6.4.1. These exopolysaccharides are especially used in stirred dairy products, to enhance the viscosity of the continuous phase in which the protein gel particles are present. In the end, the right combination of starter bacteria (type and inoculation level) and incubation conditions are thus needed, to produce fermented dairy products of the texture expected by the consumer.

The use of lactose by lactic acid bacteria during fermentation not only affects flavor and texture, but is also relevant for people that are lactose-intolerant. Because during fermentation, lactose is broken down, and fermented dairy products are naturally low in lactose. In addition, symptoms of lactose intolerance have been shown to be reduced when consuming fermented dairy products, even more than could be expected from only lactose removal (Alm, 1982). Possibly, the presence of live lactic acid bacteria in the product contributes to a further breakdown of the lactose in the digestive tract, although evidence for this is lacking (Vesa et al., 1996). This is the reason why in practice, lactose intolerance consumers tend to prefer the consumption of fermented dairy products (see also Chapter 2: Digestion, absorption, metabolism, and physiological effects of lactose and Chapter 3: Lactose intolerance and other related food sensitivities).

6.6.2 Role of lactose in quality perception of dried dairy products

The role of lactose in the product perception by consumers of reconstituted milk powder does not differ from the aspects mentioned in Section 6.6.2. What makes milk powder special is that its low water activity and long shelf life may enhance the extent of the Maillard reaction, making milk powder sensitive to the browning, flavor formation, and nutritional decline associated with it, although the underlying chemistry is similar to the liquid dairy products (see Sections 6.3.4 and 6.6.2). Again, just like for lactose-free liquid products, lactose-hydrolyzed dairy powders are also more sensitive to the Maillard reaction.

6.6.2.1 Caking

Besides the role of lactose on sensory perception of the product, it also has an influence on the physical stability of the product. Lactose, being the continuous phase of the powder particle, also determines its behavior. One of the most apparent effects of this is the caking of milk powder and other dairy powders (e.g., infant formula), which is related to the glass transition temperature of lactose (Chuy & Labuza, 1994), as also explained in Section 6.5.3. Due to the lower glass transition temperature of glucose, dairy powders in which the lactose is hydrolyzed are more sensitive to caking, and should thus be dried to a very lower water content and be stored to prevent any moisture from entering the package (Shrestha, Howes, Adhikari, & Bhandari, 2007).

Consumers have a major influence on this caking, as the rate and extent of caking strongly depends on the moisture content of the powder. While manufacturers usually package the product to prevent any moisture from entering, the consumer may leave the package open, inducing caking.

6.6.2.2 Lactose in dairy-based ingredients

Lactose is present in many dairy-based ingredients that are used in nondairy food products. Because lactose is the main components of the dry matter of whey, many whey-based ingredients contain lactose. This way, lactose can end up in many food products, which is relevant for lactose-intolerant consumers (see Section 6.6.3 and Chapter 3: Lactose intolerance and other related food sensitivities). Isolated lactose can also be used in food products. Due to its neutral taste and low sweetness, lactose is an easy to apply sugar. When replacing other sugars by lactose, for example, in bakery or confectionary applications, both sweetness and the level of the extent of Maillard reaction can be reduced (Edwards, 1984). Lactose-derived components can also be present as a bioactive ingredient in food products (Schaafsma, 2008), as explained in more detail in Section 6.5.3.

6.6.3 Lactose consumption level

The habitual intake of lactose for an average Western consumer, consuming on average 250 g of liquid dairy products, is approximately 10 g per day. However, with contribution of fermented dairy products, which naturally contain less lactose (see Section 6.4.1), the actual lactose intake from this average level of dairy product intake will be somewhat lower than that.

For lactose-intolerant consumers, there is now a much wider range of products available on the market today, that are lactose-free (see Section 6.4.2), even though a completely lactose-free diet is not needed, as a low dose of lactose can be tolerated by these consumers. The intake of approximately 10–12 g of lactose seems to be the threshold for causing symptoms in lactose-intolerant consumers (for more details, see Chapter 3: Lactose intolerance and other related food sensitivities). As described in Chapter 3, Lactose intolerance and other related food sensitivities, the level of self-reported lactose intolerance is much higher than the level of clinically proven lactose intolerance. As these consumers with self-reported lactose intolerance generally limit their intake of dairy products, lactose-free dairy products are more popular than would be estimated based on real rates of lactose intolerance. This has led dairy companies to develop lactose-free dairy products, even in countries with low levels of lactose intolerance, like Finland (Jelen & Tossavainen, 2003), even though their consumption is not related to a reduction in perceived lactose intolerance-related symptoms (Suarez, Savaiano, & Levitt, 1995), as discussed in more detail in Chapter 3, Lactose intolerance and other related food sensitivities.

The lactose intake does not come from traditional dairy products alone. The European Union has a total milk production of about 165 billion kg of milk, of which 37% is used for cheese production. This means that approximately 60 billion kg of whey is produced annually, containing 3 billion kg of lactose, which can be isolated for use in food, feed, and pharmaceutical applications. With 500 million inhabitants, that translates to about 15 g per capita per day, which is more than the habitual consumption of lactose through liquid dairy product consumption. Although part of this lactose will be exported, it does indicate that the level of lactose intake through liquid dairy products is somewhat in the same range as the intake through other sources. Especially for consumers that are lactose-intolerant (Chapter 3: Lactose intolerance and other related food sensitivities), this lactose present in nondairy products may be more challenging to prevent, as it is present in a very wide range of food and pharmaceutical products.

6.7 Conclusion and areas for future research

This chapter has provided an overview of lactose throughout the dairy chain, from its biosynthesis in the mammary gland, through its role in consumer products and its further use in dairy-based ingredients, and finally its role in the consumer perspective on dairy products. At different stages of the dairy production chain, knowledge about lactose has been found to be important, but at the same time, certain knowledge is still missing. In this final section, an overview is given for the topics dealt with in this chapter, shortly concluding on what is known, followed by discussion of the areas for potential future research on lactose in the dairy chain.

6.7.1 Primary production

In the mammary gland, lactose is synthesized from glucose. The biosynthesis pathway and the other roles of glucose in milk synthesis by the mammary gland are well known. Lactose synthesis is mainly studied for its important role in driving milk yield. However, the interrelations between the biosynthesis pathways of the different major milk components, fat, protein, and lactose, are much less studied. It is clear that a difference in energy balance, probably through its effects on glucose metabolism, affects lactose synthesis and milk yield. Changing glucose status (either by infusion of glucose or insulin), however, impacts not only lactose synthesis, but also the synthesis of the other milk components. This may be due to energy partition in the mammary gland. Energy available through glucose can be used for multiple processes in the mammary gland: as energy supply, as precursor for lactose synthesis, as precursor for synthesis of oligosaccharides, and as precursor for glycerol for triglyceride production. Understanding the mechanisms behind the effects of glucose metabolism on synthesis of all milk components is an area that deserves further study. Such knowledge could be useful for optimizing dairy cow management (e.g., feeding, dry period management) to decrease the risk for metabolic disorders, and the ensuing generally increased risks for many animal diseases.

6.7.2 Consumer products

Lactose plays a role in the quality of many dairy products. It is crucial for dairy fermentation, where lactose is both energy source for the lactic acid bacteria and substrate for specific flavor components. Where fermented are naturally low in lactose, completely lactose-free products can be made (Section 6.4.2), which are in an increasing demand. These products can be made by removing or hydrolyzing lactose, with most products using a combination of both these processes. Making lactose-free dairy products has much broader consequences for products quality than one would initially assume. With regard to the increased sweetness of the monosaccharides formed compared to lactose, this is usually solved by removing part of the lactose prior to hydrolysis. However, the glucose formed in this processing has different chemical and physical characteristics, which can impact product quality. As mentioned in Sections 6.6.2 and 6.6.3, this can lead to increased sensitivity to the Maillard reaction. As it is difficult to reduce the Maillard reaction itself, adapted temperatures during processing and/or storage (Evangelisti, Calcagno, Nardi, & Zunin, 1999) are possible solutions for this issue. However, a proper understanding of the mechanisms for the exact differences, for a wide range of different dairy products, is currently lacking. Optimization of processes to reduce the Maillard reaction may lead to lactose-free dairy products that are more similar to their lactose-containing counterparts.

6.7.3 Dairy-based ingredients

Similar to dairy products, dairy-based ingredients are also influenced by lactose. As ingredients are usually dried, and lactose is a major component of the dry matter, lactose determines most of the physical and chemical characteristics. For dairy-based, and especially lactose-based, ingredients, most research is focused on adding functionality. Several lactose-based ingredients are used due to the inability of our body to digest these components, leading to treatment options for constipation, as well as for microbiota stimulation, as described in Section 6.5.3 (see Chapter 4: Lactose—a conditional prebiotic? for a discussion on lactose as possible prebiotic). An important area of research is the use of specific dairy components for infant formula, to better mimic breast milk. Research focuses on all milk components, the lipid, protein, and carbohydrate fractions of human milk. In the area of complex carbohydrates, currently galacto-oligosaccharides are much used, either alone or in combination with fructo-oligosaccharides, in infant formula to stimulate the healthy development of the infant's microbiota. However, the oligosaccharides in human milk are much more complex than these simple straight chain oligosaccharides, with both branched structures and more different carbohydrate units being used in human milk oligosaccharides (Bode, 2012). With recent technological developments, it has become possible to create more complex oligosaccharides on an industrial scale. These processes are still based on lactose as an acceptor to which other sugars are enzymatically bound

(Sprenger, Baumgärtner, & Albermann, 2017). This may be an important future direction for application of complex sugars in dairy products like infant formula, which may be even better at mimicking the functions of human milk oligosaccharides.

References

Abraham, S., & Chaikoff, I. L. (1959). Glycolytic pathways and lipogenesis in mammary glands of lactating and nonlactating normal rats. *Journal of Biological Chemistry, 234*, 2246–2253.

Adhikari, K., Dooley, L. M., Chambers, E., & Bhumiratana, N. (2010). Sensory characteristics of commercial lactose-free milks manufactured in the United States. *Lebensmittel-Wissenschaft & Technologie, 43*, 113–118. Available from https://doi.org/10.1016/j.lwt.2009.06.017.

Aider, M., & Halleux, D. de (2007). Isomerization of lactose and lactulose production: review. *Trends in Food Science & Technology, 18*(7), 356–364. Available from https://doi.org/10.1016/j.tifs.2007.03.005.

Akersk, M. R., Bauman, D. E., Capuco, A. V., Goodman, G. T., & Tucker, A. H. (1981). Prolactin regulation of milk secretion and biochemical differentiation of mammary epithelial cells in periparturient cows. *Endocrinology, 109*, 23–30. Available from https://doi.org/10.1210/endo-109-1-23.

Alm, L. (1982). Effect of fermentation on lactose, glucose, and galactose content in milk and suitability of fermented milk products for lactose intolerant individuals. *Journal of Dairy Science, 65*, 346–352. Available from https://doi.org/10.3168/jds.S0022-0302(82)82198-X.

Anderson, S. M., Rudolph, M. C., McManaman, J. L., & Neville, M. C. (2007). Key stages in mammary gland development. Secretory activation in the mammary gland: It's not just about milk protein synthesis!. *Breast Cancer Research, 9*, 204. Available from https://doi.org/10.1186/bcr1653.

Antonowicz, I., & Lebenthal, E. (1977). Developmental pattern of small intestinal enterokinase and disaccharidase activities in the human fetus. *Gastroenterology, 72*, 1299–1303.

Atkinson, R. L., Kratzer, F. H., & Stewart, G. F. (1957). Lactose in animal and human feeding: A review. *Journal of Dairy Science, 40*, 1114–1132. Available from https://doi.org/10.3168/jds.S0022-0302(57)94603-9.

Auldist, M., Coats, S., Rogers, G., & McDowell, G. (1995). Changes in the composition of milk from healthy and mastitic dairy cows during the lactation cycle. *Australian Journal of Experimental Agriculture, 35*, 427. Available from https://doi.org/10.1071/EA9950427.

Auldist, M. J., Coats, S., Sutherland, B. J., Mayes, J. J., McDowell, G. H., & Rogers, G. L. (1996). Effects of somatic cell count and stage of lactation on raw milk composition and the yield and quality of Cheddar cheese. *Journal of Dairy Research, 63*, 269. Available from https://doi.org/10.1017/S0022029900031769.

Bode, L. (2012). Human milk oligosaccharides: Every baby needs a sugar mama. *Glycobiology, 22*, 1147–1162. Available from https://doi.org/10.1093/glycob/cws074.

Cant, J. P., Trout, D. R., Qiao, F., & Purdie, N. G. (2002). Milk synthetic response of the bovine mammary gland to an increase in the local concentration of arterial glucose. *Journal of Dairy Science*, *85*, 494–503.

Chuy, L. E., & Labuza, T. P. (1994). Caking and stickiness of dairy-based food powders as related to glass transition. *Journal of Food Science*, *59*, 43–46. Available from https://doi.org/10.1111/j.1365-2621.1994.tb06893.x.

de Vos, W. M., & Vaughan, E. E. (1994). Genetics of lactose utilization in lactic acid bacteria. *FEMS Microbiology Reviews*, *15*, 217–237.

Edwards, W. P. (1984). Uses for dairy ingredients in confectionery. *International Journal of Dairy Technology*, *37*, 122–125. Available from https://doi.org/10.1111/j.1471-0307.1984.tb00515.x.

Evangelisti, F., Calcagno, C., Nardi, S., & Zunin, P. (1999). Deterioration of protein fraction by Maillard reaction in dietetic milks. *Journal of Dairy Research*, *66*, 237–243.

Ewaschuk, J. B., Naylor, J. M., & Zello, G. A. (2005). D-Lactate in human and ruminant metabolism. *Journal of Nutrition*, *135*, 1619–1625. Available from https://doi.org/10.1093/jn/135.7.1619.

Fanaro, S., Boehm, G., Garssen, J., Knol, J., Mosca, F., Stahl, B., & Vigi, V. (2007). Galacto-oligosaccharides and long-chain fructo-oligosaccharides as prebiotics in infant formulas: A review. *Acta Paediatrica*, *94*, 22–26. Available from https://doi.org/10.1111/j.1651-2227.2005.tb02150.x.

Fitzpatrick, J. J., Descamps, N., O'Meara, K., Jones, C., Walsh, D., & Spitere, M. (2010). Comparing the caking behaviours of skim milk powder, amorphous maltodextrin and crystalline common salt. *Powder Technology*, *204*, 131–137. Available from https://doi.org/10.1016/j.powtec.2010.07.029.

Foster, K. D., Bronlund, J. E., & (Tony) Paterson, A. H. J. (2006). Glass transition related cohesion of amorphous sugar powders. *Journal of Food Engineering*, *77*, 997–1006. Available from https://doi.org/10.1016/j.jfoodeng.2005.08.028.

Gänzle, M. G., Haase, G., & Jelen, P. (2008). Lactose: Crystallization, hydrolysis and value-added derivatives. *International Dairy Journal*, *18*, 685–694. Available from https://doi.org/10.1016/j.idairyj.2008.03.003.

Greenberg, N. A., & Mahoney, R. R. (1981). Immobilisation of lactase (beta-galactosidase) for use in dairy processing: A review. *Process Biochemistry*, *16*, 2–8.

Haase, G., & Nickerson, T. A. (1966). Kinetic reactions of alpha and beta lactose. I. Mutarotation. *Journal of Dairy Science*, *49*, 127–132. Available from https://doi.org/10.3168/jds.S0022-0302(66)87811-6.

Harju, M., Kallioinen, H., & Tossavainen, O. (2012). Lactose hydrolysis and other conversions in dairy products: Technological aspects. *International Dairy Journal*, *22*, 104–109. Available from https://doi.org/10.1016/j.idairyj.2011.09.011.

Harju, M. E., & Heikkila, H. O. (1990). Process of recovering lactose from whey. *Separation & Purification Reviews*, *41*, 251–266.

Harutoshi, T. (2013). Exopolysaccharides of lactic acid bacteria for food and colon health applications. In D. J. M. Kongo (Ed.), *Lactic acid bacteria—R & D for food, health and livestock purposes*. IntechOpen. Available from: https://www.intechopen.com/books/lactic-acid-bacteria-r-d-for-food-health-and-livestock-purposes/exopolysaccharides-of-lactic-acid-bacteria-for-food-and-colon-health-applications. https://doi.org/10.5772/50839.

Heck, J. M. L., van Valenberg, H. J. F., Dijkstra, J., & van Hooijdonk, A. C. M. (2009). Seasonal variation in the Dutch bovine raw milk composition. *Journal of Dairy Science*, *92*, 4745–4755. Available from https://doi.org/10.3168/jds.2009-2146.

Hettinga, K. (2014). Study used wrong assumption about galactose content of fermented dairy products. *British Medical Journal*, *349*, g7000. Available from https://doi.org/10.1136/bmj.g7000.

Holt, C. (1983). Swelling of Golgi vesicles in mammary secretory cells and its relation to the yield and quantitative composition of milk. *Journal of Theoretical Biology*, *101*, 247–261. Available from https://doi.org/10.1016/0022-5193(83)90339-9.

Hunziker, O. F., & Nissen, B. H. (1926). Lactose solubility and lactose crystal formation. *Journal of Dairy Science*, *9*, 517–537. Available from https://doi.org/10.3168/jds.S0022-0302(26)93924-6.

Jawad, R., Elleman, C., Vermeer, L., Drake, A. F., Woodhead, B., Martin, G. P., & Royall, P. G. (2012). The measurement of the β/α anomer composition within amorphous lactose prepared by spray and freeze drying using a simple 1H-NMR method. *Pharmaceutical Research*, *29*, 511–524. Available from https://doi.org/10.1007/s11095-011-0575-6.

Jelen, P., & Tossavainen, O. (2003). Low lactose and lactose-free milk and dairy products—prospects, technologies and applications. *Australian Journal of Dairy Technology*, *58*, 161–165.

Jenness, R., Regehr, E. A., & Sloan, R. E. (1964). Comparative biochemical studies of milk. II. Dialyzable carbohydrates. *Comparative Biochemistry and Physiology*, *13*, 339–352.

Jonker, J. S., Kohn, R. A., & Erdman, R. A. (1998). Using milk urea nitrogen to predict nitrogen excretion and utilization efficiency in lactating dairy cows. *Journal of Dairy Science*, *81*, 2681–2692. Available from https://doi.org/10.3168/jds.S0022-0302(98)75825-4.

Jouppila, K., & Roos, Y. H. (1994). Glass transitions and crystallization in milk powders. *Journal of Dairy Science*, *77*, 2907–2915. Available from https://doi.org/10.3168/jds.S0022-0302(94)77231-3.

Kandler, O. (1983). Carbohydrate metabolism in lactic acid bacteria. *An tonie van Leeuwenhoek*, *49*, 209–224.

Kosikowski, F. V., & Wierzbicki, L. E. (1972). Lactose hydrolysis of raw and pasteurized milks by *Saccharomyces lactis* lactase. *Journal of Dairy Science*, *56*, 146–148. Available from https://doi.org/10.3168/jds.S0022-0302(73)85129-X.

Lee, W. J., & Lucey, J. A. (2004). Structure and physical properties of yogurt gels: Effect of inoculation rate and incubation temperature. *Journal of Dairy Science*, *87*, 3153–3164. Available from https://doi.org/10.3168/jds.S0022-0302(04)73450-5.

Mackle, T. R., Dwyer, D. A., Ingvartsen, K. L., Chouinard, P. Y., Lynch, J. M., Barbano, D. M., & Bauman, D. E. (1999). Effects of insulin and amino acids on milk protein concentration and yield from dairy cows. *Journal of Dairy Science*, *82*, 1512–1524. Available from https://doi.org/10.3168/jds.S0022-0302(99)75378-6.

Mahoney, R. R. (1997). Lactose: Enzymatic modificationIn P. F. Fox (Ed.), Berlin, Heidelberg: Springer.

Martins, S. I. F., Jongen, W. M., & van Boekel, M. A. J. (2000). A review of Maillard reaction in food and implications to kinetic modelling. *Trends in Food Science and Technology*, *11*, 364–373. Available from https://doi.org/10.1016/S0924-2244(01)00022-X.

Mendoza, M. R., Olano, A., & Villamiel, M. (2005). Chemical indicators of heat treatment in fortified and special milks. *Journal of Agriculture and Food Chemistry*, *53*, 2995–2999. Available from https://doi.org/10.1021/jf040406l.

Menzies, K. K., Lefevre, C., Macmillan, K. L., & Nicholas, K. R. (2009). Insulin regulates milk protein synthesis at multiple levels in the bovine mammary gland. *Functional & Integrative Genomics*, *9*, 197–217. Available from https://doi.org/10.1007/s10142-008-0103-x.

Morrissey, P. A. (1985). Lactose: Chemical and physicochemical properties. In P. F. Fox (Ed.), *Elsevier Applied Science*. London, UK: Publishers.

Murphy, M. R. (1992). Water metabolism of dairy cattle. *Journal of Dairy Science*, *75*, 326–333. Available from https://doi.org/10.3168/jds.S0022-0302(92)77768-6.

Newton, A. E., Fairbanks, A. J., Golding, M., Andrewes, P., & Gerrard, J. A. (2012). The role of the Maillard reaction in the formation of flavour compounds in dairy products—not only a deleterious reaction but also a rich source of flavour compounds. *Food & Function*, *3*, 1231. Available from https://doi.org/10.1039/c2fo30089c.

Nguyen, D.-A. D., & Neville, M. C. (1998). Tight junction regulation in the mammary gland. *Journal of Mammary Gland Biology and Neoplasia*, *3*, 233–246. Available from https://doi.org/10.1023/A:1018707309361.

Nickerson, T. A. (1962). Lactose crystallization in ice cream. IV. Factors responsible for reduced incidence of sandiness. *Journal of Dairy Science*, *45*, 354–359. Available from https://doi.org/10.3168/jds.S0022-0302(62)89398-9.

O'Bryan, C. A., Crandall, P. G., Ricke, S. C., Ndahetuye, J. B., & Ndahetuye, J. B. (2015). *Lactic acid bacteria (LAB) as antimicrobials in food products*. Elsevier.

Ohlsson, J. A., Johansson, M., Hansson, H., Abrahamson, A., Byberg, L., Smedman, A., ... Lundh, Å. (2017). Lactose, glucose and galactose content in milk, fermented milk and lactose-free milk products. *International Dairy Journal*, *73*, 151–154. Available from https://doi.org/10.1016/j.idairyj.2017.06.004.

Osman Swar, M. (2011). Donkey milk-based formula: A substitute for patients with cow's milk protein allergy. *Sudanese Journal of Paediatrics*, *11*, 21–24.

Rao, M. V., & Dutta, S. M. (1978). Lactase activity of microorganisms. *Folia Microbiologica (Praha).*, *23*, 210–215.

Schaafsma, G. (2008). Lactose and lactose derivatives as bioactive ingredients in human nutrition. *International Dairy Journal*, *18*, 458–465. Available from https://doi.org/10.1016/j.idairyj.2007.11.013.

Schutz, Y. (2011). Protein turnover, ureagenesis and gluconeogenesis. *International Journal of Vitamin and Nutrition Research*, *81*, 101–107. Available from https://doi.org/10.1024/0300-9831/a000064.

Settachaimongkon, S., Nout, M. J. R., Antunes Fernandes, E. C., Hettinga, K. A., Vervoort, J. M., van Hooijdonk, T. C. M., ... Van Valenberg, H. J. F. (2014). Influence of different proteolytic strains of *Streptococcus thermophilus* in co-culture with *Lactobacillus delbrueckii* subsp. *bulgaricus* on the metabolite profile of set-yoghurt. *International Journal of Food Microbiology*, 177. Available from https://doi.org/10.1016/j.ijfoodmicro.2014.02.008.

Seyis, I., & Aksoz, N. (2003). Production of lactase by *Trichoderma* sp. *Food Technology and Biotechnology*, *42*, 121–124.

Sharp, P. F. (1938). Relation between lactose and ash content of the milk of different mammals. *Journal of Dairy Science*, *21*, 127–128. Available from https://doi.org/10.3168/jds.S0022-0302(38)95626-7.

Sharpe, M. E. (1979). Lactic acid bacteria in the dairy industry. *International Journal of Dairy Technology*, *32*, 9–18. Available from https://doi.org/10.1111/j.1471-0307.1979.tb01402.x.

Shrestha, A. K., Howes, T., Adhikari, B. P., & Bhandari, B. R. (2007). Water sorption and glass transition properties of spray dried lactose hydrolysed skim milk powder. *Lebensmittel-Wissenschaft & Technologie*, *40*, 1593–1600. Available from https://doi.org/10.1016/j.lwt.2006.11.003.

Silalai, N., & Roos, Y. H. (2010). Roles of water and solids composition in the control of glass transition and stickiness of milk powders. *Journal of Food Science*, *75*, E285–E296. Available from https://doi.org/10.1111/j.1750-3841.2010.01652.x.

Sprenger, G. A., Baumgärtner, F., & Albermann, C. (2017). Production of human milk oligosaccharides by enzymatic and whole-cell microbial biotransformations. *Journal of Biotechnology*, *258*, 79–91. Available from https://doi.org/10.1016/j.jbiotec.2017.07.030.

Stacey, A., Schnieke, A., Kerr, M., Scott, A., McKee, C., Cottingham, I., ... Colman, A. (1995). Lactation is disrupted by alpha-lactalbumin deficiency and can be restored by human alpha-lactalbumin gene replacement in mice. *Proceedings of the National Academy Sciences of the United States of America*, *92*, 2835–2839. Available from https://doi.org/10.1073/pnas.92.7.2835.

Steeneveld, W., Schukken, Y. H., van Knegsel, A. T. M., & Hogeveen, H. (2013). Effect of different dry period lengths on milk production and somatic cell count in subsequent lactations in commercial Dutch dairy herds. *Journal of Dairy Science*, *96*, 2988–3001. Available from https://doi.org/10.3168/jds.2012-6297.

Stinnakre, M. G., Vilotte, J. L., Soulier, S., & Mercier, J. C. (1994). Creation and phenotypic analysis of alpha-lactalbumin-deficient mice. *Proceedings of the National Academy Sciences of the United States of America*, *91*, 6544–6548. Available from https://doi.org/10.1073/pnas.91.14.6544.

Stoop, W. M., Bovenhuis, H., & van Arendonk, J. A. M. (2007). Genetic parameters for milk urea nitrogen in relation to milk production traits. *Journal of Dairy Science*, *90*, 1981–1986. Available from https://doi.org/10.3168/jds.2006-434.

Suarez, F. L., Savaiano, D. A., & Levitt, M. D. (1995). A comparison of symptoms after the consumption of milk or lactose-hydrolyzed milk by people with self-reported severe lactose intolerance. *New England Journal of Medicine*, *333*, 1–4. Available from https://doi.org/10.1056/NEJM199507063330101.

Terré, M., Devant, M., & Bach, A. (2007). Effect of level of milk replacer fed to Holstein calves on performance during the preweaning period and starter digestibility at weaning. *Livestock Science.*, *110*, 82–88. Available from https://doi.org/10.1016/j.livsci.2006.10.001.

Thomas, M. E. C., Scher, J., Desobry-Banon, S., & Desobry, S. (2004). Milk powders ageing: Effect on physical and functional properties. *Critical Reviews in Food Science and Nutrition*, *44*, 297–322. Available from https://doi.org/10.1080/10408690490464041.

Torres, D. P. M., do, M., Gonçalves, P. F., Teixeira, J. A., & Rodrigues, L. R. (2010). Galacto-oligosaccharides: Production, properties, applications, and significance as prebiotics. *Comprehensive Reviews in Food Science and Food Safety*, *9*, 438–454. Available from https://doi.org/10.1111/j.1541-4337.2010.00119.x.

Troise, A. D., Bandini, E., De Donno, R., Meijer, G., Trezzi, M., & Fogliano, V. (2016). The quality of low lactose milk is affected by the side proteolytic activity of the lactase used in the production process. *Food Research International*, *89*, 514–525. Available from https://doi.org/10.1016/j.foodres.2016.08.021.

Troy, H. C., & Sharp, P. F. (1930). Alpha and beta lactose in some milk products. *Journal of Dairy Science*, *13*, 140–157. Available from https://doi.org/10.3168/jds.S0022-0302(30)93513-8.

Urashima, T., Asakuma, S., & Messer, M. (2007). In J. P. Kamerling, G. J. Boons, Y. C. Lee, A. Suzuki, N. Taniguchi, & A. G. J. Voragen (Eds.), *Milk oligosaccharides*. Oxford, UK: Elsevier.

Urbach, G. (1995). Contribution of lactic acid bacteria to flavour compound formation in dairy products. *International Dairy Journal*, *5*, 877–903. Available from https://doi.org/10.1016/0958-6946(95)00037-2.

Van Boekel, M. (2001). Kinetic aspects of the Maillard reaction: A critical review. *Molecular Nutrition and Food Research*, *45*, 150–159.

Van Boekel, M. A. J. S. (1998). Effect of heating on Maillard reactions in milk. *Food Chemistry*, *62*, 403–414. Available from https://doi.org/10.1016/S0308-8146(98)00075-2.

van Boekel, M. A. J. S. (2006). Formation of flavour compounds in the Maillard reaction. *Biotechnology Advances*, *24*, 230–233. Available from https://doi.org/10.1016/j.biotechadv.2005.11.004.

van Hoffen, E., Ruiter, B., Faber, J., M'Rabet, L., Knol, E. F., Stahl, B., ... Garssen, J. (2009). A specific mixture of short-chain galacto-oligosaccharides and long-chain fructo-oligosaccharides induces a beneficial immunoglobulin profile in infants at high risk for allergy. *Allergy*, *64*, 484–487. Available from https://doi.org/10.1111/j.1398-9995.2008.01765.x.

van Knegsel, A. T. M., Remmelink, G. J., Jorjong, S., Fievez, V., & Kemp, B. (2014). Effect of dry period length and dietary energy source on energy balance, milk yield, and milk composition of dairy cows. *Journal of Dairy Science*, *97*, 1499–1512.

Vesa, T. H., Marteau, P., Zidi, S., Briet, F., Pochart, P., & Rambaud, J. C. (1996). Digestion and tolerance of lactose from yoghurt and different semi-solid fermented dairy products containing *Lactobacillus acidophilus* and bifidobacteria in lactose maldigesters—is bacterial lactase important? *European Journal of Clinical Nutrition.*, *50*, 730–733.

Walstra, P., & Jenness, R. (1984). *Dairy chemistry and physics*. New York: John Wiley & Sons.

Walstra, P., Wouters, J. T. M., & Geurts, T. J. (2006). *Dairy science and technology* (2nd ed.). Taylor and Francis.

Walton, G. E., van den Heuvel, E. G. H. M., Kosters, M. H. W., Rastall, R. A., Tuohy, K. M., & Gibson, G. R. (2012). A randomised crossover study investigating the effects of galacto-oligosaccharides on the faecal microbiota in men and women over 50 years of age. *British Journal of Nutrition*, *107*, 1466–1475. Available from https://doi.org/10.1017/S0007114511004697.

Warren, C. D., Chaturvedi, P., Newburg, A. R., Oftedal, O. T., Tilden, C. D., & Newburg, D. S. (2001). Comparison of oligosaccharides in milk specimens from humans and twelve other species. *Advances in Experimental Medicine and Biology*, *501*, 325–332.

Wheelock, J. V., Rook, J. A. F., & Dodd, F. H. (1965). The relationship in the cow between the osmotic pressure of milk and of blood. *Journal of Dairy Research*, *32*, 79. Available from https://doi.org/10.1017/S0022029900018392.

Young, J. W. (1977). Gluconeogenesis in cattle: Significance and methodology. *Journal of Dairy Science*, *60*, 1–15. Available from https://doi.org/10.3168/jds.S0022-0302(77)83821-6.

Yun, J. H., Kwon, I. K., Lohakare, J. D., Choi, J. Y., Yong, J. S., Zheng, J., ... Chae, B. J. (2005). Comparative efficacy of plant and animal protein sources on the growth performance, nutrient digestibility, morphology and caecal microbiology of early-weaned pigs. *Asian-Australasian Journal of Animal Science*, *18*, 1285–1293. Available from https://doi.org/10.5713/ajas.2005.1285.

Zadow, J. G. (1986). Lactose hydrolysed dairy products. *Food Technology in Australia*, *38*, 460–462, 471, 12 refs.

Zijlstra, R. T., Whang, K. Y., Easter, R. A., & Odle, J. (1996). Effect of feeding a milk replacer to early-weaned pigs on growth, body composition, and small intestinal morphology, compared with suckled littermates. *Journal of Animal Science*, *74*, 2948–2959.

Zunft, H.-J., & Schulze, J. (1990). Does mutarotation influence lactose digestion? Experimental investigations and a mathematical model. *Computer Methods and Programs in Biomedicine*, *32*, 287–295. Available from https://doi.org/10.1016/0169-2607(90)90111-L.

Zydney, A. L. (1998). Protein separations using membrane filtration: New opportunities for whey fractionation. *International Dairy Journal*, *8*, 243–250. Available from https://doi.org/10.1016/S0958-6946(98)00045-4.

Glossary

α- or β-protons protons that are bonded to the anomeric center of a saccharide.

α-amylases α-amylase or alpha-amylase is an enzyme that hydrolyzes α-1,4 glycosidic bonds of large, α-linked polysaccharides, such as starch and glycogen, to yield glucose and maltose, isomaltose and dextrins in humans and other mammals. Humans produce pancreatic and salivary α-amylases. It is also present in plant seeds containing starch and is secreted by many fungi.

α-lactalbumin a protein that regulates the production of lactose in the milk of nearly all mammals. It is only found in the mammary glands and milk, and becomes the rate-limiting factor in lactose biosynthesis.

α-lactose one of the two anomeric forms of lactose. The regular form of lactose, the monohydrate crystal, consists of this anomer.

β4galactosyltransferase a glycosyltranferase that catalyzes the transfer of a galactose and found in many tissues in the human body.

β-galactosidase a glycoside hydrolase enzyme that catalyzes the hydrolysis of β-galactosides into monosaccharides through the breaking of a β-glycosidic bond.

β-lactoglobulin major whey protein.

β-lactose one of the two anomeric forms of lactose. The prevalent form in commercial anhydrous lactose.

ε-amino group amino groups that is the side chain of an amino acid.

A2 dairy proteins most cow's milk contains both A2 and A1 β casein proteins, representing 80% of proteins in milk, but some dairy cows produce only the A2 β casein protein. There have been claims that A1 β casein is associated with adverse gastrointestinal symptoms by interacting with intestinal opioid receptor while A2 β casein does not produce symptoms.

aberrant crypt foci (ACF) aberrant crypt foci form before colorectal polyps and are one of the earliest changes seen in the colon that may lead to cancer. ACF are, as opposed to normal epithelial cells, apoptosis resistant.

acidosis in acidosis, the pH of the blood is abnormally low. There is too much acid in the body resulting from the accumulation of acid (either by increased production of metabolic acids, such as lactic acid, or disturbances in the ability to excrete acid via the kidneys) or from the depletion of alkaline reserves. Acidosis is associated with diabetic ketoacidosis, lung disease, and severe kidney disease. The opposite condition of acidosis is alkalosis in which there is a too high pH of the blood due to excess base or insufficient acid in the body.

adaptation a two-stage process whereby random genetic changes affect the characteristics or behaviors of an organism, conferring a fitness advantage, and natural selection favors these genetic changes to increase their frequency in a population.

adenoma adenoma is a benign tumor of glandular tissue.

adhesion interaction between particles of different compounds.

adult lactase deficiency or adult hypolactasia temporary or permanent insufficient concentration of the enzyme lactase in the small intestine to hydrolize lactose, due to lactase gene downregulation, injury, illness, aging, or other factors. Undigested lactose passes into the large intestine where colonic bacteria ferment the disaccharide.

AFM atomic force microscopy: interaction between a surface and typically a needle is measured. The result is a 3D plot (image) of forces between surface and needle.

agonists a substance that initiates a physiological response when combined with a receptor.

albuterol also known as salbutamol, used as inhalation medication as bronchodilator.

allele an allele is a variant form of a gene, located at a specific position on a chromosome.

alveoli small cavity in the mammary gland in which milk is stored between milkings. Contracts during milking in response to oxytocin to eject the milk.

Amadori products intermediate component in the Maillard reaction between sugar and protein that is formed relatively early during the Maillard reaction.

amorphous lactose amorphous lactose is found mainly in two types of lactose. In typical spray-dried lactose (SuperTab 11SD) it is present at a level of ~10%, and it results from the rapid drying of the portion of dissolved lactose processed through the spray drier. In milled lactose the amorphous portion depends on the degree of milling, but as a guide a commonly used grade for wet granulation (say Pharmatose 200M) contains ~2% amorphous lactose at the time of manufacture. Exposure of amorphous α-lactose to atmospheric moisture causes its recrystallization to lactose monohydrate.

anhydrous a (lactose) form lacking water.

anhydrous crystals crystals that do not contain water in their structure.

anomeric form an anomer is an epimer at the hemiacetal/acetal carbon in a cyclic saccharide, an atom called the anomeric carbon.

antagonists substance that interferes with or inhibits the physiological action of another.

antiinflammatory substance(s) that interfere(s) with the biochemical process that leads to features of inflammation (e.g., swelling, erythema, and pain).

antimicrobial peptides (AMPs) also known as host defense peptides (HDPs), a conserved component of the innate immune response in all organisms, including plants, animals, and humans. Most AMPs have the ability to kill microbial pathogens directly, whereas others act indirectly by modulating the host defense systems.

antineoplastic effects substances that inhibit or prevent the growth or development of malignant cells.

antisolvents a term used to describe solvents that mix with another solvent but causes the crystallization or precipitation of dissolved compounds from the dissolving solvent.

API active pharmaceutical ingredient, the drug substance.

archeological pertaining to archeology, the study of human activity in prehistory and history through the analysis of material remains.

Aridol a drug or API used in bronchial provocation test.

atomic force microscope see AFM.

ATP adenosine triphosphate used as source of energy in biological systems.

autosomal dominant a genetic trait encoded on an autosome (nonsex chromosome) where one copy of the allele is sufficient to confer the trait.

a_w water activity, a measure of the amount of water that is available to chemical reactions.

bacterial overgrowth a condition where the normal number of bacteria (≤103 colony-forming units) in the upper small intestine is increased due to abnormal motor function of the small intestine.

bifidobacteria one of the major genera of bacteria that make up the colonic flora in humans and other mammals.

blood glucose test the measurement of blood glucose in response to an oral carbohydrate load to determine the host ability to digest and absorb the carbohydrate. Most commonly used in the glucose tolerance test for diabetes and still used in assessing the ability to digest lactose.

bronchial provocation test a test to provocate asthma symptoms.

brush border enzymes digestive enzymes that are expressed by the small intestinal mucosa and embedded in the brush border.

brush border membrane a brush border (striated border or brush border membrane) is the microvilli-covered luminal surface of an epithelial cell in certain absorptive tissues of the body, for example, the small intestine of the gastrointestinal tract.

budesonide a drug or API used in asthma treatment.

caking the irreversible aggregation of particles to a coherent solid (e.g., lumps).

carbohydrates biomolecules consisting of carbon, hydrogen, and oxygen. Carbohydrates include sugars, starches, cellulose, and other compounds. They are often referred to as saccharides and can be subclassified as monosaccahrides, disaccharides, oligosaccharides, and polysaccharides.

carcinogenesis also called oncogenesis or tumorigenesis, is the initiation of cancer formation, whereby normal cells are transformed into cancer cells.

cardiovascular relating to the circulatory system, which comprises the heart and blood vessels.

cataracts a medical condition in which the lens of the eye becomes progressively opaque, resulting in blurred vision.

cataracts (galactosemia II) related to the harmful influence of galactitol. In galactosemic cataracts, osmotic swelling of the lens epithelial cells (LEC) occurs.

cathelicidin an example of an antimicrobial peptide (AMP).

cation exchange columns columns for chromatographic separation, where separation of molecules is based on their charge.

ceramide a family of waxy lipid molecules. A ceramide is composed of sphingosine and a fatty acid. Bound to lactose, glucose, or galactose, it is an important substrate of lactase-phlorizin hydrolase. Different ceramides combine with amino acids and function in cell–cell interactions mediating immune and epithelial functions. In adults excess ceramides contribute to type 2 diabetes.

cheese ripening storage of cheese after production to further develop its texture and flavor.

chelating chemical components that are able to bind metal ions.

chromatography separation technology that can be based on many different principles (volatility, polarity, charge, etc.).

cirrhosis cirrhosis is a late stage of scarring (fibrosis) of the liver caused by many forms of liver diseases.

cis position to a gene same side of the strand of chromosome the specific gene is situated, opposed to trans which is on the opposite strand.

Cochrane the Cochrane Library is a collection of high-quality, independent scientific evidence to inform health-care decision-making.

cohesion interaction between particles of the same compound.

colonic pertaining or relating to the colon, also known as the large intestine.

colonic adaptation enhanced bacterial metabolism of lactose in the colon, usually occurring in adults with programmed loss of intestinal lactase function. Also may occur in children with secondary loss of intestinal lactase who continue to consume lactose.

colonocytes an epithelial cell of the colon.

colorectal cancer (CRC) refers to malignant transformation of cells at any site within the rectum to cecum (the beginning) of the colon.

congenital lactase deficiency a rare genetic condition whereby mutations in or around the lactase gene result in insufficient expression of lactase. Usually diagnosed in newborns and can be fatal if not treated.

congenital lactose intolerance the phenotypic consequences of congenital lactase deficiency.

convergent evolution the process of organisms independently evolving the same or similar traits in similar environments by identical genetic adaptions or different genetic changes that have the same effect on the phenotype.

cow's milk allergy(CMA) allergic (IgE, IgG, and IgA mediated) reactions to proteins in cow's milk such as casein, α-lactalbumin and β-lactoglobulin, and bovine serum albumin. CMA can cause symptoms of rash, respiratory wheezing, runny nose, and colitis with bloody diarrhea, particularly in children. It is not an allergy to lactose or an inability to digest lactose.

critical airflow velocity the threshold value of airflow for allowing powder particles to show fluidization behavior.

Crohn's disease a form of inflammatory bowel disease (IBD). It causes inflammation of the digestive tract, which can lead to abdominal pain, severe diarrhea, fatigue, weight loss, and malnutrition.

croscarmellose sodium cross-linked carboxymethyl cellulose, used as superdisintegrant in tablet dosage forms.

crystal water water that is part of the crystal structure, for example, in crystalline α-lactose monohydrate.

cutaneous anaphylaxis a localized hypersensitivity reaction in the form of a wheal and flare in the skin.

cytokine cytokines are cell signaling molecules that aid cell-to-cell communication in immune responses and stimulate the movement of cells toward sites of inflammation, infection, and trauma. Cytokines include any substances, such as interferon, interleukin, and growth factors, which are secreted by certain cells of the immune system and have an effect on other cells.

D- and L-lactic acid lactic acid can exist as two optimal isomers, D- and L-lactic acid, which are the mirrored structures of the same basic component.

DC a means of producing a tablet without granulation. The active ingredient is blended with excipients, typically at least a filler-binder, a disintegrant, and a lubricant, and then the blend is compressed on a tablet machine. The filler-binders are usually special grades of excipients (e.g., spray-dried lactose) exhibiting good flow and compaction properties to enable the DC process.

decantation removal of liquid fraction from a mixture of a solid and a liquid. Typically used to remove the liquid phase in a crystallization process.

degranulation a cellular process that releases antimicrobial cytotoxic or other molecules from secretory vesicles called granules found inside certain cells.

dietary fiber carbohydrate polymers with 10 or more monomers that are resistant to digestion and absorption in the small intestine of humans, being partially or completely fermented by colonic microbiota and have a physiological benefit to human health.

differential scanning calorimetry see DSC.

diffraction pattern the pattern of electromagnetic radiation generated by radiating a powder or a material. The radiation can be X-rays or (laser) light. With X-rays, information on the crystal structure is obtained; with laser light, information on the particle size is obtained from the diffraction pattern.

diluent material used to dilute a substance; it can be a liquid or a solid.

disaccharide carbohydrate built from two monosaccharide units, for example, lactose, sucrose, trehalose, maltose, isomaltose.

Down's syndrome trisomy 21: a congenital disorder arising from a chromosome defect, causing intellectual impairment and physical abnormalities including short stature and a broad facial profile.

downstream processes processes employed after the process of interest.

DPI dry powder inhaler: a pharmaceutical dosage form used to inhale medications in the form of dry powders.

DSC differential scanning calorimetry, a method to measure endothermic or exothermic effects, like phase transitions, as function of temperature.

DSC thermograms the output of a DSC measurement.

EHH (extended haplotype homozygosity) a measure of the extent of similarity of haplotypes extending out from a core haplotype or allele of interest. It measures the probability of two chromosomes chosen at random carrying the same haplotype, and can be used to estimate the age of an allele. In combination with information on allele frequency, EHH can be used to infer positive selection.

enterocyte general term referring to epithelial cells of the small intestine (duodenum, jejunum, ileum).

eosinophil attractant cytokines and chemokines that attract mobilized eosinophils (a variety of white blood cells) in immune responses.

epigenetic heritable changes in gene expression that do not involve changes to the DNA sequence. Often occur in response to the environmental conditions, including age, lifestyle, diet, and toxin exposure. Involve chemical modifications to DNA, such as methylation.

epithelial cells cells that line the outside of the body (e.g., skin), but also internally (e.g., in the intestine or the mammary gland).

EPS exopolysaccharides, carbohydrates produced by bacteria that cause a slimy structure to arise.

estrogenic activity functions of a substance that result in typical outcome of the female sex hormone estrogen (e.g., enlarged breast, ovulation).

evaporated (sweetened) milk milk from which water has been removed by evaporation (usually under vacuum). Sugar can be added to the resulting condensed milk to create sweetened condensed milk.

excipient a component of a drug product other than the API, which is intentionally added to the dosage form to enable processing into patient-friendly medicines, to control the rate at which the API dissolves from the dosage form, to aid drug stability and other reasons. For solid oral dosage forms, main classes of excipients include diluents or filler-binders, disintegrants, glidants, lubricants, coating materials, and stabilizing agents.

exogenous milk consumption the consumption of milk produced by another species (e.g., the consumption of cow's milk by humans).

exon any part of a gene that will be present in the final mature RNA produced by that gene after introns have been removed by RNA splicing.

exopolysaccharides see EPS.

exothermic peak a typical feature in a DSC thermogram that shows an event that delivers heat.

extended haplotype test a method to infer recent positive selection by examining the length of extended haplotypes around the gene or position of interest (see EHH).

farnesoid X receptor (FXR) a nuclear bile acid receptor highly expressed in the liver, intestine, kidney, adrenal glands, and adipose tissue. FXR is a master regulator of the synthesis and pleiotropic actions of endogenous bile acids.

fat globules small fat droplets in milk, ranging in size from less than 1 to more than 10 μm.

fatty acid carboxylic acid with a long aliphatic chain, which can be saturated or unsaturated.

FDA Food and Drug Administration (FDA). The US FDA is responsible for protecting public health by assuring the safety, efficacy, and security of, among other things, human drugs.

FDA GRAS/SCOGS list list provided by the FDA, contains substances that are generally recognized as safe. The Select Committee on GRAS Substances (SCOGS) evaluated and provided scientific insights on these substances.

fermentation a metabolic process that generates energy from sugars in the absence of oxygen.

fibroblast growth factors (FGFs) a family of growth factors which play an important role in tissue repair and regeneration.

flavonoids a major class of plant phenolics found widely in fruits and vegetables, tea, and red wine. They may reduce the risk of heart disease and stroke. Isoflavones are phytoestrogens, which have potent biological activity, which may reduce the risk of hormone-dependent disease.

FODMAPS fermentable, oligosaccharides, disaccharides, monosaccharides, and polyols.

food sensitivity or a food intolerance perceived difficulty in digesting a particular food with symptoms such as intestinal gas, abdominal pain, and/or diarrhea which are not caused by a food allergy involving the IgE/histamine system. It is often distinguished from lactose intolerance when referring to dairy products to signify other milk components such as fats and proteins producing symptoms.

functional gastrointestinal disorders (FGIDs) common disorders characterized by persistent and recurring gastrointestinal symptoms as a result of abnormal functioning of the gastrointestinal tract (e.g., increased or decreased motility and/or increased sensitivity). They are not caused by structural anomalies, tumors, masses, or biochemical abnormalities.

galacto-oligosaccharides (GOS) GOS are β-linked oligosaccharides with a degree of polymerization of 2–9 units that consist of glucose or galactose at the reducing end and are extended by β-linked galactose. Industrially produced GOS mainly have glucose (from lactose) at the reducing end.

galactose monosaccharide that is one of the two monosaccharides of which lactose is composed.

galactosemia a group of rare congenital inborn errors that affects an individual's ability to metabolize the sugar galactose properly due to deficiencies in the Leloir pathway. There are three phenotypes representing mutations in three key enzymes. Galactosemia I and III are the most severe affecting multiple organs.

galactosyltransferase enzyme that catalyzes the condensation reaction between glucose and galactose, forming lactose.

galectins a subclass of lectins (carbohydrate-binding proteins) that binds specifically to β-galactoside sugars, such as *N*-acetyllactosamine which can be bound to proteins by either N-linked or O-linked glycosylation. They play a role in modulating cell-to-cell or cell-to-matrix interactions and have important structural functions for nerves epithelia and immunity.

gas chromatography vaporizing a sample, without decomposition, and injecting it onto the head of the chromatographic column to separate the component molecules of the substance.

(inverse) gas chromatography (IGC) a way to analyze gas mixtures by flowing it over a substrate. Difference in affinity between gas and substrate is measured as difference in time it needs to travel through a column. In GC, the gas (mixture) is the subject of investigation with a known substrate, in IGC the substrate is the subject of investigation with a known gas (mixture).

G-coupled-receptors G-protein-coupled receptors constitute a large protein family of *receptors* that detect molecules outside the cell and activate internal signaling.

gene downregulation the process by which a gene decreases or stops transcription, often in response to external stimuli, reducing or stopping protein synthesis.

gene knockouts experimental animals in which a specific gene is disabled as a model to study the function of that gene.

gene promoter region a region of DNA located that initiates transcription of a particular gene located near the transcription start site of a gene and usually upstream of the gene.

gene-culture coevolution a model of the dynamic interactions between genetic and cultural evolutionary processes.

glass transition temperature the temperature at which a glass (e.g., amorphous lactose) changes in behavior from more solid-like to more liquid-like.

glassy lactose lactose in its amorphous form.

gluconeogenesis the formation of glucose from noncarbohydrate sources.

glucose a carbohydrate categorized as a "simple sugar" or monosaccharide that is metabolized in the body to produce energy, and is one of the molecules making up lactose.

GLUT: glucose transporters a large group of proteins also known as solute carriers or facilitated glucose transporters that carry glucose across cell membranes.

glycolysis the process in which glucose is broken down to produce energy.

glycoprotein a protein which has carbohydrate groups covalently linked to the polypeptide chain.

glycoside hydrolases enzymes that catalyze the hydrolysis of glycosidic bonds in complex glycoside sugars.

glycosuria a condition characterized by an excess of sugar in the urine, typically associated with diabetes or kidney disease.

glycosyltransferases glycosyltransferases are enzymes that establish natural glycosidic linkages.

haplotype a group of alleles inherited together on a single chromosome (i.e., from one parent).

helper T cells a type of T cell that helps coordinate the immune response and is particularly important in the adaptive immune system. They help the activity of other immune cells by releasing T cell cytokines, especially by activating B cells. Besides activating B cells to secrete antibodies and macrophages to destroy ingested microbes, they also help to activate cytotoxic T cells to kill infected target cells.

hepatic encephalopathy hepatic encephalopathy is a decline in brain function that occurs as a result of severe liver disease.

heterofermentative lactic acid bacteria that metabolize hexoses by the phosphoketolase pathway.

HMPC hydroxypropyl methyl cellulose, a derivative of cellulose used as excipient.

homofermentative lactic acid bacteria that metabolize hexoses by the glycolytic pathway.

homologous gene region a region of DNA with similar or identical sequences arising from a common ancestor.

homozygosity possessing two identical alleles at a genetic locus.

human milk oligosaccharides (HMOs) HMOs carry lactose at the reducing end and are elongated with galactose, *N*-acetylglucosamine, fucose, or sialic acid to generate a large structural diversity of oligosaccharides.

hydrogenated a chemical reaction between hydrogen and another component, which causes addition of hydrogen to the structure of that component.

hydrolysis the chemical breakdown of a compound by the addition of water.

hygroscopic material attracting water from the environment.

hypercalcemia above normal serum level of calcium. Too much calcium in the blood (and less in the skeletons) can weaken bones.

hypertension above normal measured blood pressure, usually considered above 130 systolic and 80 diastolic mm of mercury.

hypoglycemia low level of glucose in the bloodstream and may be associated with symptoms such as anxiety, sweating, clumsiness, trouble talking, confusion, loss of consciousness, tremor, palpitations, nausea, and pallor.

hypolactasia a deficit of lactase in the intestinal mucosa.

ICH International Council for Harmonization of technical requirements for pharmaceuticals for human use brings together the regulatory authorities and the pharmaceutical industry to discuss scientific and technical aspects of drug registration.

idiopathic ulcerative colitis one of the most common types of colitis, which is characterized by extensive ulcerations along the mucosa and submucosa of the bowel. One of the current recognized inflammatory bowel diseases (e.g., the other being Crohn's disease). "Idiopathic" means "of unknown cause."

IgE immunoglobulins E (IgE) are antibodies produced by the immune system. Immediate allergic reactions are the result of IgE.

IL-17 interleukin 17A (IL-17 or IL-17A) is a proinflammatory cytokine involved in the immune response against invading pathogens, but also associated with autoimmune diseases. This cytokine is produced by a group of T helper cells in response to their stimulation with another interleukin, IL-23.

ileal microbiota microbiota of the terminal small intestine.

inflammation a localized physical condition in which part of the body becomes reddened, swollen, hot, and often painful, especially as a reaction of the immune system to injury or infection. Recognized on pathological examination as tissue infiltrate with a variety of white cells.

inflammatory bowel diseases (IBD) a group of inflammatory conditions of the colon and small intestine. Idiopathic ulcerative colitis is limited to the mucosa and submucosa of the colon while Crohn's disease can involve the entire gastrointestinal tract and affect all layers of the intestinal wall.

innate immune system dominant system of host defense in most organisms which elicits a response to microorganisms and/or toxins in a nonspecific way.

insulin a hormone produced in the pancreas by the islets of Langerhans that regulates the amount of glucose in the blood. Its deficiency or abnormal resistance leads to diabetes. Also plays a role in regulating milk synthesis.

interferon (INF-γ) a dimerized soluble cytokine, a product of human leukocytes, that is critical for innate and adaptive immunity against viral, some bacterial, and protozoal infections.

interleukin any of various cytokines of low molecular weight that are produced by lymphocytes, macrophages, and monocytes and that function especially in regulation of the immune system, in particular cell-mediated immunity.

intraperitoneal injection (IP injection) the injection of a substance into the peritoneum (body cavity).

intron transcribed but not translated regions of a gene between exons that are removed from the primary messenger RNA transcript and are absent in the mature messenger RNA transcript.

ipratropium bromide a drug substance or API used as bronchodilator in among other things asthma treatment.

IPEC the International Pharmaceutical Excipients Council (IPEC), a global nonprofit organization, consist of five regional councils (IPEC-Americas, IPEC Europe, IPEC Japan, IPEC China, and IPEC India). The council comprises excipient manufacturers, distributors, and pharmaceutical companies that use these excipients. The Federation has as its key objectives harmonization of compendial standards and GMP guidelines.

irritable bowel syndrome (IBS) a common and often chronic syndrome affecting the large intestine. Signs and symptoms include cramping, abdominal pain, bloating, gas, and diarrhea or constipation, or both. Classically it is divided into four types depending on the dominant symptom present: IBS-D, diarrhea dominant, IBS-C, constipation dominant, IBS mixed D and C, and IBS undifferentiated.

isomerization chemical rearrangement of the atoms in a molecule.

isotope ratio mass spectrometry an analytical technique using a mass spectrometer to determine the isotopic composition molecules by measuring the mass of its constituent atoms.

ketosis a metabolic state in which ketone bodies are formed during β-oxidation of fatty acids.

Kreb's cycle the energy producing biochemical pathway in cells.

lactase the enzyme encoded by the LCT gene (also known as lactose-phlorizin hydrolase). Lactase is expressed primarily in the intestinal brush border, and is responsible for hydrolyzing lactose into it constituent monosaccharides, glucose and galactose.

lactase deficiency (LD) a reduction in intestinal lactase caused by a variety of factors such as genetics, age, illness, injury, and disease.

lactase expression the transcription and translation of the LCT gene resulting in the production of the enzyme lactase.

lactase nonpersistence the major phenotype in adult humans resulting from the downregulation of the lactase gene sometime during or after weaning leading to decreased levels of intestinal lactase.

lactase persistence the continued expression of the gut enzyme lactase into adulthood in sufficient quantities to permit lactose digestion, cleaving the disaccharide lactose into its component monosaccharides, glucose and galactose.

lactase β-D-galactoside galactohydrolase β-galactosidase (β-Gal; β-D-galactoside galactohydrolase; EC 3.2.1.23; also called lactase) catalyzing the hydrolysis of β-galactosides. Often found in microorganisms, plants, and animals. In mammals it is referred to as lactase.

lactase-phlorizin hydrolase a glucoside hydrolase, part of the β-galactosidase family of enzymes, that hydrolizes lactose.

lactic acid an organic acid. In milk, it can be produced by bacteria when breaking down lactose.

lactic acid producing bacteria facultative anaerobic Gram-positive bacteria that produce lactic acid as a main product of metabolism. Although many genera of bacteria produce lactic acid as a primary or secondary end-product of fermentation, the term lactic acid bacteria is conventionally reserved for genera in the order Lactobacillales, which includes mainly *Lactobacillus*, *Leuconostoc*, *Pediococcus*, *Lactococcus*, and *Streptococcus.*

lactitol sugar alcohol that is used as noncaloric sweetener.

***Lactobacillus* spp.** a genus of lactic acid bacteria that convert sugars to lactic acid. These non-spore forming bacteria are present in a variety of locations in the human body, including the digestive system.

lactobionic acid a sugar acid that can be made by oxidizing lactose.

lacto-*N*-biose sugar unit that can be attached to lactose to produce milk oligosaccharides.

lactosamine sugar unit that can be attached to lactose to produce milk oligosaccharides.

lactose a disaccharide made up of the monosaccharides, glucose and galactose, joined by a β 1–4 glycosidic bond, present in varying concentrations in the milk of virtually all mammals.

lactose blood glucose test the measurement of blood glucose in response to an oral lactose load. Most commonly used as an indirect test to ascertain lactose digestion. A failure to raise blood glucose levels beyond a threshold suggests lactose is not being digested.

lactose breath hydrogen test the most common indirect test to determine lactose digestion by measuring breath hydrogen levels following an oral lactose load. A failure to raise hydrogen levels above 20 ppm is broadly indicative of being able to hydrolyze lactose whereas formation of lactose in the colon will raise breath hydrogen levels.

lactose intolerance adverse gastrointestinal symptoms including gas, bloating, cramping, and flatus caused by the ingestion of lactose.

lactose maldigestion the failure to digest lactose into its monosaccharides, glucose and galactose, by the intestinal enzyme lactase.

lactose monohydrate the α-lactose monohydrate crystal is the predominant solid form of lactose preparations.

lactose permease enzyme mediating lactose transport in bacteria.

lactose sensitivity generalized symptoms beyond those described for lactose intolerance such as mood alterations, headaches, and fatigue attributed to lactose consumption.

lactose synthase enzyme that catalyzes the reaction in which lactose is formed from glucose and galactose.

lactulose a disaccharide consisting of galactose and fructose that is universally maldigested by mammals. It also induces lactic acid producing bacteria and has some similar clinical features to lactose.

laser diffraction a method to measure particle sizes of powders by the means of obtaining a diffraction pattern of laser light.

LCT the abbreviated gene name for lactase-phlorizin hydrolase, which encodes the enzyme lactase.

lectins lectins are carbohydrate-binding proteins and macromolecules that are highly specific for sugar moieties in other molecules. They play numerous roles in biological recognition phenomena and support multiple cellular functions. Galactose, a main product of lactose digestion, is a common ligand to lectins, than (sub) classified as galectins.

LNP the abbreviation for lactase nonpersistence. The inability to produce lactase in sufficient quantities to hydrolyze lactose in adulthood.

long-chain triacylglycerol fatty acids with 14 or more carbons, including myristic acid, palmitic acid, stearic acid, and arachidic acid. These are found in dairy fat, coconut oil, palm kernel oil, peanut oil, and other vegetable oils.

LP the abbreviation for lactase persistence; the continued expression of lactase into adulthood.

lumen the inside space of the body, surrounded by epithelial cells.

lymphocyte trafficking lymphocyte trafficking (or migration) refers to their ability to migrate from one place to another in the body. This is common, since the initiation of immune responses is centralized in secondary lymphoid organs while the effector phase takes place where needed in peripheral tissues.

lymphocytes a form of small leukocyte (white blood cell) with a single round nucleus, occurring especially in the lymphatic system.

lysozyme also known as *N*-acetylmuramide glycanhydrolase or muramidase, this glycoside hydrolase is an antimicrobial enzyme forming part of the innate immune system. It catalyzes the hydrolysis of 1,4-β-linkages between *N*-acetylmuramic acid and *N*-acetyl-D-glucosamine residues in peptidoglycan, a constituent of bacterial cell walls.

macronutrients nutritional carbohydrates, proteins, and fats consumed in a diet.

magnesium stearate a substance used as lubricant in tableting of powders.

Maillard reaction chemical reaction between reducing sugars and amino acids, which leads to, for example, flavor formation and browning of many foods upon heating.

mannitol a sugar alcohol, prepared by hydrogenation of fructose. It is also used as a medication.

mast cells an immune cell filled with basophil granules, found especially in connective tissue (between body and outside world), and releasing histamine and other substances during inflammatory and allergic reactions. Although best known for their role in allergy and anaphylaxis, mast cells play an important protective role as well, being intimately involved in wound healing, angiogenesis, immune tolerance, defense against pathogens, and blood–brain barrier function.

MCC microcrystalline cellulose. A substance prepared by the acid digestion of cellulose, a common excipient used as filler/binder.

MEC mammary epithelial cell.

membrane filtration separation of components over a membrane, often based on differences in size.

membrane separation see membrane filtration.

Mendelian genetics the study of the inheritance of traits determined by a single gene.

Mendelian randomization method a method of using measured variation in genes of known function to examine the causal effect of a modifiable exposure on disease in observational studies. The design has a powerful control for reverse causation and confounding which otherwise bedevil epidemiological studies.

mesh a unit of size, commonly utilized to distinguish sieves sizes as it relates to the amount of sieve threats per width.

Mesolithic the period between the Paleolithic and Neolithic, all three forming the Stone Age. The dates vary globally but in Europe it is approximately the period from 15,000 to 6000 years ago.

metabolic status balance between energy available and energy used.

metabolic syndrome the metabolic syndrome is essentially a cluster of biochemical and physiological abnormalities and metabolic complications of obesity that occurring together, increase the risk of heart disease, stroke, and diabetes. The criteria listed include abdominal obesity, determined by increased waist circumference, raised triglycerides, reduced high-density lipoprotein, elevated blood pressure, and raised plasma glucose.

metabolites components that are precursors, intermediates, or end products of metabolic pathways.

metabolome the total spectrum of metabolites present within an organism, cell, or tissue.

methylation the process by which a methyl group (CH_3) covalently bonds to DNA, often to a cytosine residue, which can alter transcription. Methylation changes the expression of a gene without changing the DNA sequence itself.

microbiome the community of microorganisms including bacteria, fungi, and viruses that live in or on the human body.

microbiota an ecological community of commensal, symbiotic, and pathogenic microorganisms found in and on all multicellular organisms. Microbiota includes bacteria, archaea, protists, fungi, and viruses.

microcrystalline cellulose see MCC.

micronized milled to such a size that average particle size is below 3–5 μm.

micronutrients the minerals, vitamins, and other nutrient components in a diet, required in smaller amounts than macronutrients. They are involved in a range of physiological functions to maintain health including cell growth, regulation, replication, and development.

microvilli any of the minute hair-like structures projecting from the exposed surface of the cell in order to increase the surface area for absorption, secretion, cellular adhesion, or mechanotransduction.

milk a nutrient- and energy-dense liquid food produced in the mammary glands providing a matrix of macro- and micronutrients to mammalian neonates.

milk fat globule membrane (MFGM) a complex and unique structure composed primarily of lipids and proteins that surrounds milk fat globules secreted from the milk producing cells of humans and other mammals. Contains sphingolipids.

milk sensitivity symptoms produced by milk consumption such as headache, rash, mood changes, and nausea which are not caused by allergies or reactions to lactose, but may be caused by other components in milk.

molecular agonists an agonist is a chemical molecule that binds to a receptor and activates the receptor to produce a biological response. This usually refers to molecules that mimic normal functions of the receptor that are activated by natural host molecules.

monohydrate crystal in which there is a one-to-one molecular ratio of the crystalline component and water.

monosaccharides simple sugar that is the basic building block of all more complex carbohydrates.

monounsaturated fatty acids (MUFAs) a type of fatty acids defined chemically as having one double bond in the fatty acid chain with all other carbon atoms single bonded.

multilayer adsorption technique BET Brunauer, Emmett, and Teller (BET) absorption is commonly used to analyze the specific surface area of a solid material.

multiple sclerosis (MS) an autoimmune disease in which the insulating covers of nerve cells in the brain and spinal cord are damaged due to demyelination.

multivariate analysis statistical evaluation that takes many descriptors of a product or a process in account at once.

mutarotation the change of one anomer (e.g., geometric form of a sugar) into another.

***N*-acetylglucosamine (GlcNAc)** a monosaccharide derivative of glucose. It is a monomeric unit of the polymer chitin. After cellulose, it is the second most abundant carbohydrate. In mammals, GlcNAc is a component in oligosaccharides, for example, in glycoproteins.

***N*-acetyllactosamine (Gal(β 1–4)GlcNAc)** an acylaminosugar and a component in many glycoproteins. It is an organic compound containing a sugar linked to a chain through *N*-acyl group. Often a substrate for galactosidases, fucosyltransferases, and sialyltransferases.

NADPH a cofactor in basic energy metabolism.

natural selection the key mechanism of adaptive evolution whereby the differential survival and reproduction of individuals are attributed to differences in a trait (phenotype). Over time, a heritable trait can become more common or less frequent in a population if the trait confers an advantage or disadvantage to reproductive fitness.

necrotizing enterocolitis overgrowth of the small intestine of neonates by pathogenic Enterobacteriaceae, leading to intestinal inflammation and necrotic cell death of the intestinal mucosa.

Neolithic the New Stone Age from ~12,000 years ago in southwest Asia. This period is associated with a rapid change in diet and lifestyle as humans transitioned from being food-seeking to food-producing. The Neolithic transition first occurred ~12,000 years ago in southwestern Asia.

neonatal lactose intolerance failure to achieve full maturation of intestinal lactase after birth and can cause diarrhea.

neoplastic functions the functions of membranes in neoplastic (abnormal growing) cells.

nephrocalcinosis a disorder that occurs when too much calcium is deposited in the kidneys.

NMR nuclear magnetic resonance is used to analyze atoms that have a nuclear spin. Molecules and environment of molecules can be analyzed to great detail.

NMR nuclear magnetic resonance.

nuclear magnetic resonance see NMR.

nucleotide phosphodiesterase pyrophosphatase 7 a dimeric enzyme that catalyzes the hydrolysis of phosphate diester bonds.

nutrient a food component used by an organism for survival, growth, and reproduction.

oligosaccharide saccharides containing 2–10 monosaccharides.

osmolarity a measure for the number of molecules dissolved in a solution.

osmotic diarrhea diarrhea caused by carbohydrates that are not hydrolyzed and absorbed in the small intestine and, therefore, as osmotic active substances hold/draw too much water into the bowels.

osmotic pressure the pressure that can move water through a membrane from lower to higher osmolarity, to equilibrate the osmolarity on both sides of the membrane.

osteoporosis a medical condition in which the bones become brittle and fragile from loss of tissue, typically as a result of hormonal changes, or deficiency of calcium or vitamin D.

Ostwald ripening the growth of larger crystals on the expense of smaller crystals.

PABA para-aminobenzoic acid or PABA is an organic compound that can be synthesized indirectly in the human body by gut bacteria. Common dietary sources of PABA include kidney and liver, mushrooms, and whole grains.

Paleolithic the Old Stone Age; this period in human prehistory spans from ~3.3 million to ~11,500 years ago, characterized by successive stone tool industries, range expansion, and new hunting and food processing technologies.

partitioning the differential diffusive distribution of a component over multiple phases.

PAT process analytical technology is a term used to describes the technologies and requirements on analytical techniques in processing environment.

pathobiont any potentially pathological (disease-causing) organism which, under normal circumstances, lives as a symbiont.

PCA principal component analysis: a multivariate analysis tool.

perceived sandiness sensoric sensation as if sand is present in a food product. Can be caused by the presence of large (>10 μm) crystals (or other particles).

permeate the liquid passing through the membrane when applying membrane filtration.

pH a measure for the acidity of a solution, defined by the logarithm of the number of free protons (H^+).

pharmaceutical grade lactose lactose that complies to regulatory requirements of pharmaceutics.

phlorizin a flavonoid derived from the bark of the apple tree and converted to phloretin which has effects on serum glucose and is the second major substrate of lactase-phlorizin hydrolase enzyme encoded by the LCT gene.

phlorizin-glycosyl-*N*-acylsphingosine glucohydrolase chemical name of glycosylceramidase also known by other names in common use include *phlorizin* hydrolase, phloretin-glucosidase, *glycosyl* ceramide *glycosylhydrolase*, cerebrosidase, *phloridzin* β-glucosidase, lactase-*phlorizin* hydrolase, and *phloridzin* glucosidase.

physical properties the properties of a material that are determined using physical analyses (e.g., viscosity).

polycose a trademark for an easily digestible and rapidly absorbed nutritional supplement for oral and tube feeding containing glucose polymers.

polymorphic forms a specific crystal form of a compound. Many compounds do have multiple polymorphic forms that can be pure forms, but can also be a complex mixture of forms of the same (pure) compound.

polymorphism the existence of two or more genetic variants in a population that occur at significant frequencies.

polysaccharides polymeric carbohydrate molecules composed of long chains of monosaccharide units bound together by glycosidic bonds which can be branched or linear in structure.

polyunsaturated fatty acids (PUFAs) a type of fatty acids that includes the essential fatty acids (omega-3 and omega-6 fatty acids) among other fatty acids (like conjugated fatty acids); defined chemically as having two or more double bonds in the fatty acid chain.

postcrystallization all process following the crystallization process.

pre- and pro-lactase intracellular precursors to mature lactase which are cleaved and after folded in the endoplasmic reticulum.

prebiotic index the PI equation is based on changes in key bacterial groups during fermentation. The bacterial groups are bifidobacteria, lactobacilli, clostridia, and bacteroides.

prebiotics compounds that encourage the growth and metabolic activity of beneficial microorganisms, particularly in the gut.

precursor adenomas tubular adenoma is a type of colonic polyp and a precursor of colorectal cancer.

pre–pro-lactase lactase-mRNA is translated by a membrane-bound ribosome into a polypeptide called pre–pro-lactase. During translation, the 1927 long amino acid sequence is fed into the ER (endoplasmic reticulum), but remains anchored in the lipid bilayer of the ER. The pre–pro-lactase precursor is than processed in several steps to gain matured active lactase.

primary adult onset hypolactasia declining levels of intestinal lactase, usually caused by age-related downregulation of lactase, but not caused by injury, illness, or damage to the small intestine.

probiotics live microorganisms that, when administered in adequate amounts, confer a health benefit on the host.

protolacteal fluids fluids that were secreted from glands (most likely sweat glands near hair follicles) in species ancestral to mammals, and generally believed to be the evolutionary precursor to milk.

recurrent adenomas adenomas that regrow between interval times of colonoscopy.

reducing sugar a sugar that can form an open ring structure that is chemically reactive (e.g., in the Maillard reaction).

retentate the liquid not passing through the membrane, and thus being retained, when applying membrane filtration.

rumen the first digestive chamber in the digestive tract of ruminant animals. Feed is fermented by bacteria in the rumen.

saccharolytic fermentation fermentation by saccharolytic bacteria, that is, bacteria capable of splitting up and fermenting carbohydrates and producing volatile short chain fatty acids (SCFAs). Probiotics are saccharolytic, meaning they use carbohydrates (fiber) for energy.

Saccharomycetaceae sp. family of fungi comprising the typical yeasts that reproduce by budding and ferment carbohydrates.

salbutamol sulfate see albuterol.

secondary hypolactasia reduced intestinal lactase levels caused by injury, disease, infection, medication, or surgery damaging the intestinal villi. An individual may produce lactase, but a secondary factor prevents sufficient levels of lactase being bound in the small intestine.

secondary lactase deficiency small bowel diseases leading to loss of intestinal villous surface where lactase usually resides can develop lactose maldigestion. Examples of diseases include celiac (gluten enteropathy), Crohn's disease of small bowel, parasitic infestation with giardia lamblia, radiation enteritis, and others.

secondary lactase insufficiency this form of lactose maldigestion occurs when small intestinal lactase production decreases after an illness, injury, or surgery involving your small intestine.

selective pressures factors causing differential fitness between individuals within a species. It is the driving force of evolution by natural selection.

self-reported lactose intolerance (SRLI) personal belief without tests of lactose digestion that lactose containing foods will induce symptoms of lactose intolerance and/or sensitivity.

SGLT: sodium-dependent glucose cotransporters (or sodium-glucose linked transporter, SGLT) are a family of energy-dependent sodium/glucose cotransporters found in the intestinal mucosa (enterocytes) of the small intestine.

short chain fatty acids (SCFAs) end products of carbohydrate fermentation by bacteria. Main SCFAs are acetate, propionate, and butyrate.

sodium starch glycolate cross-linked carboxymethyl potato starch, commonly used as a disintegrant for tablets and capsules.

sphingolipids important polar lipids forming part of the milk fat globule membrane. Coupled with different molecules these have functions on cell functions and communications.

supersaturation a solution of a component at a concentration that is higher than its normal maximum solubility.

T3 triiodothronine; active form of thyroid hormone.

T4 thyroxine; inactive thyroid hormone.

texture the sensory perception of a product structure, different from flavor.

Tg glass transition temperature.

Th1 Th1 cells are a subset of helper-inducer T-lymphocytes which synthesize and secrete interleukin-2 gamma-interferon, interleukin-2, and tumor necrosis factor (TNF)-β. In contrast, Th2 cells produce IL-4, IL-5, IL-6, IL-10, and IL-13. While Th1 cells tend to generate responses against intracellular parasites such as bacteria and viruses, Th2 cells produce immune responses against helminths and other extracellular parasites. Due to their ability to kill antigen-presenting cells and their lymphokine-mediated effector activity Th1 cells are associated with vigorous delayed-type hypersensitivity reactions.

Th17 Th17 cells, by virtue of their production of IL-17 and IL-17F, are generally thought to be proinflammatory and play an important role in host defense against infection, by recruiting neutrophils and macrophages to infected tissues.

thyroid hormones produced by the thyroid gland utilizing iodine. They control many body metabolic functions including temperature regulation. The active form of the thyroid hormone, thyroxine, is triiodothyronine.

thyroxine thyroxine (T4) is the main hormone secreted into the bloodstream by the thyroid gland. It is the inactive form and most of it is converted to an active form called triiodothyronine (T3) by organs such as the liver and kidneys.

tight junctions junctions between cells whose function is to prevent leakage.

T_m melting temperature.

transcription the copying of a DNA sequence into an RNA molecule. The first step in gene expression which begins with the production of an RNA molecule from a gene (DNA).

transgalactosylation transgalactosylation is the process by which the enzyme β-galactosidase hydrolyzes lactose.

Treg the regulatory T cells (Tregs), formerly known as suppressor T cells, are a subpopulation of T cells that modulate the immune system, maintain tolerance to self-antigens, and prevent autoimmune disease.

triglycerides lipids composed of a glycerol backbone with three fatty acids.

triiodothyronine active form of thyroid hormone; T3.

UDP uridine diphosphate, abbreviated as UDP, is a nucleotide diphosphate. It is an ester of pyrophosphoric acid with the nucleoside uridine. UDP is an important factor in glycogenesis.

UDP-Gal uridine diphosphate galactose is a nucleoside diphosphate sugar which can be epimerized into UDP-glucose for entry into the mainstream of carbohydrate metabolism. In lactose synthesis, it provides a source of galactose.

ulcerative colitis a disease that causes inflammation and sores ulcers in the lining of the large intestine (colon). It is one of the inflammatory bowel diseases along with Crohn's disease.

ultrafiltration membrane filtration using a membrane with micrometer size pores that retains proteins but that allows smaller solutes (e.g., lactose and salts) to pass through.

urinary galactose tests a rarely used indirect test to assess lactose digestion by measuring galactose levels in urine, lactose/galactose, and galactose/creatinine ratios following an oral lactose load. Lactase nonpersistent individuals will have lower urinary galactose levels than lactase-persistent individuals who can hydrolyze lactose.

UVB light intermediate wavelength ultraviolet (UV) radiation. Ultraviolet can be classified into three types: UVA, UVB, and UVC. Only UVA and UVB pass through the earth's ozone layer.

water activity see a_w.

whey powder whey from which the water has been removed by drying.

X-ray diffraction X-ray diffraction gives information of the crystal form of a substance.

Index

Note: Page numbers followed by "*f*" and "*t*" refer to figures and tables, respectively.

M

N

O

P

CPI Antony Rowe
Eastbourne, UK
June 18, 2019